Frontiers in Sustainability

Sazada Siddiqui • Mohd Tariq
Zishan Ahmad Wani • Pankaj Kumar
Editors

Frontiers in Sustainability

Biodiversity Conservation in a Changing Climate

Springer

Editors
Sazada Siddiqui
College of Science, Department of Biology
King Khalid University
Abha, Saudi Arabia

Mohd Tariq
Faculty of Allied, Health and Life Sciences
Sumandeep Vidyapeeth Deemed to be University
Vadodara, Gujarat, India

Zishan Ahmad Wani
Department of Environmental Science and Engineering
SRM University-AP
Neerukonda, Andhra Pradesh, India

Pankaj Kumar
Department of Earth and Environmental Science, Parul Institute of Applied Sciences
Parul University
Vadodara, Gujarat, India

ISBN 978-3-032-19739-9 ISBN 978-3-032-19740-5 (eBook)
https://doi.org/10.1007/978-3-032-19740-5

This Springer imprint is published by the registered company Springer Nature Switzerland AG
The registered company address is: Gewerbestrasse 11, 6330 Cham, Switzerland

Preface

Frontiers in Sustainability: Biodiversity Conservation in a Changing Climate is a comprehensive collection of the most recent breakthroughs in the field of sustainability science. This book is a necessary tool for academics, professionals, and government officials who are committed to tackling the complex issues of biodiversity conservation vis-à-vis climate change. Every chapter will provide an in-depth exploration of particular technical advancements and scientific findings that have the potential to revolutionize our approach to biodiversity conservation and mitigating climate change. The editors have carefully selected and organized contributions from prominent specialists around the world, guaranteeing that the content is both authoritative and widely applicable. Concrete examples of real-world scenarios and practical uses are emphasized, showcasing the measurable effects of theoretical investigations. Moreover, this book highlights the significance of cooperation and incorporation among many sectors to accomplish sustainable objectives. *Frontiers in Sustainability* serves as a guiding light for innovation and advancement in the area by discussing both the present state of environmental research and the upcoming trends that will influence the future. This book is essential for individuals who are dedicated to creating a positive impact on the world by implementing sustainable practices.

About the Volume

This volume attempts to provide the latest, crucial, and rich knowledge and expert studies on the paradigms, ideas, procedures, techniques, applications, and case studies pertaining to the *Frontiers in Sustainability*: *Exploring Cutting-Edge Research in Environmental Science and Technology*. The present volume comprises 13 chapters contributed by global subject experts. It is believed that this new volume on *Frontiers in Sustainability: Biodiversity Conservation in a Changing Climate* will be of the utmost relevance to academics, researchers, and students working in the realm of biodiversity conservation and climate change. Besides, it will enable the policy makers and managerial experts across the globe to draft policies and strategies for biodiversity conservation and the mitigation of changing climatic scenarios.

Dr. Sazada Siddiqui would like to thank all her colleagues and friends in the Department of Biology, College of Science, King Khalid University, Abha, Saudi Arabia. I am highly grateful to King Khalid University for providing administrative and technical support.

Abha, Saudi Arabia	Sazada Siddiqui
Vadodara, Gujarat, India	Mohd Tariq
Neerukonda, Andhra Pradesh, India	Zishan Ahmad Wani
Vadodara, Gujarat, India	Pankaj Kumar

The original version of the book has been revised. A correction to this book can be found at https://doi.org/10.1007/978-3-032-19740-5_14.

Contents

Exploring the Concept and Significance of Biodiversity in Modern Times

Apurva Sharma, Snigdha Singh, Pankaj Kumar, and Vedpathak Mangesh Mohan

Abstract The Earth's biological resources are essential for human existence and socioeconomic advancement. Furthermore, it has offered boundless services to humanity, either directly or indirectly, for ages. Biological variety has several optional values, making it a worldwide asset of immense significance for present and future generations. Conversely, the danger to species and ecosystems has reached unprecedented levels, as species extinction from human activities persists alarmingly. Humanity relies on robust ecosystems that sustain significant groups and cultures. The effective operation of ecosystems delivers essential goods and services crucial for human health and survival, encompassing nutrition and food security, clean air and fresh water, medicinal resources, cultural and spiritual values, and support for local livelihoods and economic development. Furthermore, it may assist in mitigating illnesses and stabilizing the environment. The historical Rio Conventions on biodiversity, environmental protection, and development commenced in 1992, including the Earth Summit, the Convention on Biological Diversity, the United Nations Framework Convention on Climate Change, and the United Nations Convention to Combat Desertification, collectively striving to preserve functional ecosystems for the benefit of humanity.

Keywords Biodiversity · Ecosystems · Conservation planning · DNA banks · Biotechnology

A. Sharma (✉) · S. Singh · P. Kumar
Department of Earth and Environmental Science, Parul Institute of Applied Sciences, Parul University, Vadodara, Gujarat, India

V. M. Mohan
Department of Environmental Science, School of Earth Sciences, Punyashlok Ahilyadevi Holkar Solapur University, Solapur, Maharashtra, India

S. Siddiqui et al. (eds.), *Frontiers in Sustainability*,
https://doi.org/10.1007/978-3-032-19740-5_1

1 Introduction

Biodiversity refers to the diversity and variability of living species across all levels, from genes to ecosystems. Biodiversity refers to the variety among organisms that originate from many sources, including terrestrial, marine, and aquatic habitats and ecological complexes they include; it encompasses diversity within species, between species, and within ecosystems (Lausch et al., 2016). Ecosystem, species, and genetic diversity form the three levels of organization in the concept of Biodiversity (Fig. 1). Alpha, beta, and gamma diversity are three key measures of biodiversity used in ecology to describe species richness and distribution at different spatial scales. Alpha diversity refers to the diversity within a single habitat or area, beta diversity measures the difference in species composition between habitats, and gamma diversity represents the overall diversity across a larger landscape or region. Biodiversity enhances human health and well-being. Their disappearance undermines this base. Biodiversity delivers ecosystem services that include goods and services essential for the functioning of ecosystems vital for our food and freshwater; assists in climate regulation; mitigates floods and diseases; offers recreational advantages; and contributes to aesthetic and spiritual enrichment. Biodiversity enhances local livelihoods, medicinal resources (both traditional and contemporary), and economic advancement. Human health fundamentally relies on ecosystem services facilitated by biodiversity and its goods and services (Patz et al., 2012).

Biodiversity, particularly the types of wild plants, has many positive effects on human health. For everyday food and sustenance, local populations may continue to gain from plant variety and sustainable ecosystem management. Numerous worldwide concerns, including food security, climate change, conservation biology, and human health, are pertinent to research on how humans interact with plants (Clark et al., 2014; Quave & Pieroni, 2015). Because of its ability to regulate fires to preserve forest regions, Indigenous peoples' traditional ecological knowledge (TEK) has garnered much attention in recent decades (Clark et al., 2014). The preservation of local resources and the provision of local models for sustainable living via

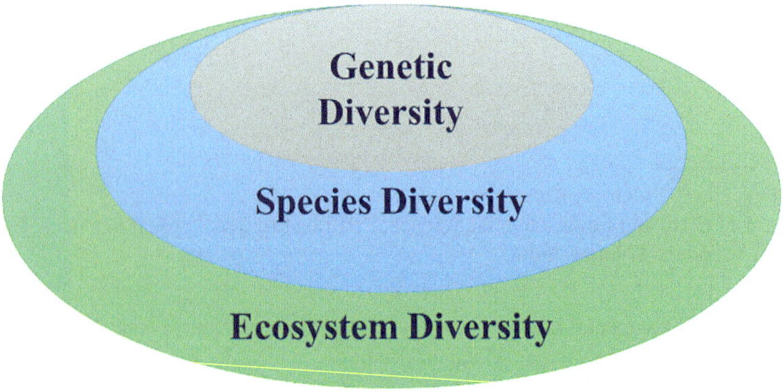

Fig. 1 Levels of organization of the concepts of biodiversity (NBC, 2025)

multi-ecosystem services depend heavily on extensive agricultural management (Atreya & Shrestha, 2020; Eloy et al., 2019; Mauerhofer et al., 2018; Tattoni et al., 2017). Additionally, TEK has recently drawn more attention from the ecological science community, which views it as having great potential to help develop better strategies for protecting species diversity in protected areas and reviving local resource use (Uchida et al., 2021).

2 Understanding the Importance of Biodiversity: Ecosystems and Economy

Our food, biomaterials, biofuels, pollination, biocontrol, genetic resources, cultural values, and several other unintended advantages are all provided by the biological creatures that power Earth's biogeochemistry, which has a significant impact on environmental conditions at both local and global scales. The indispensable ecosystem services obtained from nature are directly or indirectly dependent on the biological diversity of a location (Fig. 2). On a fundamental level, the biogeochemistry of an ecosystem is primarily determined by the cumulative mass of these species and their collective biological activities. Nevertheless, a startling variety of creatures often make up this bulk. Understanding how life's remarkable variety relates to biogeochemical or ecosystem functioning is a key difficulty for contemporary research, even though the biological mechanisms that underlie biogeochemistry are typically well characterized (Chivian, 2002). A summary of the importance of biodiversity for our ecosystem is described in Table 1.

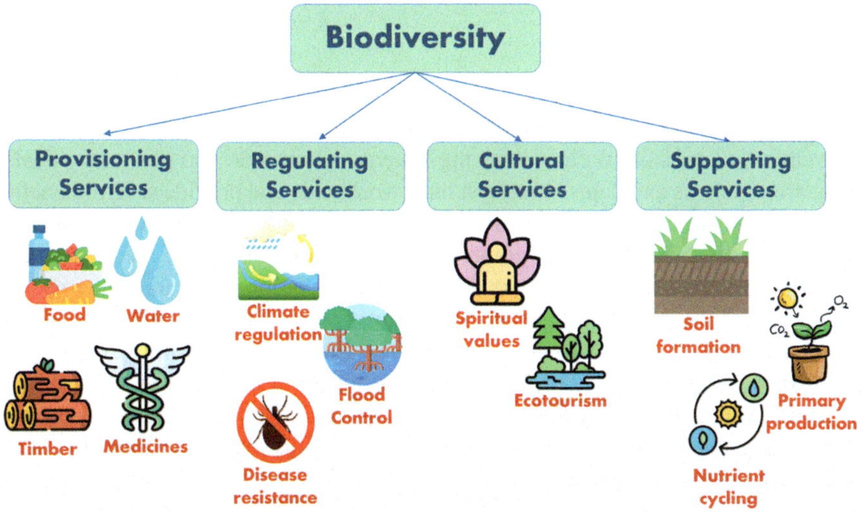

Fig. 2 Importance of biodiversity for ecosystem services

Table 1 Important aspects of biological diversity

Aspect	Importance	Examples
Ecological stability	Maintains ecosystem functions like nutrient cycling, soil formation, and climate regulation.	Forests regulate rainfall patterns; wetlands filter pollutants.
Economic value	Food, fuel, fiber, medicines, and industrial products are essential for economies.	Fisheries, timber industry, medicinal plants (e.g., Artemisinin for malaria).
Source of genetic resources	Supplies genetic material for improving crops, livestock, and disease resistance.	Wild relatives of rice and wheat are used in breeding programs.
Ecosystem services	Supports pollination, seed dispersal, water purification, and pest control services.	Bees pollinate crops; mangroves prevent coastal erosion.
Climate regulation	Plays a role in carbon sequestration and regulating greenhouse gases.	The Amazon rainforest acts as a carbon sink.
Cultural and aesthetic value	Contributes to cultural identity, spirituality, recreation, and tourism.	Sacred groves in India; ecotourism in African savannas.
Scientific and educational value	Offers opportunities for research, education, and innovation.	Discovery of penicillin from fungi; biodiversity studies informing conservation policies.
Resilience against environmental changes	Enhances ecosystem adaptability to disturbances such as climate change, pests, and diseases.	Diverse coral reefs recover from bleaching events faster than less diverse ones.
Food security	Provides a diverse range of foods critical for nutrition and sustenance.	Traditional crops like quinoa and millet are adapted to harsh conditions.
Medical and pharmaceutical resources	Many drugs are derived from natural compounds found in biodiversity.	Rosy periwinkle produces anti-cancer compounds (vincristine, vinblastine).

2.1 Wetlands

Wetland ecosystems support human health and well-being in several ways: flood prevention, water supply, clean drinking water, nourishment, and pharmaceuticals. Irrigation systems and dams have been built worldwide and provide several benefits to society, such as local employment, tourism, fisheries, food and fodder for the local population, increased agricultural production, and small-scale electricity generation for home and industrial use. Because they function as natural filtering systems and filter out silt, heavy metals, and other impurities from water, wetlands are referred to as the "kidneys of nature." They also collect and hold onto rainfall, progressively releasing it to plant roots, and transferring it into aquifers and surface streams. All of these advantages come free of charge from natural ecosystems. Conserving biodiversity not only sustains our economy and creates employment, but it also saves money. (Patz et al., 2012).

2.2 *Forests*

Forests are the most essential resource on Earth, providing many necessary ecological services in nature from which human beings benefit. Forests naturally spread their seeds and pollen grains. By relocating to areas with more hospitable conditions, forest species may preserve their genetic diversity and long-term adaptability to change across time and space. However, this ability may have been diminished by anthropogenic landscape alterations, such as changes in land use and gene pools, and population fragmentation may harm populations' genetic and reproductive condition (Thompson et al., 2009).

Changes in gene frequency and seed and pollen distribution may co-occur and interact during adaptation. To maintain within-population levels of genetic diversity and avoid genetic drift and loss of genetic diversity that can result from inbreeding within small, isolated, or fragmented tree populations, for example, dispersal frequently encourages gene flow among highly fragmented tree populations (Farwig et al., 2008b).

A relatively quick response to an environmental challenge is made possible by high levels of genetic variability within a larger, local population or gene pool of a particular tree species (such as typical populations seen in the boreal or temperate biomes). The gene pool may be reduced to favor genotypes that are most resilient to disturbances like toxic chemicals, pest infestations, or other forms of interspecific competition, climate change, or soil water and nutrient conditions as a result of differential survival brought on by natural selection pressures (Thompson et al., 2009).

2.3 *Economic Value of Biodiversity*

Pharmaceutical bioprospecting has gradually expanded in high-biodiversity regions like coral reefs and tropical rainforests, as several necessary medications have been found in natural products. Natural resources were used to create recent biotechnology breakthroughs, including the capacity to sequence and clone genes (Chivian & Bernstein, 2008). Maintaining genetic variety will become more and more important for pharmaceutical bioprospecting, among other things, as illnesses recur and many modern infections become more resistant to antibiotics. Numerous animal species, especially mammals, are also beneficial models in biomedical research for physiological processes, including adapting to harsh and unfavorable environments, which may affect human health. The ability of hibernating bears to recycle urea and maintain bone density, for instance, may have direct implications for renal dialysis patients or the prevention of osteoporosis. Another emerging field with potential benefits to human health is biocontrol, which employs biological substitutes for chemical pesticides, such as bacteria, fungi, viruses, and insects (Patz et al., 2012).

3 The Threat of Biodiversity Loss: Causes and Consequences

Along with the appropriation and acceleration of human activities, such as the improper use and management of agricultural land, forest destruction, overgrazing, poverty, legal frameworks, governmental policies, and institutional structures, ecosystem degradation is a natural process. One of the world's most significant environmental issues today is ecosystem deterioration. Twenty billion hectares of land have deteriorated globally, accounting for 22% of the total cropland, prairie, and forest area. Due to global degradation, about 5 million hectares of farmland cannot produce crops yearly. In 20 years, a third of the world's arable land will be gone if current trends continue. Ecosystem degradation is the cause of the notable drop in land productivity. Due to environmental degradation, worldwide agricultural productivity has decreased by 11.9–13.4% during the last 50 years (Daily, 1995; UNEP, 1997).

Furthermore, land degradation has led to several environmental issues that seriously threaten human survival and sustainable development, including the filling up of rivers and lakes, changes in the soil's organic carbon reserve, the loss of biodiversity, and the disappearance of special habitats. Therefore, preventing land degradation, restoring and rebuilding degraded ecosystems, and taking concrete steps to make the sustainable use of land resources a reality are the hot spots for government agencies, non-governmental organizations, scientific communities, and the general public in order to ensure food safety and maintain a healthy environment (Hai et al., 2007). It has been acknowledged that the introduction of new species is a significant factor contributing to the global decline in biodiversity (Catford, 2017; Strayer & Dudgeon, 2010), although it is clear that creatures that live in freshwater environments are more at risk (Moorhouse & Macdonald, 2015).

China's lake area has shrunk by 1.3 million hectares since the early 1950s. Mining caused the disappearance of almost 1000 lakes. Mining reduced the lake area in the Yangtze River's downstream zone from 25,828 km^2 in 1949 to 10,493 km^2 now, or 59.4% of the original lake area. Among them, Dongtinghu Lake, formerly the biggest lake, has shrunk from 4300 km^2 in 1950 to fewer than 2270 km^2. There are presently just 741 lakes in the Hubei province, down from over 1000 in the past, and the water surface has decreased to around three-quarters. Over the last 56 years, the wetland area has decreased by more than half. Because of habitat degradation, species that are bred in ecosystems are rapidly becoming extinct in order to create farmland for food (Guo et al., 2001). Approximately 10–15% of plants worldwide were thought to be endangered, but in China, the percentage of endangered plants and endangered species rose to 15–20% and 4000–5000, respectively. The distribution regions of valuable wild species, such as high-nose antelopes, Yangtze River dolphins, wild elephants, bears, and northeast tigers, have decreased. Their numbers have drastically declined, according to research conducted over the last 30 years (China Forestry Publishing House, 2020).

4 Conservation Imperatives for Sustaining Biodiversity

As evidenced by the growing global area covered by protected areas and the corresponding decline in biodiversity, traditional conservation policies primarily focused on establishing protected areas (Calado et al., 2014), with more or less stringent land use restrictions may have been ineffective and generally inadequate. There are at least three ways that science might impact environmental conservation efforts. First, a corpus of scientific theory and practice may provide some building blocks for policymaking. Second, when asked to help implement policies and conventions and explain the social and economic ramifications of using different approaches or scenarios, science can provide answers (this role is best filled when science is integral to the process, not simply called in for peer review or when technical or political problems emerge). Third, science can and should be used to assess how well political processes accomplish declared biodiversity objectives. The efficiency of research in achieving these three goals will be improved by a systematic framework for conservation planning (Butchart et al., 2010; Fernandes et al., 2017; Gómez-Baggethun & Ruiz-Pérez, 2011; Mora & Sale, 2011; Wiens et al., 2009).

5 Biotechnological Approaches for Biodiversity Conservation

Human DNA sample and data collection have significantly expanded in the last several years. Future medicines, clinical care, and genetic research depend on this activity. Modern scientific techniques should be used for conservation to preserve collections and maximize their utility. These genetic repositories support advancements in human health and offer valuable insights into the broader spectrum of Biodiversity, helping us understand genetic variation, species evolution, and the conservation of biological diversity.

5.1 Molecular Marker

A genetic marker is a gene or DNA sequence linked to a trait. It has a known position on a chromosome. It may be defined as a variation resulting from a mutation or change in the visible genetic locus. A genetic marker may be a lengthy DNA sequence, like mini and microsatellites, or a small one, such as the sequence around a single-base-pair alteration (single-nucleotide polymorphism) (Al-Samarai & Al-Kazaz, 2015). The breeder's primary goal is to select animals with exceptional genetic potential to be parents for the next generation. The phenotype of an animal for a specific attribute was employed as a selection strategy in the first effort to enhance animals. Morphological markers—external animal traits such as udder form, coat color, body shape, skin texture, and anatomical features—are used in this application (Gizaw, 2011).

Because morphological markers depend on subjective assessments and descriptions, the results drawn from their application are often incorrect when used to assess farm animal genetics. In order to examine the genetic resources, another kind of marker is represented by cytological markers that are part of several criteria, including chromosomal karyotypes, bandings, repetitions, translocations, deletions, and inversions (Nam et al., 2020). Genetic variety results from chromosomal mutations. These mutations were indicators to pinpoint the gene's precise position on a particular chromosome. Cytological markers enable us to examine the genetic diversity of domesticated animals by comparing the number and shape of chromosomes between domesticated animals and their wild counterparts. Because of its high qualities and quick, easy, and affordable method, cytological markers are still often employed to help determine the origin and categorization of species. Biochemical indicators, such as blood types and isozymes, comprise the third marker category. These indicators stand for biochemical characteristics that protein electrophoresis might examine. Genetic variation within species and evolutionary links between species were examined using the variations in the amino acid content of soluble proteins and isozymes. Since proteins and isozymes are not genetic materials, the use of these markers was restricted. Since they result from gene expression, environmental influences may impact them. Thus, the molecular markers become the focus of researchers' direct attention. The most accurate molecular indicators are those based on the individual's genome's nucleotide sequence alterations (Al-Samarai & Al-Kazaz, 2015).

5.1.1 Marker-Assisted Selection (MAS)

One of the most important methods for enhancing animal performance is selection. Two kinds of data may be used to estimate the best linear unbiased prediction (BLUP), which integrates pedigrees and phenotypes to provide estimated breeding values (EBVs). Marker-assisted selection (MAS) is a novel method that uses DNA markers to provide a third kind of data. DNA in linkage equilibrium with a quantitative trait locus (QTL) may be the basis for the MAS (LE-MAS). The term "LE" describes molecular markers in linkage disequilibrium with QTL (LDMAS), which are genotype frequencies at one locus distinct from genotype frequencies at the second locus. The term "LD" describes the non-random pairing of alleles between two loci, or the mutation selection that causes the QTL effect (Gene-MAS). The cattle business uses all three MAS kinds (Al-Samarai & Al-Kazaz, 2015).

5.2 DNA Banks

In general, in order to establish successful conservation methods, more plant conservationists are using DNA technology. The DNA bank is a practical, long-term, and simple technique for preserving genetic resources for biodiversity. Unlike conventional seed or field gene banks, DNA banks reduce the danger of exposing genetic information in natural environments. It maintains the DNA's stability in cold storage and requires a small sample size. Genetic procedures are required to introduce the stored genetic material since DNA cannot be extracted from whole plants (Sara Peris).

New resources known as plant DNA banks have been developed with much promise for describing and using biodiversity. Because they serve as vital national and international resources, other botanical collections, genetic resource collections, herbaria, botanic gardens, and seed banks are highly prized (Waldren et al., 1999). Historically, live plants kept in seed banks, field gene banks, botanic gardens, or in situ reserves have served as the foundation for collections used in biotechnology and biodiversity assessments or based on dried plants managed in herbarium collections (Maxted et al., 2000; Miller & Nyberg, 1995). In order to manage the world's genetic resources, solve the biodiversity catastrophe, and realize their full potential, DNA collections are now crucial everywhere. DNA technologies are being used more and more by plant conservationists to provide answers to fundamental botanical queries that might aid in developing effective conservation plans (Karp et al., 1996; Kelleher et al., 2005; Lowe & Harris, 2004).

Herbarium specimens' physical, chemical, and anatomical traits may be analyzed to determine genetic variation within and across species. None of them, however, can be said to have the same potential as DNA. Ultimately, the complete genetic makeup of the target organism is represented by genomic DNA samples. Many crops spread vegetatively, including potatoes, sweet potatoes, yams, cassava, and fruit tree species (Futema et al., 2014).

DNA preservation and in situ conservation techniques offer an integrated conservation approach. An alternate method of conserving vulnerable and vegetatively propagated plant species is using in vitro gene banks, which store plant material in a nutritional medium under controlled circumstances (Bhat et al., 1995; Fay, 1994; Sharma & Chandel, 1996).

It is now feasible to provide a germplasm preservation method that combines the potential for fast clonal replication and disease eradication, uniquely using in vitro procedures (Grout et al., 1977). Field gene banks are labor-intensive and risk losing germplasm to illness, stress, or natural disasters. Another potential ex situ conservation method currently in the experimental phases is the cryogenic preservation of seeds or vegetative material (Rajasekharan & Sahijram, 2015).

5.2.1 DNA Banking and Its Beneficial Effects

- Study is essential to health care, and access to biological materials, including DNA, is necessary for contemporary study. The possible advantages justify the creation of DNA banks. However, the potential for abuse necessitates appropriate management and safeguarding the subjects' interests.
- Biological samples in anonymous collections were first gathered without identities, making it difficult to connect them to their original origins. Although bio-

logical materials in anonymised collections were originally identifiable, they can no longer be linked to their original origins since they have been permanently stripped of any identities.

- For research reasons, biological components in identifiable collections are not identified. Nevertheless, a code may be used to link them to their origins. The biological materials in identifiable collections are identified by identifiers such as name, patient number, or exact pedigree location.
- All forms of DNA banking need informed consent. Ethics committee monitoring is required for research use to guarantee a suitable balance between risks and benefits.
- People may request their agreement for wider usage since it is hard to predict all the possible research uses for a collection. If so, there is no need to contact them again. However, subjects should be allowed to communicate if they want to withdraw. (Godard et al., 2004).

6 Role of Biotechnology in Evaluating Genetic Diversity

In vitro plant tissue culture involves aseptically cultivating plant cells, tissues, organs, embryos, protoplasts, or seeds on a nutritional medium under regulated temperature, humidity, photoperiod, light, and conditions. Scientific knowledge and practical expertise help plant tissue culture thrive.

Potent yet straightforward tissue culture enables researchers to grow and alter plants in a sterile environment, making it useful for plant breeding. Rapid clonal propagation or micropropagation; disease elimination; anther/pollen culture for haploid plants; protoplast isolation for hybrid production; embryo culture; botanical substance production; cell selection and mutation; studies in fundamental plant anatomy, development, and nutrition; plant biotechnology; somatic embryogenesis; synthetic seed production; callus culture; and flower cult (Kyte et al., 2013).

Plant embryos need a growth medium to develop and grow in vitro. Young embryos require a more complicated medium to grow (Bridgen et al., 2018; Haslam & Yeung, 2011).

7 Unconventional Biotechnological Methods

Unconventional biotechnological methods have emerged as powerful tools for conserving biological diversity, especially in the face of habitat loss, climate change, and species extinction. Unlike traditional approaches, these methods involve advanced scientific techniques such as micropropagation, cloning, somatic embryogenesis, cryopreservation, and plant tissue culture. They offer innovative ways to preserve genetic resources, restore endangered species, and ensure the long-term sustainability of ecosystems. The following section highlights some of the key unconventional biotechnological methods used in biodiversity conservation. They have been discussed in this section and summarized in Table 2.

7.1 Micropropagation and Cloning

Micropropagation produces mass plant propagules from any plant component or cell in vitro. Micropropagation propagules multiple stock plants in microenvironments. Conserving endangered plants often involves micropropagation and cloning from explants. Plant propagation in vitro has great promise for developing high-quality medications and preserving medicinal plants. Shoot development began

Table 2 Biotechnological methods in biodiversity conservation (Pathak & Abido, 2014)

Method	Description	Applications	Examples
Micropropagation	Rapid asexual multiplication of plants under controlled conditions using tissue culture.	Mass production of rare or endangered plant species; reforestation programs.	Micropropagation of orchids, bamboo, and medicinal plants like *Rauvolfia*.
Cloning	Production of genetically identical organisms from a single parent cell or organism.	Preservation of valuable or endangered genetic traits; revival of extinct species (theoretically).	Cloning of endangered animals like the gaur (Bos gaurus) and attempts with the *Pyrenean ibex*.
Somatic embryogenesis	Formation of embryos from somatic (non-reproductive) cells without fertilization.	Regeneration of plants for conservation and breeding programs.	Somatic embryos developed in crops like coffee, cocoa, and oil palm.
Organogenesis	Formation of new organs (shoots, roots) from tissues or cells in vitro.	Restoration of endangered plants and plant genetic improvement.	Regeneration of shoots and roots in potato and tomato tissue culture.
Cryopreservation	Ultra-low temperature storage of biological samples (cells, tissues, seeds) for long-term conservation.	Conservation of endangered species' germplasm; preserving genetic diversity.	Cryopreservation of seeds in the Svalbard Global Seed Vault; freezing of animal embryos.
Plant tissue culture	Plant cells, tissues, or organs are grown under sterile conditions on a nutrient-culture medium.	Rapid multiplication of elite and endangered plant varieties; genetic improvement studies.	Conservation of rare medicinal plants like *Taxus baccata* and *Withania somnifera*.
DNA banking	Storage of DNA samples for future research, breeding, and conservation efforts.	Securing genetic material for endangered species and facilitating genetic diversity studies.	DNA banks preserve genetic material from species like tigers, elephants, and rare plants.

directly from explant nodes or indirectly via callus-mediated de-differentiation of shoot initials in the in vitro plant regeneration/micropropagation procedure. Next, shoot initials undergo elongation and development to create miniature plantlets with well-developed shoots and root systems for soil transplantation (Castellanos-Hernández et al., 2009; Sadeq et al., 2014). Micropropagation uses plant pieces as explants to rapidly, season-independently propagate, maintain, and store rare and endangered plants (Chandra et al., 2010; Sarasan et al., 2006).

7.2 Somatic Embryogenesis and Organogenesis

Somatic embryogenesis via differentiation of a single somatic cell or tissue to regenerate several plants concurrently is a typical approach in plant tissue culture. Regenerating endangered plants for conservation via somatic embryogenesis and organogenesis from explant cultures is the most common method (Sadeq et al., 2014). An explant culture in an adequate medium regenerates complete plants through direct or indirect somatic embryogenesis. In direct somatic embryogenesis, plants grow straight from explants without callus induction (mass of unorganized cells) and callus de-differentiation, as in indirect somatic embryogenesis. Induction of somatic embryos from cultured explants in specialized culture conditions and their germination into complete plantlets requires numerous processes and plant growth regulators (Sadeq et al., 2014).

7.3 Cryopreservation

One important biotechnological approach for ex situ plant conservation is cryopreservation, which is suitable for the long-term preservation of plant genetic material. Cryopreservation is particularly beneficial for the conservation of rare, endangered, and threatened plant species (Dussert et al., 1997; Paunescu, 2009; Zhao et al., 2008). Long-term storage of cultivated plant components needs ultra-cold storage. Basic procedures and laboratory facility needs are covered here. Most cryopreservation experimental systems (cell suspensions, calli, shoot tips, embryos, etc.) include substantial cellular water levels. This makes them vulnerable to freezing. Thus, artificial dehydration is crucial. However, conventional methods use freeze-induced dehydration, whereas contemporary methods use vitrification (Zhao et al., 2008). Subjecting plant cell and tissue cultures to supraoptimal temperature with or without cryoprotectants stops them from proliferating or metabolizing. The classical freezing process involves pre-growth, cryoprotection, gradual chilling (0.5–2 °C/min) to −40 °C, fast immersion in liquid nitrogen (LN), storage, thawing, and recovery. It requires technical competence to freeze, store, thaw, and reculture live plant cells during cryopreservation against cryogenic damage. Maintaining cryogenic cultures in LN at −196 °C or in the vapor phase at −135 °C preserves tissue viability after rewarming. Plant species are cryopreserved via vitrification, dehydration, and encapsulation (Stacey & Day, 2007). Molecular marker analysis shows cryopreservation maintains genetic integrity (Liu et al., 2008). Cold-stored tissue offers a safer, cleaner, disease-free genetic supply for international exchange (Feng et al., 2011).

7.4 Plant Tissue Culture

In vitro plant tissue culture (PTC) propagates plants quickly, season-independently, and sterilely. Cloning plant material and creating disease-free, clean plant stock is highly effective. PTC is a sunrise technique that boosts agriculture and industry. Plant tissue culture may develop any plant cell into a whole plant using totipotency. Various culturing techniques employ various organs (Chawla & Singh, 2009). The approach for different types of culture increases plant output by quickly regenerating plants without seeds or employing seeds with poor germination rates (Malabika et al., 2017).

8 Genetically Modified Foods

Genetic manipulation improves seeds and hybrids instead of selective breeding. Biotechnology tweaks a few genes (often just one) and leaves the rest unchanged, unlike selective breeding, which created green revolution seed variants. Backcrossing may introduce new genetic modifications into all conventional crop kinds. Genetically modifying all seed kinds maintains agricultural biodiversity with just a modest alteration in seed genetic structure. However, the magnitude of biodiversity losses due to GM crops relies on local GM crop adoption rather than a generic GM crop (Amini et al., 2014).

9 Significance of Biodiversity in Modern Times

Biodiversity holds immense significance in this era as it majorly influences ecological balance, our economic development, and also the well-being of all living forms (Bennett, 2017; Harrison et al., 2014; Reichstein et al., 2014; Viglizzo et al., 2015). Thc essential ecosystem services depend on it and provide us with increased resilience to the changing environments. Key roles and importance of biodiversity in sustaining life and human Development is shown in Table 3.

9.1 Ecological Importance of Biodiversity

The health and functioning of ecosystems largely depend on Biodiversity. Various processes responsible for providing us with ecosystem services, such as the different provisioning services, soil formation, pollination, nutrient cycling, climate regulation, etc., depend on Biodiversity (Marshall et al., 2025). All the diversity in flora and fauna has a unique contribution to the functioning of the ecosystem, thereby making it more productive and improving the overall health of the environment. Ecological stability and resilience are aided significantly in biologically diverse regions, as ecosystems rich in species can withstand disturbances better. If an event negatively affects a particular species, the others can fulfill their ecological role in a highly diverse ecosystem (Ahmed & Sarma, 2022). Rich biodiversity also serves as

Table 3 Key roles and importance of biodiversity in sustaining life and human development

Aspect	Significance	Examples
Ecological stability	Biodiversity ensures the stability and resilience of ecosystems, helping them withstand disturbances like climate change and pollution.	Diverse forests resist pests better than monocultures; coral reefs recover better with diverse species.
Ecosystem services	Biodiverse ecosystems provide vital services like air and water purification, soil fertility, climate regulation, and pollination.	Wetlands filter water, bees pollinate crops, and forests regulate carbon dioxide.
Food security	Genetic diversity among crops and livestock ensures food variety, nutrition, and resistance to diseases and changing climate conditions.	Rice varieties resistant to floods and drought-tolerant maize.
Medicinal resources	Many modern medicines are derived from plant, animal, and microbial biodiversity; untapped species may hold cures for diseases.	Penicillin (fungus); anti-cancer drug Taxol (Pacific yew tree).
Economic benefits	Biodiversity supports industries like agriculture, pharmaceuticals, biotechnology, and ecotourism, contributing to national economies.	The coffee industry relies on diverse plant-pollinator networks in ecotourism in the Amazon rainforest.
Cultural and spiritual values	Biodiversity shapes traditions, religions, folklore, and identity, contributing to cultural heritage and human well-being.	Sacred groves in India; totem animals in indigenous cultures.
Scientific research and innovation	Studying biodiversity aids in scientific discoveries, ecosystem modeling, biotechnology innovations, and climate change solutions.	Biomimicry: Velcro inspired by burrs; climate-resilient crops.
Climate Change Mitigation and adaptation	Diverse ecosystems store carbon and help communities adapt to changing climatic conditions.	Mangroves protect against storm surges; forests sequester carbon.
Ethical responsibility	Humans have a moral duty to protect other life forms, acknowledging their intrinsic value and right to exist.	Conservation efforts like the World Wildlife Fund (WWF) campaigns.

a buffer against any environmental changes. When subjected to shocks like climatic fluctuations, invasive species, human interventions, etc., diverse ecosystems have a better chance of survival than areas with low diversity.

9.2 *Economic Value of Biodiversity*

Various economic activities are directly or indirectly supported by the accessibility of biological diversity in the regions (Fig. 3). Agricultural activities depend widely on genetic resources for breeding resilient and high-yielding crops (Silova & Desimoni, 2025). Diversity in crop varieties offers important characteristics like drought resistance, pest tolerance, nutritional improvement, salt tolerance, etc., which are necessary for sustainable food production. Biodiversity ensures the availability of multiple species in highly yielding fisheries and forestry sectors (Carpenter et al., 2012). Overexploitation of one species can lead to the collapse of an ecosystem and eventually lead to food insecurity as the world's population increases. The discovery of different pharmaceuticals, which are applied in various life-saving medicines, including antibiotics and anti-cancer drugs, can also be attributed to the diversity among microorganisms, plants, and animals. The potential for future discoveries also relies on our capability to conserve biodiversity (McKenzie et al., 2025).

Biodiversity is also responsible for promoting the ecotourism industry, upon which there is a heavy reliance for the livelihood of the local population. The economic value of the different ecosystem services, including clean water, pollination, air purification, soil formation, etc., which were underestimated in traditional

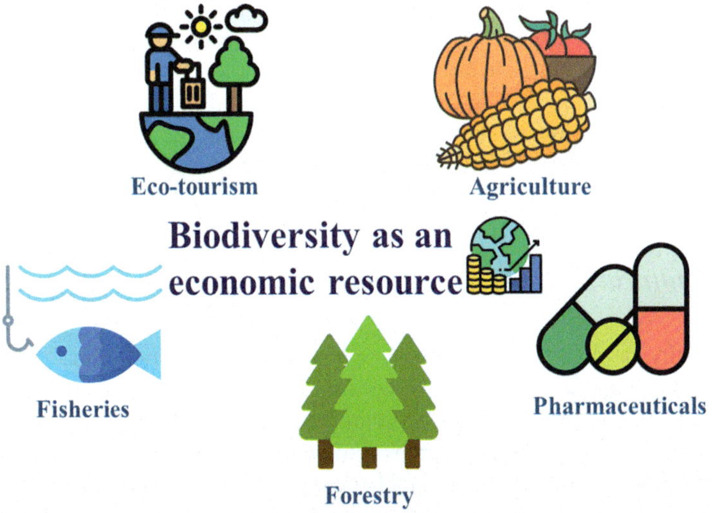

Fig. 3 Biodiversity: a foundation for economic sectors

financial assessments, is being evaluated by nations now and has a considerable value in the economies (Duraiappah et al., 2014).

9.3 Biodiversity and Food Security

The future of food production can only be secured by preserving genetic diversity. Improving crop varieties to withstand stressful climate change conditions, diseases, and pests relies heavily on the breeder's access to a diverse genetic material (Lambin et al., 2014). The increasing volatility in the climate makes it even more essential to maintain a broad genetic base for an uninterrupted supply of foods (Gibb et al., 2025). A diverse variety of plants and animals is responsible for fulfilling the dietary requirements of humans residing in different corners of the world, which may not have access to the usual resources. Therefore, conserving biodiversity is about preserving species and ensuring a healthy and resilient food system for humanity (Asaad et al., 2017).

9.4 Biodiversity and Climate Change

The role of biodiversity in climate change mitigation and adaptation is pivotal and intricate. Conserving forests, grasslands, peatlands, and wetlands significantly enhances carbon sequestration and reduces atmospheric greenhouse gases. Therefore, preserving Biodiversity has emerged as an effective strategy for combating climate change (Lohbeck et al., 2016). A biodiverse ecosystem offers a range of microclimates and habitats that enable species to adjust to changing climatic conditions (Kenter et al., 2011). The overall resilience of an ecosystem is enhanced with a biodiverse ecosystem, as mangrove forests protect coastal areas from storm surges. At the same time, diverse agricultural systems are more capable of withstanding climate-induced stresses. Thus, preserving Biodiversity is integral to building a climate-resilient future (Bannar-Martin et al., 2018).

9.5 Cultural, Ethical, and Aesthetic Significance

Biodiversity holds profound cultural, spiritual, and aesthetic values. Indigenous communities around the world have deep-rooted connections to their natural environments. Their knowledge systems, traditions, and cultural identities are intimately tied to the Biodiversity of their regions (Emmett Duffy, 2009). Conserving biodiversity is essential for preserving cultural heritage and indigenous rights. In many societies, biodiversity has religious and spiritual significance. Sacred groves, revered species, and traditional ecological knowledge illustrate the moral and spiritual dimensions of biodiversity conservation. Moreover, the aesthetic appreciation of nature, manifested through art, literature, and recreation, depends on the richness of biological diversity. Beyond cultural and aesthetic values lies an ethical responsibility to protect the Earth's biological wealth. As stewards of the planet, humans have a moral obligation to ensure the survival of other species and maintain the integrity of ecosystems for future generations (Tengö et al., 2014).

9.6 Scientific and Technological Contributions

Technological innovation and scientific exploration have been highly dependent on biological diversity as our understanding of life and biological processes aids in developing scientific principles (López-Feldman & Chávez, 2017). For example, biomimicry-technological innovation inspired by nature—has led to breakthroughs in engineering, materials science, and medicine. Genetic resources from diverse organisms are harnessed for developing new crops, medicines, and industrial products. Looking ahead, biodiversity holds immense potential for future technological developments. From the discovery of new antibiotics to the development of climate-resilient crops and bio-based industries, the biological wealth of the planet is a critical asset for addressing global challenges and advancing sustainable development (Farwig et al., 2008a).

10 Conclusion

Ecological equilibrium, human well-being, and sustainable development depend on biodiversity. This chapter examined biodiversity's ecological, economic, cultural, and scientific relevance, showing how interconnected humans and nature are. Our daily lives depend on biodiversity, which provides ecological services like pollination, water purification, climate management, and genetic resources for agriculture and medicine. However, habitat degradation, pollution, invasive species, and climate change are increasing biodiversity loss, necessitating immediate action. Integrated ecological and human techniques must be used to preserve biodiversity outside protected areas. Molecular markers, DNA banking, micropropagation, and cryopreservation provide new ways to conserve and restore endangered species. Successful conservation requires understanding traditional ecological knowledge and enabling fair benefit sharing across communities. In the face of global difficulties, biodiversity becomes a natural treasure and a key answer for climate resilience, food security, and sustainable economies. As stewards of Earth, we must protect its biological riches ethically and practically. A comprehensive, science-informed approach, inclusive policies, and public participation are needed to stop biodiversity loss and foster recovery for future generations. Protecting biodiversity is essential to humanity's future.

Acknowledgments The authors are thankful to the Department of Earth and Environmental Science, Parul Institute of Applied Sciences, Parul University, Vadodara for their support during the preparation of this chapter.

References

Ahmed, M., & Sarma, P. (2022). Small farmers and agricultural transformation: A sociological study in three villages of India. *Alexandria Science Exchange Journal, 43*(2), 325–342. https://doi.org/10.21608/ASEJAIQJSAE.2022.245692

Al-Samarai, F. R., & Al-Kazaz, A. A. (2015). Molecular markers: An introduction and applications. *European Journal of Molecular Biotechnology, 9*, 118–130. https://www.researchgate.net/publication/307717534_Molecular_Markers_an_Introduction_and_Applications

Amini, S., Sharafi, S., Komeili, H. R., & Tabaee, N. A. (2014). Effect of biotechnology on biodiversity. *International Journal of Farming and Allied Sciences*. www.ijfas.com

Asaad, I., Lundquist, C. J., Erdmann, M. V., & Costello, M. J. (2017). Ecological criteria to identify areas for biodiversity conservation. *Biological Conservation, 213*, 309–316. https://doi.org/10.1016/J.BIOCON.2016.10.007

Atreya, P. N., & Shrestha, J. (2020). Biodiversity of neglected and underutilized fruits of Nepal: A review. *Fundamental and Applied Agriculture, 5*(4), 470–483. https://doi.org/10.5455/FAA.122860

Bannar-Martin, K. H., Kremer, C. T., Ernest, S. K. M., Leibold, M. A., Auge, H., Chase, J., Declerck, S. A. J., Eisenhauer, N., Harpole, S., Hillebrand, H., Isbell, F., Koffel, T., Larsen, S., Narwani, A., Petermann, J. S., Roscher, C., Cabral, J. S., & Supp, S. R. (2018). Integrating community assembly and biodiversity to better understand ecosystem function: The community assembly and the functioning of ecosystems (CAFE) approach. *Ecology Letters, 21*(2), 167–180. https://doi.org/10.1111/ELE.12895

Bennett, V. J. (2017). Effects of road density and pattern on the conservation of species and biodiversity. *Current Landscape Ecology Reports 2017 2:1, 2*(1), 1–11. https://doi.org/10.1007/S40823-017-0020-6

Bhat, S. R., Chandel, K. P. S., & Malik, S. K. (1995). Plant regeneration from various expiants of cultivated piper species. *Plant Cell Reports, 14*(6), 398–402. https://doi.org/10.1007/BF00238605

Bridgen, M. P., Van Houtven, W., & Eeckhaut, T. (2018). *Plant tissue culture techniques for breeding* (pp. 127–144). Springer. https://doi.org/10.1007/978-3-319-90698-0_6

Butchart, S. H. M., Walpole, M., Collen, B., Van Strien, A., Scharlemann, J. P. W., Almond, R. E. A., Baillie, J. E. M., Bomhard, B., Brown, C., Bruno, J., Carpenter, K. E., Carr, G. M., Chanson, J., Chenery, A. M., Csirke, J., Davidson, N. C., Dentener, F., Foster, M., Galli, A., et al. (2010). Global biodiversity: Indicators of recent declines. *Science, 328*(5982), 1164–1168. https://doi.org/10.1126/SCIENCE.1187512/SUPPL_FILE/BUTCHART_SOM.PDF

Calado, H., Fonseca, C., Vergílio, M., Costa, A., Moniz, F., Gil, A., & Dias, J. A. (2014). Small Islands conservation and protected areas. *Revista de Gestão Costeira Integrada, 14*(2), 167–174.

Carpenter, S. R., Folke, C., Norström, A., Olsson, O., Schultz, L., Agarwal, B., Balvanera, P., Campbell, B., Castilla, J. C., Cramer, W., DeFries, R., Eyzaguirre, P., Hughes, T. P., Polasky, S., Sanusi, Z., Scholes, R., & Spierenburg, M. (2012). Program on ecosystem change and society: An international research strategy for integrated social–ecological systems. *Current Opinion in Environmental Sustainability, 4*(1), 134–138. https://doi.org/10.1016/J.COSUST.2012.01.001

Castellanos-Hernández, O. A., Rodríguez-Sahagún, A., Acevedo-Hernández, G. J., Rodríguez-Garay, B., Cabrera-Ponce, J. L., & Herrera-Estrella, L. R. (2009). Transgenic Paulownia elongata S. Y. Hu plants using biolistic-mediated transformation. *Plant Cell, Tissue and Organ Culture, 99*(2), 175–181. https://doi.org/10.1007/S11240-009-9590-2/METRICS

Catford, J. A. (2017). Hydrological impacts of biological invasions. In *Impact of Biological Invasions on Ecosystem Services* (pp. 63–80). https://doi.org/10.1007/978-3-319-45121-3_5

Chandra, S., Bandopadhyay, R., Kumar, V., & Chandra, R. (2010). Acclimatization of tissue cultured plantlets: From laboratory to land. *Biotechnology Letters, 32*(9), 1199–1205. https://doi.org/10.1007/S10529-010-0290-0/METRICS

Chawla, H. S., & Singh, A. K. (2009). Intellectual property rights in biotechnology. In *Plant tissue culture and moelcular markers*. https://books.google.co.in/books?hl=en&lr=&id=iEGmRvdlV38C&oi=fnd&pg=PA51&dq=Chawla,+2009+plant+tissue+culture&ots=ZawYnjxPhN&sig=AS9otNOfqSl_P23xFgkIJpfY-1E&redir_esc=y#v=onepage&q&f=false

China Forestry Publishing House. (2020). *National report on sustainable forest management*.

Chivian, E. (2002). *Biodiversity: Its Importance to Human Health A Project of the Center for Health and the Global Environment Harvard Medical School under the auspices of the World Health Organization, the United Nations Development Programme, and the United Nations Environment Programme*. www.med.harvard.edu/chge

Chivian, E., & Bernstein, A. (2008). *Sustaining life: How human health depends on biodiversity*. https://books.google.co.in/books?hl=en&lr=&id=jIc3AwAAQBAJ&oi=fnd&pg=PP1&dq=Biodiversity+Chivian+and+Bernstein+2008&ots=ScnxV3749K&sig=5tH3yX1ActN62Bdn3iueHasmVBI#v=onepage&q=Biodiversity%20Chivian%20and%20Bernstein%202008&f=false

Clark, N. E., Lovell, R., Wheeler, B. W., Higgins, S. L., Depledge, M. H., & Norris, K. (2014). Biodiversity, cultural pathways, and human health: A framework. *Trends in Ecology & Evolution, 29*(4), 198–204. https://doi.org/10.1016/J.TREE.2014.01.009

Daily, G. C. (1995). Restoring value to the world's degraded lands. *Science (New York, N.Y.), 269*(5222), 350–354. https://doi.org/10.1126/SCIENCE.269.5222.350

Duraiappah, A. K., Asah, S. T., Brondizio, E. S., Kosoy, N., O'Farrell, P. J., Prieur-Richard, A. H., Subramanian, S. M., & Takeuchi, K. (2014). Managing the mismatches to provide ecosystem services for human Well-being: A conceptual framework for understanding the new commons. *Current Opinion in Environmental Sustainability, 7*, 94–100. https://doi.org/10.1016/J.COSUST.2013.11.031

Dussert, S., Chabrillange, N., Engelmann, F., Anthony, F., & Hamon, S. (1997). Cryopreservation of coffee (*Coffea Arabica* L.) seeds: Importance of the precooling temperature. *Cryo-Letters, 18*, 269–276.

Eloy, L., Schmidt, I. B., Borges, S. L., Ferreira, M. C., & dos Santos, T. A. (2019). Seasonal fire management by traditional cattle ranchers prevents the spread of wildfire in the Brazilian Cerrado. *Ambio, 48*(8), 890–899. https://doi.org/10.1007/S13280-018-1118-8/METRICS

Emmett Duffy, J. (2009). Why biodiversity is important to the functioning of real-world ecosystems. *Frontiers in Ecology and the Environment, 7*(8), 437–444. https://doi.org/10.1890/070195

Farwig, N., Braun, C., & Böhning-Gaese, K. (2008a). Human disturbance reduces genetic diversity of an endangered tropical tree, Prunus africana (Rosaceae). *Conservation Genetics, 9*(2), 317–326. https://doi.org/10.1007/S10592-007-9343-X/METRICS

Farwig, N., Sajita, N., & Böhning-Gaese, K. (2008b). Conservation value of forest plantations for bird communities in western Kenya. *Forest Ecology and Management, 255*(11), 3885–3892. https://doi.org/10.1016/J.FORECO.2008.03.042

Fay, M. F. (1994). In what situations is in vitro culture appropriate to plant conservations? *Biodiversity and Conservation, 3*(2), 176–183. https://doi.org/10.1007/BF02291887/METRICS

Feng, C., Yin, Z., Ma, Y., Zhang, Z., Chen, L., Wang, B., Li, B., Huang, Y., & Wang, Q. (2011). Cryopreservation of sweetpotato (Ipomoea batatas) and its pathogen eradication by cryotherapy. *Biotechnology Advances, 29*(1), 84–93. https://doi.org/10.1016/J.BIOTECHADV.2010.09.002

Fernandes, G. W., Vale, M. M., Overbeck, G. E., Bustamante, M. M. C., Grelle, C. E. V., Bergallo, H. G., Magnusson, W. E., Akama, A., Alves, S. S., Amorim, A., Araújo, J., Barros, C. F., Bravo, F., Carim, M. J. V., Cerqueira, R., Collevatti, R. G., Colli, G. R., da Cunha, C. N., D'Andrea, P. S., et al. (2017). Dismantling Brazil's science threatens global biodiversity heri-

tage. *Perspectives in Ecology and Conservation, 15*(3), 239–243. https://doi.org/10.1016/J.PECON.2017.07.004
Futema, M., Plagno, V., Li, K. W., Whittall, R. A., Neil, H. A. W., Seed, M., Bertolini, S., Calandra, S., Descamps, O. S., Graham, C. A., Hegele, R. A., Karpe, F., Durst, R., Leitersdorf, E., Lench, N., Nair, D. R., Soran, H., van Bockxmeer, F. M., & Humphries, S. E. (2014). Whole exome sequencing of familial hypercholesterolaemia patients negative for LDLR/APOB/PCSK9 mutations. *Journal of Medical Genetics, 51*(8), 537–544. https://doi.org/10.1136/JMEDGENET-2014-102405
Gibb, R., Redding, D. W., Friant, S., & Jones, K. E. (2025). Towards a 'people and nature' paradigm for biodiversity and infectious disease. *Philosophical Transactions of the Royal Society B: Biological Sciences, 380*(1917). https://doi.org/10.1098/RSTB.2023.0259
Gizaw, S. (2011). *Characterization and conservation of indigenous sheep genetic resources: A practical framework for developing countries*. ILRI (aka ILCA and ILRAD). https://books.google.co.in/books?hl=en&lr=&id=yosH24rBKygC&oi=fnd&pg=PR4&dq=biodiversity+Gizaw+et.al,+2007&ots=ISEQZV9cAq&sig=KseC1P1b4p_RqGqtpAasDaJ90cE&redir_esc=y#v=onepage&q&f=false
Godard, B., Schmidtke, J., Cassiman, J. J., & Aymé, S. (2004). Data storage and DNA banking for biomedical research: Informed consent, confidentiality, quality issues, ownership, return of benefits. A professional perspective. European Journal of Human Genetics 2003 11:2*, 11*(2), S88–S122. https://doi.org/10.1038/sj.ejhg.5201114
Gómez-Baggethun, E., & Ruiz-Pérez, M. (2011). Economic valuation and the commodification of ecosystem services. *Progress in Physical Geography, 35*(5), 613–628. https://doi.org/10.1177/0309133311421708
Grout, B. W. W., Westcott, R. J., & Henshaw, G. G. (1977). Scanning electron microscope studies of multiple shoot production by meristem tip cultures of solarium × curtilobum. *Annals of Botany, 41*(6), 1113–1116. https://doi.org/10.1093/OXFORDJOURNALS.AOB.A085401
Guo, Y., Somerfield, P. J., Warwick, R. M., & Zhang, Z. (2001). Large-scale patterns in the community structure and biodiversity of freeliving nematodes in the Bohai Sea, China. *Journal of the Marine Biological Association of the United Kingdom, 81*(5), 755–763. https://doi.org/10.1017/S0025315401004568
Hai, R., Zhian, L., Weijun, S., Zuoyue, Y., Shaolin, P., Chonghui, L., Mingmao, D., & Jianguo, W. (2007). Changes in biodiversity and ecosystem function during the restoration of a tropical forest in south China*, 50*(2), 277–284. https://doi.org/10.1007/s11427-007-0028-y
Harrison, P. A., Berry, P. M., Simpson, G., Haslett, J. R., Blicharska, M., Bucur, M., Dunford, R., Egoh, B., Garcia-Llorente, M., Geamănă, N., Geertsema, W., Lommelen, E., Meiresonne, L., & Turkelboom, F. (2014). Linkages between biodiversity attributes and ecosystem services: A systematic review. *Ecosystem Services, 9*, 191–203. https://doi.org/10.1016/J.ECOSER.2014.05.006
Haslam, T. M., & Yeung, E. C. (2011). Zygotic embryo culture: An overview. *Methods in Molecular Biology, 710*, 3–15. https://doi.org/10.1007/978-1-61737-988-8_1
Karp, A., Seberg, O., & Buiatti, M. (1996). Molecular techniques in the assessment of botanical diversity. *Annals of Botany, 78*(2), 143–149. https://doi.org/10.1006/ANBO.1996.0106
Kelleher, C. T., Hodkinson, T. R., Douglas, G. C., & Kelly, D. L. (2005). Species distinction in Irish populations of Quercus petraea and Q. Robur: Morphological versus molecular analyses. *Annals of Botany, 96*(7), 1237–1246. https://doi.org/10.1093/AOB/MCI275
Kenter, J. O., Hyde, T., Christie, M., & Fazey, I. (2011). The importance of deliberation in valuing ecosystem services in developing countries—Evidence from the Solomon Islands. *Global Environmental Change, 21*(2), 505–521. https://doi.org/10.1016/J.GLOENVCHA.2011.01.001
Kyte, D., Ives, J., Draper, H., Keeley, T., & Calvert, M. (2013). Inconsistencies in quality of life data collection in clinical trials: A potential source of bias? Interviews with research nurses and Trialists. *PLoS One, 8*(10), e76625. https://doi.org/10.1371/JOURNAL.PONE.0076625
Lambin, E. F., Meyfroidt, P., Rueda, X., Blackman, A., Börner, J., Cerutti, P. O., Dietsch, T., Jungmann, L., Lamarque, P., Lister, J., Walker, N. F., & Wunder, S. (2014). Effectiveness

and synergies of policy instruments for land use governance in tropical regions. *Global Environmental Change, 28*(1), 129–140. https://doi.org/10.1016/J.GLOENVCHA.2014.06.007

Lausch, A., Bannehr, L., Beckmann, M., Boehm, C., Feilhauer, H., Hacker, J. M., Heurich, M., Jung, A., Klenke, R., Neumann, C., Pause, M., Rocchini, D., Schaepman, M. E., Schmidtlein, S., Schulz, K., Selsam, P., Settele, J., Skidmore, A. K., & Cord, A. F. (2016). Linking earth observation and taxonomic, structural and functional biodiversity: Local to ecosystem perspectives. *Ecological Indicators, 70*, 317–339. https://doi.org/10.1016/J.ECOLIND.2016.06.022

Liu, G., Zhou, H., Li, Y., Li, G., Cui, L., Liu, W., & Cao, Y. (2008). Evaluation of the viability and osteogenic differentiation of cryopreserved human adipose-derived stem cells. *Cryobiology, 57*(1), 18–24. https://doi.org/10.1016/J.CRYOBIOL.2008.04.002

Lohbeck, M., Bongers, F., Martinez-Ramos, M., & Poorter, L. (2016). The importance of biodiversity and dominance for multiple ecosystem functions in a human-modified tropical landscape. *Ecology, 97*(10), 2772–2779. https://doi.org/10.1002/ECY.1499

López-Feldman, A., & Chávez, E. (2017). Remittances and natural resource extraction: Evidence from Mexico. *Ecological Economics, 132*, 69–79. https://doi.org/10.1016/J.ECOLECON.2016.10.010

Lowe, A., & Harris, S. (2004). *Ecological genetics: Design, analysis, and application - Andrew Lowe, Stephen Harris*. Paul Ashton - Google Books. https://books.google.co.in/books?hl=en&lr=&id=W3UkvPORFRcC&oi=fnd&pg=PR5&ots=00PoHSYEOu&sig=2zgNQbRgcRu5p3m1LNb-lCXJVTc&redir_esc=y#v=onepage&q&f=false

Malabika, Pathak, R., Bahrain, O., & Pathak, M. R. (2017). Highly efficient plant regeneration of African daisy (Osteospermum hybrida) through somatic embryogenesis of leaf derived callus. *IOSR Journal of Biotechnology and Biochemistry (IOSR-JBB), 3*(6), 25–30. https://doi.org/10.9790/264X-03062529

Marshall, E., Keem, J. L., Penman, T. D., & Di Stefano, J. (2025). Simulating fuel management for protecting regional biodiversity under climate change. *Journal of Environmental Management, 373*, 123731. https://doi.org/10.1016/J.JENVMAN.2024.123731

Mauerhofer, V., Ichinose, T., Blackwell, B. D., Willig, M. R., Flint, C. G., Krause, M. S., & Penker, . M. (2018). Underuse of social-ecological systems: A research agenda for addressing challenges to biocultural diversity. *Land Use Policy, 72*, 57–64. https://doi.org/10.1016/J.LANDUSEPOL.2017.12.003

Maxted, N., Erskine, W., Robertson, L. D., & Asthana, A. N. (2000). Are our germplasm collections museum items? In *Linking research and marketing opportunities for pulses in the 21st century. Current plant science and biotechnology in agriculture* (pp. 589–602). Springer. https://doi.org/10.1007/978-94-011-4385-1_56

McKenzie, E. J., Jones, M., Seega, N., Siikamäki, J., & Vijay, V. (2025). Science and technical priorities for private sector action to address biodiversity loss. *Philosophical Transactions of the Royal Society B: Biological Sciences, 380*(1917). https://doi.org/10.1098/RSTB.2023.0208

Miller, A. G., & Nyberg, J. A. (1995). *Collecting herbarium vouchers*. Royal Botanic Garden.

Moorhouse, T. P., & Macdonald, D. W. (2015). Are invasives worse in freshwater than terrestrial ecosystems? *Wiley Interdisciplinary Reviews: Water, 2*(1), 1–8. https://doi.org/10.1002/WAT2.1059

Mora, C., & Sale, P. F. (2011). Ongoing global biodiversity loss and the need to move beyond protected areas: A review of the technical and practical shortcomings of protected areas on land and sea. *Marine Ecology Progress Series, 434*, 251–266. https://doi.org/10.3354/MEPS09214

Nam, V. T., Le Bich, P., Hang, N. N., Linh, L. H., Ly, H. T., Hue, T., Ha, N. H., Hanh, H. H., & Hien, L. T. T. (2020). Molecular markers for analysis of plant genetic diversity. *Journal of Biotechnology, 18*, 589–608.

NBC. (2025). *What is biodiversity?* National Biodiversity Centre. https://nbc.gov.bt/what-is-biodiversity/

Pathak, M. R., & Abido, M. S. (2014). The role of biotechnology in the conservation of biodiversity. *Journal of Experimental Biology and Agricultural Sciences, 2*. https://www.researchgate.net/publication/265126076_The_Role_of_Biotechnology_in_the_Conservation_of_Biodiversity

Patz, J., Corvalan, C., Horwitz, P., & Campbell-Lendrum, D. (2012). *Our planet, our health, our future: Human health and the Rio conventions: Biological diversity, climate change and desertification*. https://www.researchgate.net/publication/269334236_Our_Planet_Our_Health_Our_Future_Human_health_and_the_Rio_Conventions_biological_diversity_climate_change_and_desertification

Paunescu, A. (2009). Biotechnology for endangered plant conservation: A critical overview. *Romanian Biotechnological Letters, 14*(1), 4095–4103. http://cms.iucn.org/about/work/programmes/species/red_list

Quave, C. L., & Pieroni, A. (2015). A reservoir of ethnobotanical knowledge informs resilient food security and health strategies in the Balkans. *Nature Plants, 1*(2), 1–6. https://doi.org/10.1038/nplants.2014.21

Rajasekharan, P. E., & Sahijram, L. (2015). In vitro conservation of plant germplasm. In *Plant biology and biotechnology: Volume II: Plant genomics and biotechnology* (pp. 417–443). https://doi.org/10.1007/978-81-322-2283-5_22

Reichstein, M., Bahn, M., Mahecha, M. D., Kattge, J., & Baldocchi, D. D. (2014). Linking plant and ecosystem functional biogeography. *Proceedings of the National Academy of Sciences of the United States of America, 111*(38), 13697–13702. https://doi.org/10.1073/PNAS.1216065111/SUPPL_FILE/PNAS.1216065111.SAPP.PDF

Sadeq, M. A., Pathak, M. R., Salih, A. A., Abido, M., & Abahussain, A. (2014). Somatic embryogenesis and organogenesis for regeneration of endangered multipurpose desert plant *Leptadenia pyrotechnica* Forsk. Decne in the Kingdom of Bahrain. *American Journal of Plant Sciences, 2014*(15), 2342–2353. https://doi.org/10.4236/AJPS.2014.515248

Sarasan, V., Cripps, R., Ramsay, M. M., Atherton, C., McMichen, M., Prendergast, G., & Rowntree, J. K. (2006). Conservation in vitro of threatened plants - Progress in the past decade. *In Vitro Cellular and Developmental Biology - Plant, 42*(3), 206–214. https://doi.org/10.1079/IVP2006769/METRICS

Sharma, N., & Chandel, K. P. S. (1996). In vitro conservation of Orchis latifolia: A threatened, medicinal terrestrial orchid. *Indian Journal of Plant Genetic Resources, 9*(1), 109–113. https://www.indianjournals.com/ijor.aspx?target=ijor:ijpgr&volume=9&issue=1&article=012

Silova, I., & Desimoni, V. (2025). Promoting food security and biodiversity restoration: Insights from Kenyan youth climate change activists. *Youth, 5*(1), 16. https://doi.org/10.3390/YOUTH5010016

Stacey, G. N., & Day, J. G. (2007). Long-term ex situ conservation of biological resources and the role of biological resource centers. *Methods in Molecular Biology (Clifton, N.J.), 368*, 1–14. https://doi.org/10.1007/978-1-59745-362-2_1

Strayer, D. L., & Dudgeon, D. (2010). Freshwater biodiversity conservation: Recent progress and future challenges. *Journal of the North American Benthological Society, 29*(1), 344–358. https://doi.org/10.1899/08-171.1

Tattoni, C., Ianni, E., Geneletti, D., Zatelli, P., & Ciolli, M. (2017). Landscape changes, traditional ecological knowledge and future scenarios in the Alps: A holistic ecological approach. *Science of the Total Environment, 579*, 27–36. https://doi.org/10.1016/J.SCITOTENV.2016.11.075

Tengö, M., Brondizio, E. S., Elmqvist, T., Malmer, P., & Spierenburg, M. (2014). Connecting diverse knowledge systems for enhanced ecosystem governance: The multiple evidence base approach. *Ambio, 43*(5), 579–591. https://doi.org/10.1007/S13280-014-0501-3/FIGURES/2

Thompson, I., Mackey, B., Mcnulty, S., & Mosseler, A. (2009). *Forest resilience, biodiversity, and climate change* (Technical Series, No. 43) (pp. 1–67). Secretariat of the Convention on Biological Diversity. https://research.fs.usda.gov/treesearch/36775

Uchida, K., Blakey, R. V., Burger, J. R., Cooper, D. S., Niesner, C. A., & Blumstein, D. T. (2021). Urban biodiversity and the importance of scale. *Trends in Ecology & Evolution, 36*(2), 123–131. https://doi.org/10.1016/J.TREE.2020.10.011/ASSET/8D31C099-2CD2-4F79-96BF-C31AC6F65385/MAIN.ASSETS/B2.SML

UNEP. (1997). *UNEP launches World Atlas of desertification | Meetings coverage and press releases*. https://press.un.org/en/1997/19971009.unep11.html

Viglizzo, E. F., Nosetto, M. D., Jobbágy, E. G., Ricard, M. F., & Frank, F. C. (2015). The ecohydrology of ecosystem transitions: A meta-analysis. *Ecohydrology, 8*(5), 911–921. https://doi.org/10.1002/ECO.1540

Waldren, S., Weisler, M. I., Hather, J. C., & Morrow, D. (1999). The non-native vascular plants of Henderson Island, South Central Pacific Ocean. *Atoll Research Bulletin, 463*(459). https://doi.org/10.5479/si.00775630.463.1

Wiens, J., Sutter, R., Anderson, M., Blanchard, J., Barnett, A., Aguilar-Amuchastegui, N., Avery, C., & Laine, S. (2009). Selecting and conserving lands for biodiversity: The role of remote sensing. *Remote Sensing of Environment, 113*(7), 1370–1381. https://doi.org/10.1016/J.RSE.2008.06.020

Zhao, Z. G., Li, W. M., Chen, Z. C., You, Y., & Zou, P. (2008). Hematopoiesis capacity, immunomodulatory effect and ex vivo expansion potential of mesenchymal stem cells are not impaired by cryopreservation. *Cancer Investigation, 26*(4), 391–400. https://doi.org/10.1080/07357900701788049

Biodiversity and Its Services; Role in Human Welfare and Ecosystem Stability

Zareena Batool, Shujaul Mulk Khan, Abdullah Abdullah, Qurat Ul Ain, Nazir Mohammad, Muhammad Shakeel Khan, Maria Shah, Jawad Hussain, and Shakil Ahmed Zeb

Abstract Biodiversity encompasses a variety of life on Earth and supports the functioning of ecosystems that deliver essential services for human existence and quality of life. It has been recognized that biodiversity plays a critical role in maintaining ecological balance, improving ecosystem resilience, and supporting human well-being by providing various ecosystem services. Ecosystem services are benefits from ecosystems that directly or indirectly enhance humans. The ecosystem services concept was popularized by the Millennium Ecosystem Assessment (MEA), which categorized these services into four broad types, namely provisioning services, regulating services, supporting services, and cultural services. Provisioning services include services that fulfill material needs like food, medicine, fodder, fuel, fiber, biochemical, drinking water, etc. Regulating services include process services such as climate, carbon sequestration, erosion, diseases, pests, and water regulation, etc. These regulating services reduce the effects caused by both anthropogenic and natural disasters. Supporting services include a process that enables the provision of all other services, such as pollination, nutrient cycling, and soil formation. Cultural services enhance aesthetics, spirituality, recreation, and cultural heritage value. However, over the past few decades, humans have been changing ecosystems more rapidly than in any comparable period in history. Biodiversity is at extreme risk due to habitat loss, climate change, over-exploitation, pollution, invasive species, and unsustainable use. These induce ecosystem services, diminishing their ability to support human needs. It is essential to identify the importance of ecosystem services, especially their complicated relationship with

Z. Batool · Q. U. Ain · N. Mohammad · M. S. Khan · M. Shah · J. Hussain · S. A. Zeb
Department of Plant Sciences, Quaid-i-Azam University, Islamabad, Pakistan

S. M. Khan (✉)
Department of Plant Sciences, Quaid-i-Azam University, Islamabad, Pakistan

Pakistan Academy of Sciences, Islamabad, Pakistan

A. Abdullah
Department of Plant Sciences, Quaid-i-Azam University, Islamabad, Pakistan

Kunming Institute of Botany, Kunming, Yunnan, China

S. Siddiqui et al. (eds.), *Frontiers in Sustainability*,
https://doi.org/10.1007/978-3-032-19740-5_2

social and ecological systems. Additionally, it also focuses on how the trade-offs of this interaction affect human well-being in the present. Understanding the complex relations between biodiversity and ecosystem services is important for developing holistic approaches to effective environmental management. Building worldwide collaboration and educating the public are essential to addressing these challenges effectively. This chapter emphasizes the critical role of biodiversity in sustaining the ecosystem's health and providing various ecosystem services while underscoring the major challenges and providing effective conservation strategies to ensure ecological balance and sustainable development for present and future generations.

Keywords Millennium Ecosystem Assessment · Conservation · Supporting services · Medicinal plants

1 Introduction

The term biodiversity represents the variation found in living things across their habitats, differences between species, and within ecosystems (Rawat & Agarwal, 2015). The Parties to the Convention on Biological Diversity (CBD) explain biodiversity as the "variability among all living organisms from all sources which include terrestrial, marine, and other aquatic ecosystems and their respective ecological ecosystems where species diversity comprises both within-species, between-species and ecosystem diversity." As part of the CBD definition, it has several distinct components. The definition specifically highlights variability as an essential feature, but it also interprets this as a measurement of variation extent rather than deterministic features, suggesting spatial or temporal elements. This inclusive definition therefore supports multiple kinds of diversity assessment (including abundance and phylogenetic approaches). The definition of the CBD includes three distinct levels of variability, starting from population genetics measures within species groups and extending to species diversity across taxa and ecosystem variability across major biomes or vegetation types. Additionally, the theory includes ecological variability, which emerges from the role of species within complex ecosystems. This recognizes the interactive nature of biodiversity through ecological processes. The proposed framework omits consideration of abundance-based metrics, which are relevant for describing ecosystem services despite their connection to biodiversity (Mace, 2005).

Humans always have intrinsic relationship with the ecosystem in which they live and drive numerous benefits known as ecosystem services (Ndayizeye et al., 2020). Ecosystem services can be defined as the environmental processes, functions, or characteristics that directly and indirectly enhance human well-being. The Ecosystem services concept was popularized by the Millennium Ecosystem Assessment (MEA), which categorized these services into four broad categories, i.e., provisioning, supporting, cultural, and regulating services. Provisioning services include services that fulfill material needs like food, medicine, fodder, fuel, fiber, biochemical, drinking water, etc. Regulating services include process services such as climate, erosion, diseases, pests, and water regulation, etc. These regulating services reduce the effects caused by both anthropogenic and natural disasters. Supporting services include a process that enables the provision of all other services. Cultural services include services that enhance aesthetics, spirituality, recreation, and cultural heritage value (MEA, 2005) (Fig. 1). These services are received by humans in many ways across space and time, often without acknowledging that these are freely provided by nature. It has become a widely adopted framework to inform policymakers and land managers in understanding the connection between human well-being and ecosystem functions. There is growing recognition of the need to include the evaluation of sociocultural criteria in the evaluation of ecosystem services, specifically to ensure the success of conservation projects (Ndayizeye et al., 2020). This concept highlights the strong bonds between humans and essentials with biodiversity, recognizing the importance of plants that support the survival and growth of species (Cheminal et al., 2022). The Convention on Biological Diversity (CBD) emphasizes the significance of recognizing and demonstrating these benefits to ensure that the policymaking process considers the significance of natural resources and biodiversity (Thomaz, 2023). It is essential to identify the importance of ecosystem services, especially their complicated relationship with social and ecological systems. Additionally, it also focuses on how the exchange of

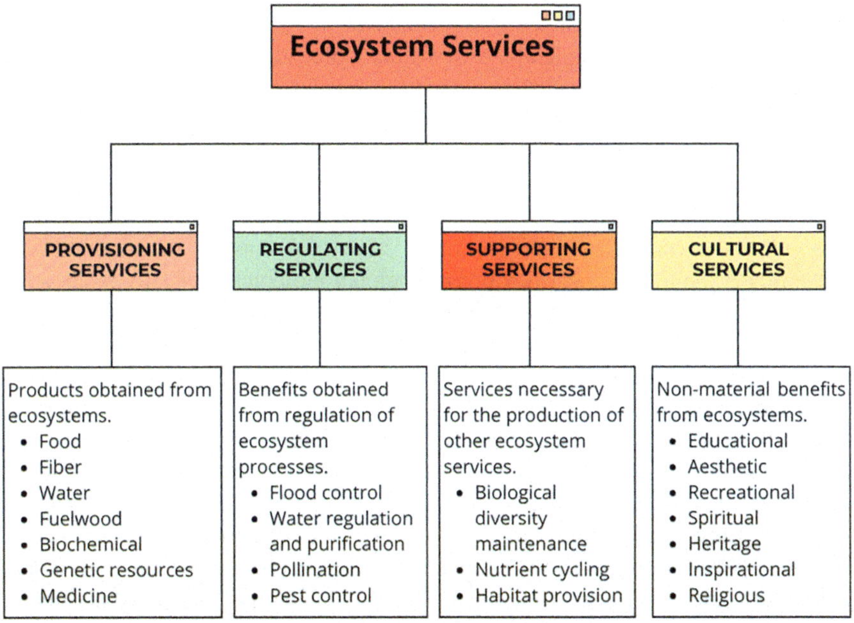

Fig. 1 Classification of ecosystem services

this interaction cause impact on human welfare in current scenario (García-Pardo et al., 2022).

The Ecosystem services framework distinguishes that humans and diversity of plants in their culture and economy are characterized as a vital part of the ecosystem. This interrelated connection between human actions and the ecosystem forms a sustainability growth that highlights the dependency of humans on their environment. It emphasizes the need to evaluate human and ecosystem prosperity together. For the assessment of the relative significance of ESs among a community and to ensure their maintenance and management, it is necessary to determine and identify ecosystem services. Understanding the connection between ecosystem services and the services for people is crucial for protecting resources, especially those vital for populations at risk of global change (Maroyi, 2022). A sustainable society is only achieved when both the conditions of humans and the ecosystem improve mutually or favorably (Cheminal et al., 2020).

2 Importance of Biodiversity in Providing Ecosystem Services

All life forms on Earth, including human beings, rely on ecosystem services to survive. These include a variety of services such as the supply of food and energy, protection against extreme environmental conditions, better air and water quality, and the provision of vast cultural and aesthetic values. When ecosystems that supply these services experience functional breakdown, these valuable services become either degraded or lost completely. A well-functioning ecosystem has two core components: the diversity of species that inhabit it and the ecological interaction among those species, alongside their relationships to environmental elements. Ecosystem services and biological diversity are inextricably linked. Biodiversity maintains the vital ecological functions that sustain ecosystem services while presenting itself as either a different ecosystem service, such as providing genetic materials for drug development, or directly providing intrinsic value to humans due to its unique range of species (Lipton et al., 2018). Currently, global biological diversity is declining due to habitat fragmentation and deforestation, which pose significant threats to endangering ecosystem stability and functionality. Ecosystem functions operate through biological mechanisms that control energy and fluxes of matter at diverse spatial and temporal scales. Biodiversity represents a wide variety of living organisms that serve as the critical drivers of these functions. Scientific research and debate continue to study the precise mechanism through which biodiversity affects ecosystem functioning (Lohbeck et al., 2016). All benefits nature provides to humanity are included under the classification of ecosystem services (ESs). Most Ecosystem Services derive from Biodiversity, which serves as the principal foundation of the majority of ESs, while their functioning and services depend directly on it. Ecosystems provide diverse services for human welfare while maintaining biodiversity as their foundation (Huang et al., 2022). The primary way by which Human society receives its benefits from ecosystems is through the functional properties of healthy ecosystems that produce goods and services. Biodiversity plays a fundamental role in ecosystem services by directly producing goods and sustaining ecosystem functions (EFs) (Harrison et al., 2014).

Sometimes it is possible to identify the same areas of high significance for both biodiversity conservation and delivery zones for ecosystem services, which presents potential integrated management possibilities. The recognition of this concept at the policy level has resulted in a greater focus on whole-ecosystem approaches to protect biodiversity. Nevertheless, researchers continue to debate how well ES frameworks work for biodiversity conservation while acknowledging our limited understanding of B–ES linkages. The relationship between biodiversity and individual ecosystem service provision shows inconsistent results when measured on a single linear positive scale. Research findings show that B–ES relationships appear in three distinct ways: (i) exhibit mixed relationships, (ii) take on different forms and shapes (e.g., nonlinear relationships), or (iii) do not exist at all. Scientists have discovered mixed B–ES relationships among individual ES because biodiversity affects specific services differently in diverse ecological settings. Moreover, differences in the B-ES interactions among individuals may lead to both synergies and trade-offs between numerous ES (Duncan et al., 2015). The vital importance of biodiversity to support ecosystem services makes forest degradation lead to considerable impacts, which increase susceptibility to natural and anthropogenic disturbance. These disturbances become more frequent and severe, leading to biodiversity decline, which ultimately diminishes forest resilience to climate extremes and pests and pathogens and invasive species, and other disturbance factors. Consequently, forest ecosystems will diminish their ability to provide ecosystem services because of this decline (Vilà & Hulme, 2017).

The provision of specific ecosystem services at various scales is thought to be positively impacted by biological diversity at species and population levels and is closely linked to ecosystem functioning. Simultaneously, both biodiversity and ecosystem functioning are shaped by interactions between individuals or species, which directly rely on the availability and conditions of habitats. For instance, increases in community and habitat area improve the ecosystem services that improve water flow (i.e., flood protection) and water quality (i.e., water purification) (Harrison et al., 2014). Additionally, biodiversity is believed to contribute to the temporal stability of ecosystem services, which is particularly relevant in ecosystems that are supposed to serve multiple functions. Thus, there is growing concern regarding how biodiversity loss would affect ecosystem integrity as well as human well-being and livelihood. Several initiatives promote the ecosystem service concept, which tries to integrate natural and social systems to provide a more comprehensive strategy for decision-making in this context where biodiversity is being linked to human well-being (Díaz et al., 2015). The Millennium Ecosystem Assessment (MEA, 2005), which examined the status of ecosystems and their potential to support human well-being, came to the conclusion that the majority of wetlands worldwide had been destroyed or degraded during the twentieth century. This highlights the urgent need for integrative frameworks that take into account the dynamics of entire systems.

The central framework for this assessment was based on a simple conceptual guiding principle:

Biodiversity ? ecosystem functioning ? ecosystem services ? human well ? being

where each arrow denotes a causal relationship, and ecosystem services (hence after ES) can be thought of as ecosystem function that eventually benefits humans. This framework is based on the assumption that a greater level of biodiversity enhances ecosystem functioning to some degree, which raises ES and, depending on the factors involved, may ultimately improve human well-being (Fig. 2). Many studies have tried to make an explicit or implicit connection between the stability of ecosystem functioning and the biological composition of ecosystems, as indicated by biodiversity proxies. These studies assumed that such relationships might play an important function in the delivery of ES. In general terms, scientists have assumed that if biodiversity in ecosystems and functioning are associated, then ES and human well-being are directly related (Pinto et al., 2014).

Ecosystem services emerged as a policy tool for preserving biodiversity largely as a result of the influence of the Convention on Biological Diversity's global strategic plan 2011–2020 (Aichi biodiversity targets), before there was a clear scientific consensus regarding the interdependence of ecosystem services and biodiversity had been fully established. Still today, there is no agreement on the nature of the connections and how they work, despite the abundance of evidence demonstrating a

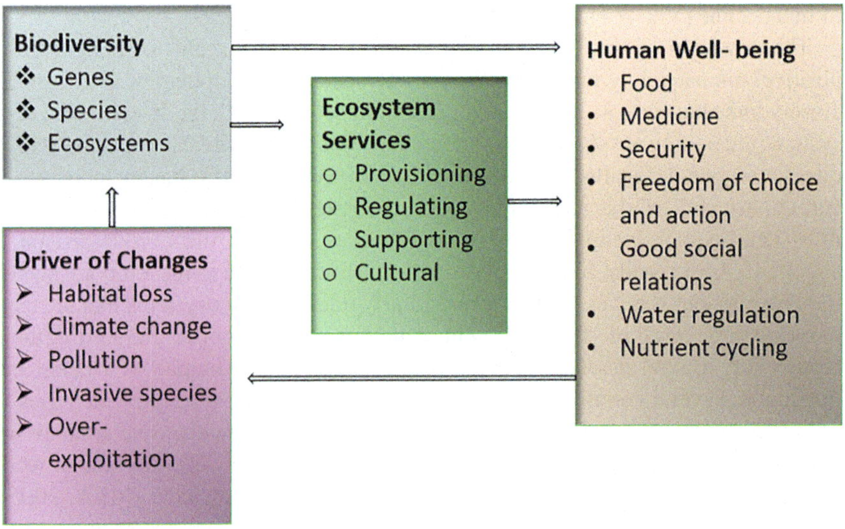

Fig. 2 Conceptual framework showing the relationship between biodiversity, ecosystem services, human well-being, and drivers of change to biodiversity

positive relationship between biodiversity, ecosystem functions, and the delivery of specific ecosystem services (Liquete et al., 2016).

2.1 Provisioning Services of Biodiversity

Provisioning services encompass the materials and energy extracted from ecosystems. Many of these services, which are among the most utilized by humans, include food, fiber, water, and medicinal products (Zeb et al., 2026). Human wellness depends on the reliable supply of food, water, and other vital provisioning services (Everard, 2018). All people, whether in urban or rural areas, wealthy or poor, and living in the Global South or the Global North, depend to some extent on the provision of ecosystem services, including food, water, medicine, timber, and fiber (Shackleton, 2021).

The most important bond between ecosystem services and human well-being arises from ecosystems providing goods through their provisioning services. Ecosystem provision delivers products to people, including food products, fibers, various fuels from peat, as well as wood and non-woody biomass, along with water supplies from lakes, rivers, and aquifers. Plantation forests and agroecosystems are examples of highly controlled ecosystems that may supply these things. Therefore, human activities are focused on providing services. Ecosystems and people will be at risk when humans take too many provisioning services from nature. The sustainability of ecosystems provides vital benefits that sustain human well-being by supplying clean air for breathing, drinking water, medicines, industrial materials, and agricultural products. Ecosystem service decline and depletion cause adverse impacts on human life quality. The decline of ecosystem provisioning services drives producers to use more inputs while they attempt to compensate for reduced outputs from these ecosystems. The human health risk will also grow because of provisioning ecosystem service degradation and loss (Deng et al., 2013).

2.1.1 Food

Food provision forms the basic link between human health and biodiversity. The human population mainly obtained the food either by farming practice or through harvesting the naturally occurring wild population (Batool et al., 2025). In the widest sense, biodiversity is the source of all the existing food we eat today and has the potential to produce novel foods in the future. There are strong connections between food supply and biodiversity. All the food species, whether they are hunted or collected from a population (Ullah et al., 2024), through minimal management, or produced in highly intensive production systems, exist within ecosystems, and within that ecosystem, their productivity is influenced by the activity of other components of biodiversity (Wilby et al., 2009).

Biodiversity plays an important role in Food security (Abdullah et al., 2021), and the broader provision of essential goods and services. This is especially true for many indigenous populations' livelihoods, as they have limited availability of external production inputs (Thorn et al., 2020).

Biodiversity is essential to the survival of the world's food supplies, which sustain all living forms, including humans and animals. All food-related industries and services are built on it. Human food forms that come from biodiversity resources include meat, nuts, vegetables, fruits, and food additives like flavorings, colorants, and preservatives. The source of these foods can be farmed or wild. Approximately 12,500 out of 300,000 flowering plant species are thought to be edible to humans; however, a significantly greater number may occasionally be used, including roughly 200 species that have been domesticated for food. But more than 75% of the food that humans eat now comes from 12 different plant species, either directly or indirectly (bananas, beans, cassava, maize, millet, potatoes, rice, sorghum, soybean, sugar cane, sweet potatoes, and wheat). Although it is more challenging to list the diversity of animals that are used for food, many of them provide goods for consumption (such as meat and milk). Animals that are frequently used by humans directly or indirectly include insects (moths, bees, wasp, beetles), crustaceans (lobsters, crabs, and shrimp), mollusks (bivalves, gastropods, and squid), echinoderms (sea urchins, and sea cucumbers) and vertebrates (fish, amphibians, reptiles, birds, and mammals). Many rural communities rely on wild-harvested meat for 30–80% of their protein intake, primarily due to the lack of cultivated alternative protein sources (Muluneh, 2021).

2.1.2 Medicine

Medicinal plants not only provide access to affordable medicine (Ullah et al., 2025), but they also offer a creative and alternative treatment with amazing potential for creating jobs and revenue. Plant products are employed as raw materials in the creation of modern medicine in addition to being utilized in traditional healthcare (Mir et al., 2021). Only a small portion of all known species have had their pharmacological potential investigated to date, making biodiversity conservation crucial for drug discovery and development because their loss eliminates as-yet-unidentified genetic and natural chemical resources. A potentially significant medication is thought to be lost every 2 years, due to biodiversity loss, and 21% (~15,000) of known medicinal plant species are in danger of going extinct. Similarly, the majority of known antibiotics (such as streptomycin and penicillin) and antifungal drugs (like Griseofulvin) were discovered by soil fungi. About 7% (26,000) of the known vascular plant species are thought to have been used medicinally (Linhares et al., 2023). More than 30% of novel drugs produced worldwide are derived from natural compounds found in plants and microorganisms that have been identified by science and/or traditional knowledge. For example, more than 70% of cancer treatments are derived from naturally occurring substances (Newman & Cragg, 2020).

2.2 Regulating Services of Biodiversity

The regulating services of biodiversity encompass functions like climate regulation, soil erosion, waste treatment, purification of water, pollination, and pest regulation (Mohri et al., 2013). Several ecological processes contribute to the operation of the ecosystem, particularly the regulation of the environment in which people live and work, and this leads to the development of regulating services. This regulation establishes the average and variance under such circumstances (Quijas & Balvanera, 2013). One of the most important ways to combat climate change is to maintain climate regulation ecosystem services (CRES), but changes in land use and land cover (LULC) have a significant impact on the provision of these services. Land use and land cover change, particularly deforestation and agricultural practices, are responsible for about 30% of global carbon emissions. These factors have led researchers to look for land management strategies that can provide improved climate regulation services; this is known as "climate-smart" landscape management (Campos et al., 2022). The implementation of agricultural diversification methods leads to biodiversity improvements, which enable essential supporting and regulating ecosystem services that ultimately determine crop production levels. Both crop and noncrop diversification effectively increased the delivery of pest management and pollination services. Water regulation refers to water quality and quantity (Tamburini et al., 2020).

2.2.1 Climate Regulation

Climate regulation deals with greenhouse gas dynamics, but carbon sequestration involves the long-term storage of carbon (Tamburini et al., 2020). The rising temperatures and altered precipitation patterns of climate change have caused substantial ecological disruption throughout the world's ecosystems. Climate change consequences have driven international organizations to adopt the Paris Agreement (Rockström et al., 2017) as a means to control its effects on biosphere health and human welfare. Climate regulation stands as one of the essential ecosystem services, yet forests function as their core components. The regulation of climate happens through forest ecosystems that trap carbon, which decreases global warming (Ali et al., 2023). Scientists use the term carbon sequestration to describe the long-term process where carbon dioxide and various other carbon elements are physically trapped for extended periods to minimize atmospheric carbon levels. Carbon sequestration is the storage of carbon, while climate regulation is the dynamics of greenhouse gases (Tamburini et al., 2020). In grasslands, the mechanisms of carbon sequestration, carbon storage as soil organic matter, and greenhouse gas fluxes are closely related. It is commonly known that intensifying grassland management through higher nutrient inputs, particularly nitrogen, results in increased carbon sequestration. However, increased emissions of greenhouse gases other than CO_2 could offset the climate mitigation impact of improved management. Due to the rising effects of climate change, global temperature and precipitation have been changing, which has ultimately disrupted entire ecosystems worldwide. The serious consequences of climate change for the biosphere and human welfare have prompted international cooperation in the creation of potential countermeasures, such as the Paris Agreement (Rockström et al., 2017). One of the most significant ecosystem services of forests and forestry is climate regulation. Numerous ecosystem services are produced by healthy forests, including the regulation of climate through carbon sequestration (a method for capturing and storing carbon dioxide and other forms of carbon over an extended period to slow down global warming) (Vauhkonen & Packalen, 2018).

2.2.2 Erosion Prevention

Recently, there has been an effort to comprehend how biodiversity affects ecosystem services, including physical services like erosion management. Root biomass output can be enhanced by plant diversification, according to studies. Soil erosion prevention is a proposed ecosystem service offered by biodiversity since it is commonly believed that higher root biomass in different systems gives soils stability (Allen et al., 2016). By stabilizing the soil and lowering water runoff, permanent vegetation prevents soil erosion. If permanent grasslands are not overgrazed or poorly maintained, they can significantly reduce soil erosion. Grassland vegetation's ability to prevent erosion is closely linked to other benefits like soil fertility, carbon sequestration (Ali et al., 2022), and water supply and management (Bengtsson et al., 2019). Forests efficiently control soil losses from ecosystems by shielding the soil from wind and water erosion. Forests have a particularly significant influence given the lengthy time scales involved in soil development (Altieri et al., 2018).

2.3 Supporting services of Biodiversity

Supporting services of biodiversity include processes that enable the provision of all the other services, such as pollination, nutrient cycling, primary production, habitat maintenance, and soil formation. Supporting services are the necessary ecological processes that maintain other ecosystem services. Supporting services are needed for the building of all other ecosystem services, such as habitat provisioning and soil formation (Guo et al., 2010). Unlike provisioning, regulating, or cultural services, their effect on human well-being is typically indirect or long-term. Although not directly observable, they support services such as food and water supply. For example, soil development enables agriculture, and photosynthesis sustains oxygen levels, but their impacts take place over long periods. Some services, such as erosion prevention, can be both supporting and regulating, depending on the speed at which their effects are realized (Oguh et al., 2021).

2.3.1 Pollination

Pollination, aided by insects, birds, mammals, wind, and water, is crucial for crop productivity and the genetic diversity of crops and wild plants (Rao, 2018). Biodiversity declines, especially in pollinators, may have negative effects on crop production and decrease the types of nutrients supplied in human diets (Sharma & Birman, 2024). Human activities create habitat disturbances that allow invasive species to spread through the provision of dispersal pathways and increasing competition with indigenous vegetation (Jehangir et al., 2024a). Thousands of species have been deliberately or inadvertently moved outside their native ranges through human activities (Jehangir et al., 2024b). Successful conservation of biodiversity needs detailed knowledge of a region's phytogeographic patterns and total floristic composition (Ejaz et al., 2024).

2.3.2 Nutrient Cycling

Plants and soil microorganisms play a major role in nutrient cycling by decomposing organic matter and enhancing the amount of nutrients in the soil (Rao, 2018). Biodiversity is essential to ecosystem services that support human well-being. It contributes to supporting nutrient cycling and soil development (Israr et al., 2024). These services are vital for survival and economic progress (Rathoure & Kumar, 2024).

2.3.3 Soil Formation

Soils are at the nexus of global biogeochemical cycling of carbon, nutrients, and water and the highest terrestrial biodiversity. Consequently, they are the basis of key ecosystem services. Soil processes can be controlled to improve one service with a subsequent gain in benefits or, on the other hand, trade-offs against other services (Smith et al., 2015). Biodiversity underpins ecosystem services that are vital to human well-being. It has supporting soil formation that is essential for human needs (Rathoure & Kumar, 2024). Forest loss has enormously accelerated soil erosion, which affects approximately 11 million km^2 worldwide, an area roughly the size of the USA and Mexico, with high erosion levels. Annually, about 75 billion tons of soil are lost, eroding up to 40 times faster than natural formation. By the late twentieth century, one-third of arable land had already degraded due to erosion, contributing to reduced yields, economic loss, and food insecurity. Soil, a slowly renewable resource, is shaped by topography, parent material, age, biota, vegetation, and anthropogenic activities. Unsustainable agriculture quickly drains nutrients, especially in less than 10 years. In arid areas, substituting deep-rooted indigenous vegetation with shallow-rooted crops raises the water table, leading to salinization and soil degradation. Soil sustains essential ecosystem processes like water regulation, nutrient cycling, decomposition of organic matter, renewal of fertility, and structural support. It also enables freshwater recharge by increasing infiltration and decreasing runoff. Its mineralogy and texture determine nutrient retention, with clay soils retaining nutrients better than sandy soils (Sekercioglu, 2010). Ecosystems minimize soil erosion through cover and root vegetation, root architecture, soil fauna, and surface roughness, such as rocks. Dense vegetation fixes the soil, and reduce soil erosion that casued by wind and water (Rao, 2018).

2.4 Cultural Services of Biodiversity

Cultural services enhance aesthetics, spirituality, recreation, and cultural heritage value. Over the past few decades, the significance of individual wild plant and animal species in maintaining ecosystem services has received increasing attention (Brockerhoff et al., 2017). Cultural services are a vital aspect of biodiversity within ecosystems (Assandri et al., 2018). The connection between biodiversity and culture is essential across various contexts. Many individuals join conservation organizations to protect and preserve species facing extinction. Understanding how biodiversity influences human well-being and health through cultural pathways is important. A framework can be developed to link biodiversity changes and human cultural values, aiming for a safe environment and good health. (Clark et al., 2014).

The local role of biodiversity in providing cultural ecosystem services is highly recognized. However, the extent to which local species can simultaneously offer multiple cultural ecosystem services to different groups of people within the same area (Díaz & Malhi, 2022). The research explores the role of local biodiversity in providing multiple ecosystem services. This involves activities such as species naming, mapping, screening, accommodation for tourists, and other rural amenities. Plant diversity underpins several economic activities, highlighting the need to further investigate the role of biodiversity in supporting other cultural ecosystem services and inspirational values (Firn et al., 2007).

2.4.1 Recreational Activities and Biodiversity

Human activities exert a significant impact on biodiversity around the globe. Among various concerns, recreational activities pose a serious threat to biodiversity, affecting littoral areas in diverse ways and causing consistent shifts in the living environment across several regions (Kinczel et al., 2020). Coastal areas show functional diversity, considering various influencing factors. Islands extensively affected by the depletion of coastal resources highlight this impact (Riera et al., 2016). Evidence suggests a need to sustain biodiversity while maximizing both biodiversity and the recreational values of green spaces, despite practical challenges. To enhance both biodiversity and the recreational value of urban green spaces, a substantial increase in biodiversity is needed, favoring engagement in recreational activities within highly diversified environments. Activities such as hiking, camping, and tourism exemplify these recreational uses of biodiversity (Brovina & Sallaku, 2024). The living environment and the frequency of park visits significantly influence feelings toward urban biodiversity. Additionally, the living environment, education level, and age are more likely to shape perceived sensory dimensions. Therefore, it is essential to assess the quality of urban green spaces (Semeraro et al., 2021).

2.4.2 Aesthetic Values

Scenic landscapes and opportunities for wildlife viewing represent significant aesthetic values within cultural ecosystem services, contributing to human well-being (Gorodeisky, 2021). Natural scenery like Gir Forest (Gujarat) and Valley of Flowers (Uttarakhand) is valued for its beauty and ecological significance. Ancient Indian scriptures, such as the Vedas and Puranas, identified the ecological significance of trees like the banyan (Ficus benghalensis), which was considered Kalpavriksha, the wish-fulfilling tree (Rao, 2018). However, this visual value of landscapes is interconnected with biodiversity and other ecosystem services, though this relationship is not always well-supported or understood. Consequently, the feeling of landscapes, biodiversity, and ecological value holds particular importance but remains poorly comprehended. It is crucial to recognize that both current and future biodiversity faces a heightened risk of extinction (Jones & Tingley, 2022). To gain a better understanding of aesthetic value and the ecological components of biodiversity, it is beneficial to integrate the study of aesthetics with the ecological functions at both the landscape and species levels (Loreau et al., 2021). This integrated approach underscores the urgent need to engage society in conservation efforts. Such a function-based approach can foster a shift in the belief of nature and culture, leading to ecologically sound preservation practices and effective policies (Peng et al., 2021).

2.4.3 Spiritual and Inspirational Experiences

Sacred sites and artistic inspiration represent the spiritual and inspirational dimensions of biodiversity. Historically, regional plant and animal species have played symbolic and practical functions in spiritual rituals, religious rituals, mascots, national symbols, currency, and business and organization names. Integrating locally present species can reinforce people's affiliation with their place, creating a sense of place and identity attached to regional diversity (Almeida-Gomes et al., 2022). Such advantages consist of spiritual enrichment, education, contemplation, recreation, and aesthetic experiences derived from ecosystems. Natural or semi-natural habitats tend to form individual and community identities over generations. Sacred groves, for instance, conserve biodiversity and food diversity in different Indian cultures (Table 1). Leisure activities such as walking and camping enable people to connect with nature, but need to be regulated to prevent ecosystem degradation. Most religions ascribe spiritual value to ecosystems. *Citrus maxima* (pomelo) is grown in Bihar for Chhath Puja offerings, an example of the religious value

Table 1 Various ecosystem services from forests and sacred sites of India

Ecosystem services	Sites	Description	Benefits	References
Carbon sequestration	Nagarhole National Park	Reduce the greenhouse effect	37,934 Total carbon fixed	Ninan and Kontoleon (2016)
Recreation	Nagarhole National Park	Recreation	24.91 million revenues realized by the park authorities from visitors to the park	Ninan and Kontoleon (2016)
Fixing and cycling of nutrients	Corbett Tiger Reserve	Forests facilitate the breakdown of various nutrients and help them leach to nearby agricultural fields.	The average crop yield here is about 2000 kg ha^{-1}	Badola et al. (2010)
Forest as a source of fuelwood	Moist temperate forest, upper Kedarnath valley, Garhwal	Fuelwood is the major source of household energy for the rural population of India	1187.98–414.62 kg/household/year	Dhyani and Dhyani (2016)

assigned to certain plants. In the current global context, preserving species and conserving nature has become an exceedingly challenging task to undertake and sustain (Lai & Stacchezzini, 2021). Consequently, various national and international treaties have been established worldwide with the aim of biodiversity conservation. Studying ecosystems and biodiversity forms the basis of educational and scientific research related to cultural services. Researchers, academics, and educators play a crucial role in supporting environmental management and the conservation of biodiversity (Bibi et al., 2024).

3 Threats to Biodiversity and Drivers of Ecosystem Decline

The Anthropocene emerged during a period of unparalleled biological diversity in the history of the Earth (Sarrazin & Lecomte, 2016). However, as the human population expands and is increasingly transforming natural environments, they are driving biological diversity down to its lowest point since the end of the Mesozoic era, 65 million years ago (Pievani, 2014). The impacts of these exceptional fights between humans and nature are incalculable, but nowadays they are determinant (Palombo, 2021). In one sense, the disruption of biodiversity stands for the most crucial aspects of environmental change, yet as a result, its benefits are less known, primarily because the true value of biota on Earth, comprising both fauna and flora, remains mostly understudied (Bush & Payne, 2021). Global biodiversity is facing an unprecedented crisis, with its components' extinction rates accelerating due to

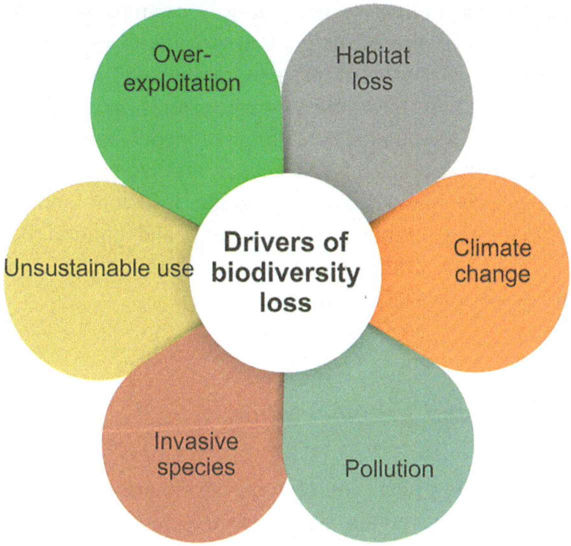

Fig. 3 The threats to biodiversity

combinations of anthropogenic activities, industrial revolution, and climate change (Shivanna, 2020). Moreover, habitat destruction, pollution, overexploitation of natural resources, and the introduction of invasive species are among the key and primary contributors to ecosystem decline (Adla et al., 2022) (Fig. 3). These threats not only disturb ecological balance but also compromise the resilience of natural systems to recover from environmental pressures (Mumby et al., 2014). Understanding the complex relationship between declining drivers and biodiversity is necessary for developing effective conservation strategies, ensuring sustainable solutions for key components negatively affecting biodiversity (Schickhoff et al., 2024).

3.1 Climate Change

Climate change is a pervasive threat to global biodiversity (Ahmad et al., 2025), and ecosystem reliability (Malhi et al., 2020). Altering precipitation patterns, temperature, and seasonal cycles of weather results in destabilizing species survival, disrupting ecological interactions, and reshaping habitats (Trew & Maclean, 2021). These tumbling effects collectively affect the structure and function of the ecosystems, diminishing their ability to deliver essential ecosystem services, such as clean water, plant pollination, and the carbon cycle, upon which human beings are mostly dependent (Pecl et al., 2017). Understanding the trajectory and magnitude of these ecological transformations is essential for enabling communities to proactively anticipate risks, improve conservation strategies, and employ adaptive measures to protect both natural and human systems of way of life (Weiskopf et al., 2020). Climate change harnesses the primary photosensitizing organisms that interlink aquatic and terrestrial food chains across geographic and time frames (Chust et al., 2014). These shifts threaten the stability of ecosystems, altering their capability to support biodiversity and keep in balance global biogeochemical cycles essential for planetary health (Lefort et al., 2015). Moreover, climate change disrupts biodiversity by altering species interactions, facilitating invasions and extreme events, diminishing provisioning, regulating, supporting, and cultural benefits (Tapale, 2024), disproportionately affecting vulnerable communities, and necessitating adaptive management strategies (Onoh et al., 2024).

3.2 Habitat Loss and Fragmentation

Conservation areas at the global and regional levels are important for preserving biodiversity; however, they remain vulnerable due to ecosystem degradation and habitat fragmentation. Studies investigating how these entities impact biodiversity within such areas typically examine localized or regional contexts, with limited focus on broader, consistent patterns (Yuan et al., 2024). The deterioration in suitable habitat leads to the fragmentation of environments characterized by isolated patches of differing eminence, existing within a barren environment (Mullu, 2016). This occurrence typically involves both the direct decrease of habitat and the subsequent fragmentation of remaining ecosystems (Lindenmayer & Fischer, 2013). To assess systematically the distinct impacts of habitat loss and fragmentation on biodiversity, researchers modeled population changes across diverse landscapes with varying composition and spatial structure (Fardila et al., 2017). These areas spanned the complete continuum of habitat configurations, from extensively fragmented to entirely endless (Kuipers et al., 2019).

3.3 Pollution

Among the major biodiversity threats, ecologists are nowadays focusing on various types of pollution, such as water pollution, air pollution, and chemical pollution, due to their significant impacts on biodiversity (Munir et al., 2016). Similarly, till recently, pollution has received constrained consideration from ecological communities as a contributor to climate change. However, increasing evidence emphasizes that 'triple crises' global change, biodiversity collapse, and environmental toxifications are intricately linked (Iyiola et al., 2022). These contests cannot be siloed; instead, they plead for holistic frameworks for research and policy actions, underscoring synergies between mitigation, conservation, and detoxication efforts (Sigmund et al., 2023). Reducing the detrimental effects of pollution on biodiversity loss needs a universal approach to addressing the anthropogenic chemical activities in ecosystems, necessitating cross-disciplinary collaboration among ecologists, ecotoxicologists, environmental chemists, and allied fields (Iyiola et al., 2022). Innovations, in reasoned techniques, system modeling, and transdisciplinary frameworks now enable systematic assessment and mitigation of this complex challenge, positioning pollution ecology at the precipices of transformative breakthroughs in understanding and interventions that can restore ecological integrity in degraded biodiversity hotspots (Terlizzi et al., 2005).

3.4 Overexploitation of Natural Resources

Overexploitation of natural resources is referred to as the unsustainable extraction and harvesting of resources and ecosystems beyond their capacity to regenerate, posing a severe threat to global biodiversity, destabilizing ecological balance, and accelerating species extinction (Cafaro et al., 2022). This repetition, driven by logging, industrial fishing, commercial hunting, and agricultural expansions, leads to a decline in important species populations, disrupts the food chain, and degrades natural niches. Additionally, overexploitation of natural resources threatens biodiversity (Khan et al., 2025), not only to destabilize ecosystems but also to erode the genetic diversity of organisms. Addressing these systematic challenges from existing frameworks like the United States Endangered Species Act and the Kunming-Montreal Global Biodiversity Framework 2022, enforce protections, while emerging strategies such as regenerative circular economies and biodiversity restoration funds (Goswami, 2024). Human activities, particularly the overexploitation of natural resources, are accelerating biodiversity loss (Hussain et al., 2025) therefore, protecting biodiversity is not a choice but an urgent imperative to preserve the ecological foundations that sustain all life forms on the Earth. (Rawat & Agarwal, 2015).

3.5 Invasion of Species

Biological invasiveness remains a substantial threat to biodiversity (Shah et al., 2025), contributing to extinction, distribution range, and habitat degradation. When these factors of exclusive dynamics act synergistically, their impact compounds, and interactions among these threats amplify challenges for field conservationists and policymakers (Mainka & Howard, 2010). The invasive species is also considered a major threat to animal and plant communities within an ecosystem, profoundly disturbing native biodiversity. Non-native organisms are proposed to extend beyond their natural ranges, proliferate rapidly, and outcompete indigenous species for certain resources, such as nutrients, water, and space. Native taxa, ecological processes, and exacerbating native global biodiversity loss are some important factors for the extinction of native organisms. The capabilities of invasive species to diminish natural habitats emphasize their status as an urgent and pervasive threat to biodiversity. Furthermore, managing invasive species is currently the basis of biodiversity conservation, imposing international collaboration to control irritated areas, introductions, and reduce their ecological, social, and economic concerns.

4 Conservation Efforts for Biodiversity and Ecosystem Services

The protection of biodiversity and ecosystem services represents a fundamental requirement for ecological sustainability. However, public tools for spatial evaluation and adaptive measures for integrated conservation stand underdeveloped. Human activities have accelerated the degradation of the ecosystem, together with biodiversity loss, causing ecosystem services and functions to be at risk. Conservation efforts include both in situ methods like designation of protected areas (for example, national parks, wildlife sanctuaries) and ex situ conservation, including botanical gardens and seed banks. Various Global frameworks, such as the Convention on Biological Diversity (CBD) and the Kunming-Montreal Global Biodiversity Framework, support conservation, sustainable resource management, and the distribution of benefits. The Aichi Targets emphasize three main priorities, including restoration, effective policy coordination, and sustainable land management (IPBES, 2019). Analyzing multiple conservation scenarios is a useful way to assess the actual trade-offs involved in nature conservation, and it can help combine social and economic objectives with the conservation of biodiversity and ecosystem services. Future research directions could include the economic valuation of ecosystem services to better understand the economic balance between the costs and benefits of biodiversity protection (Manhães et al., 2018).

We must design policies that enable the consideration of ecosystem service change in corporate environmental management systems, environmental impact assessment procedures, and conservation and rural development planning, including the design of protected areas and reserve systems.

5 Conclusion

Biodiversity plays multiple roles in sustaining ecosystem services by regulating different ecosystem processes and directly producing goods like food, medicine, timber, along with cultural and aesthetic services. However, Global biodiversity is facing an unprecedented crisis, with its components' extinction rates accelerating due to combinations of anthropogenic activities, industrial revolution, and climate change. Addressing these challenges needs integrative and effective management approaches, which identify and analyze the diverse functions of biodiversity. Future strategies must use comprehensive ecological solutions to properly measure the roles of biodiversity to ensure the ecosystem works and produces useful outputs. An integrated approach will be crucial to ensure the sustainable use of ecosystem services while safeguarding species, habitats, and ecological landscapes.

References

Abdullah, A., Khan, S. M., Pieroni, A., Haq, A., Haq, Z. U., Ahmad, Z., Sakhi, S., Hashem, A., Al-Arjani, A.-B. F., Alqarawi, A. A., & Abd_Allah, E. F. (2021). A comprehensive appraisal of the wild food plants and food system of tribal cultures in the Hindu Kush mountain range; A way forward for balancing human nutrition and food security. *Sustainability, 13*(9), 5258. https://doi.org/10.3390/su13095258

Adla, K., Dejan, K., Neira, D., & Dragana, Š. (2022). Degradation of ecosystems and loss of ecosystem services. In *One health* (pp. 281–327). Elsevier.

Ahmad, Z., Khan, S. M., Wani, Z. A., Pant, S., & Bhat, J. A. (2025). Editorial: Conserving plants in a changing climate. *Frontiers in Conservation Science, 6*. https://doi.org/10.3389/fcosc.2025.1743981

Ali, S., Khan, S. M., Siddiq, Z., Ahmad, Z., Ahmad, K. S., Abdullah, A., Hashem, A., Al-Arjani, A.-B. F., & Abd_Allah, E. F. (2022). Carbon sequestration potential of reserve forests present in the protected Margalla Hills National Park. *Journal of King Saud University Science, 34*(4), 101978. https://doi.org/10.1016/j.jksus.2022.101978

Ali, S., Khan, S. M., Ahmad, Z., Siddiq, Z., Ullah, A., Yoo, S., Han, H., & Raposo, A. (2023). Carbon sequestration potential of different forest types in Pakistan and its role in regulating services for public health. *Frontiers in Public Health, 10*. https://doi.org/10.3389/fpubh.2022.1064586

Allen, D. C., Cardinale, B. J., & Wynn-Thompson, T. (2016). Plant biodiversity effects in reducing fluvial erosion are limited to low species richness. *Ecology, 97*(1), 17–24.

Almeida-Gomes, M., de Oliveira Roque, F., Garcia, L. C., Ganci, C. C., Pacheco, E. O., Sano, N. Y., De Almeida, A. C., Bolzan, F., & Schirpke, U. (2022). Local biodiversity supports cultural ecosystem services in the Pantanal. *Wetlands, 42*(7), 69.

Altieri, V., De Franco, S., Lombardi, F., Marziliano, P. A., Menguzzato, G., & Porto, P. (2018). The role of silvicultural systems and forest types in preventing soil erosion processes in mountain forests: A methodological approach using cesium-137 measurements. *Journal of Soils and Sediments, 18*, 3378–3387.

Assandri, G., Bogliani, G., Pedrini, P., & Brambilla, M. (2018). Beautiful agricultural landscapes promote cultural ecosystem services and biodiversity conservation. *Agriculture, Ecosystems & Environment, 256*, 200–210.

Badola, R., Hussain, S. A., Mishra, B. K., Konthoujam, B., Thapliyal, S., & Dhakate, P. M. (2010). An assessment of ecosystem services of Corbett Tiger Reserve, India. *The Environmentalist, 30*(4), 320–329.

Batool, Z., Khan, S. M., Abdullah, A., Hussain, J., Khan, M. S., Hussain, I., Abbass, Z., Wu, Z.-Y., & Balslev, H. (2025). Foraging wild food plants in the dry gorges of Pakistan's high Karakorum mountains: A generational perspective. *Economic Botany, 79*(4), 529–550. https://doi.org/10.1007/s12231-025-09656-y

Bengtsson, J., Bullock, J., Egoh, B., Everson, C., Everson, T., O'connor, T., O'farrell, P., Smith, H., & Lindborg, R. (2019). Grasslands—More important for ecosystem services than you might think. *Ecosphere, 10*(2), e02582.

Bibi, C., Tabassum, S., & Roussel, Y. (2024). International environmental law: Challenges and opportunities for biodiversity conservation. *Journal of Energy and Environmental Policy Options, 7*(2), 28–35.

Brockerhoff, E. G., Barbaro, L., Castagneyrol, B., Forrester, D. I., Gardiner, B., González-Olabarria, J. R., Lyver, P. O. B., Meurisse, N., Oxbrough, A., & Taki, H. (2017). Forest bio-

diversity, ecosystem functioning and the provision of ecosystem services. *Biodiversity and Conservation, 26*, 3005–3035.
Brovina, F., & Sallaku, D. (2024). Sustainable development of forest parks for active recreation: A balance between nature conservation and physical education. *Ukrainian Journal of Forest and Wood Science, 15*(3), 165.
Bush, A. M., & Payne, J. L. (2021). Biotic and abiotic controls on the Phanerozoic history of marine animal biodiversity. *Annual Review of Ecology, Evolution, and Systematics, 52*(1), 269–289.
Cafaro, P., Hansson, P., & Götmark, F. (2022). Overpopulation is a major cause of biodiversity loss and smaller human populations are necessary to preserve what is left. *Biological Conservation, 272*, 109646.
Campos, J. C., Rodrigues, S., Sil, Â., Hermoso, V., Freitas, T. R., Santos, J. A., Fernandes, P. M., Azevedo, J. C., Honrado, J. P., & Regos, A. (2022). Climate regulation ecosystem services and biodiversity conservation are enhanced differently by climate-and fire-smart landscape management. *Environmental Research Letters, 17*(5), 054014.
Cheminal, A., Kokkoris, I. P., Strid, A., & Dimopoulos, P. (2020). Medicinal and aromatic Lamiaceae plants in Greece: Linking diversity and distribution patterns with ecosystem services. *Forests, 11*(6), 661.
Cheminal, A., Kokkoris, I. P., Zotos, A., Strid, A., & Dimopoulos, P. (2022). Assessing the ecosystem services potential of endemic floras: A systematic review on the Greek endemics of Peloponnese. *Sustainability, 14*(10), 5926.
Chust, G., Allen, J. I., Bopp, L., Schrum, C., Holt, J., Tsiaras, K., Zavatarelli, M., Chifflet, M., Cannaby, H., & Dadou, I. (2014). Biomass changes and trophic amplification of plankton in a warmer ocean. *Global Change Biology, 20*(7), 2124–2139.
Clark, N. E., Lovell, R., Wheeler, B. W., Higgins, S. L., Depledge, M. H., & Norris, K. (2014). Biodiversity, cultural pathways, and human health: A framework. *Trends in Ecology & Evolution, 29*(4), 198–204.
Deng, X., Li, Z., Huang, J., Shi, Q., & Li, Y. (2013). A revisit to the impacts of land use changes on the human wellbeing via altering the ecosystem provisioning services. *Advances in Meteorology, 2013*(1), 907367.
Dhyani, S., & Dhyani, D. (2016). Significance of provisioning ecosystem services from moist temperate forest ecosystems: Lessons from upper Kedarnath valley, Garhwal, India. *Energy, Ecology and Environment, 1*(2), 109–121.
Díaz, S., & Malhi, Y. (2022). Biodiversity: Concepts, patterns, trends, and perspectives. *Annual Review of Environment and Resources, 47*(1), 31–63.
Díaz, S., Demissew, S., Carabias, J., Joly, C., Lonsdale, M., Ash, N., Larigauderie, A., Adhikari, J. R., Arico, S., & Báldi, A. (2015). The IPBES conceptual framework—Connecting nature and people. *Current Opinion in Environmental Sustainability, 14*, 1–16.
Duncan, C., Thompson, J. R., & Pettorelli, N. (2015). The quest for a mechanistic understanding of biodiversity–ecosystem services relationships. *Proceedings of the Royal Society B: Biological Sciences, 282*(1817), 20151348.
Ejaz, U., Khan, S. M., Abdullah, S. J., Chandio, N. H., Qurat-Ul-Ain, R. A., Rizvi, Z. F., & Khalid, N. (2024). Dynamic interchange between floristic composition and industrial pollution: An ecological perspective. *Pakistan Journal of Botany, 56*(6), 2229–2243.
Everard, M. (2018). Management of provisioning services: Overview. *The Wetland Book, 1*, 997–1004.
Fardila, D., Kelly, L. T., Moore, J. L., & McCarthy, M. A. (2017). A systematic review reveals changes in where and how we have studied habitat loss and fragmentation over 20 years. *Biological Conservation, 212*, 130–138.
Firn, J., Erskine, P. D., & Lamb, D. (2007). Woody species diversity influences productivity and soil nutrient availability in tropical plantations. *Oecologia, 154*, 521–533.

García-Pardo, K. A., Moreno-Rangel, D., Domínguez-Amarillo, S., & García-Chávez, J. R. (2022). Remote sensing for the assessment of ecosystem services provided by urban vegetation: A review of the methods applied. *Urban Forestry & Urban Greening, 74*, 127636.

Gorodeisky, K. (2021). On liking aesthetic value. *Philosophy and Phenomenological Research, 102*(2), 261–280.

Goswami, D. (2024). Addressing the challenge of overexploitation of natural resources in pursuit of development. *Sprin Journal of Arts, Humanities and Social Sciences, 3*(3), 65–67.

Guo, Z., Zhang, L., & Li, Y. (2010). Increased dependence of humans on ecosystem services and biodiversity. *PLoS One, 5*(10), e13113.

Harrison, P., Berry, P., Simpson, G., Haslett, J., Blicharska, M., Bucur, M., Dunford, R., Egoh, B., Garcia-Llorente, M., & Geamănă, N. (2014). Linkages between biodiversity attributes and ecosystem services: A systematic review. *Ecosystem Services, 9*, 191–203.

Huang, Z., Qian, L., & Cao, W. (2022). Developing a novel approach integrating ecosystem services and biodiversity for identifying priority ecological reserves. *Resources, Conservation and Recycling, 179*, 106128.

Hussain, J., Khan, S. M., Khan, M. S., Saqib, Z., Abdullah, Ahmad, Z., Shah, M., Mohammad, N., Batool, Z., Sohail, A., & Liu, J. (2025). Distribution modelling and future prediction of a threatened species—Heracleum candicans Wall. ex DC.; Within the framework of biotic and abiotic interactions. *Journal of Environmental Management, 387*, 125818. https://doi.org/10.1016/j.jenvman.2025.125818

IPBES. (2019). Intergovernmental science-policy platform on biodiversity and ecosystem services. In *Summary for policy makers of the global assessment report on biodiversity and ecosystem services of the intergovernmental science-policy platform on biodiversity and ecosystem services*. IPBES Secretariat.

Israr, A., Khan, S. M., Abdullah, A., Ejaz, U., Jehangir, S., Ahmad, Z., Hashem, A., Avila-Quezada, G. D., & Abd_Allah, E. F. (2024). Fire-induced vegetation dynamics: An in-depth discourse on revealing ecological transformations of the Mahaban and surrounding forests. *Fire, 7*(1), 27. https://doi.org/10.3390/fire7010027

Iyiola, A. O., Akinrinade, A. J., & Ajayi, F. O. (2022). Effects of water pollution on biodiversity along the coastal regions. In *Biodiversity in Africa: Potentials, threats and conservation* (pp. 345–367). Springer.

Jehangir, S., Khan, S. M., Ahmad, Z., Ejaz, U., Ain, Q. U., Lho, L. H., Han, H., & Raposo, A. (2024a). Distribution of the Cannabis sativa L. in the Western Himalayas: A tale of the ecological factors behind its continuous invasiveness. *Global Ecology and Conservation, 49*, e02779.

Jehangir, S., Khan, S. M., Ejaz, U., Zahid, N., Rashid, N., Noshad, Q., Din, Z. U., & Shoukat, A. (2024b). Alien flora of Pakistan: Taxonomic composition, invasion status, geographic origin, introduction pathways, and ecological patterns. *Biological Invasions, 26*(8), 2435–2451.

Jones, G. M., & Tingley, M. W. (2022). Pyrodiversity and biodiversity: A history, synthesis, and outlook. *Diversity and Distributions, 28*(3), 386–403.

Khan, M. S., Khan, S. M., Abdullah, J. L., Wu, Z.-Y., Hussain, J., Zeb, S. A., Mohammad, N., Batool, Z., Saqib, Z., Afza, R., Manan, F., & Ali, S. (2025). Ecological assessment of Iris hookeriana across subalpine and alpine regions of the Hindu-Himalayas. *Frontiers in Forests and Global Change, 8*. https://doi.org/10.3389/ffgc.2025.1539025

Kinczel, A., Maklári, G., & Müller, A. (2020). Recreational activities and motivation among young people. *Geosport for Society, 12*(1), 53–65.

Kuipers, K. J. J., May, R. F., Graae, B. J., & Verones, F. (2019). Reviewing the potential for including habitat fragmentation to improve life cycle impact assessments for land use impacts on biodiversity. *The International Journal of Life Cycle Assessment, 24*, 2206–2219.

Lai, A., & Stacchezzini, R. (2021). Organisational and professional challenges amid the evolution of sustainability reporting: A theoretical framework and an agenda for future research. *Meditari Accountancy Research, 29*(3), 405–429.

Lefort, S., Aumont, O., Bopp, L., Arsouze, T., Gehlen, M., & Maury, O. (2015). Spatial and body-size dependent response of marine pelagic communities to projected global climate change. *Global Change Biology, 21*(1), 154–164.

Lindenmayer, D. B., & Fischer, J. (2013). *Habitat fragmentation and landscape change: An ecological and conservation synthesis*. Island Press.

Linhares, Y., Kaganski, A., Agyare, C., Kurnaz, I. A., Neergheen, V., Kolodziejczyk, B., Kędra, M., Wahajuddin, M., El-Youssf, L., & dela Cruz, T. E. (2023). Biodiversity: The overlooked source of human health. *Trends in Molecular Medicine, 29*(3), 173–187.

Lipton, D., Rubenstein, M. A., Weiskopf, S. R., Carter, S. L., Peterson, J., Crozier, L., Fogarty, M., Gaichas, S., Hyde, K. J., & Morelli, T. L. (2018). Ecosystems, ecosystem services, and biodiversity. *USGS Report*, 268–321.

Liquete, C., Cid, N., Lanzanova, D., Grizzetti, B., & Reynaud, A. (2016). Perspectives on the link between ecosystem services and biodiversity: The assessment of the nursery function. *Ecological Indicators, 63*, 249–257.

Lohbeck, M., Bongers, F., Martinez-Ramos, M., & Poorter, L. (2016). The importance of biodiversity and dominance for multiple ecosystem functions in a human-modified tropical landscape. *Ecology, 97*(10), 2772–2779.

Loreau, M., Barbier, M., Filotas, E., Gravel, D., Isbell, F., Miller, S. J., Montoya, J. M., Wang, S., Aussenac, R., & Germain, R. (2021). Biodiversity as insurance: From concept to measurement and application. *Biological Reviews, 96*(5), 2333–2354.

Mace, G. M. (2005). An index of intactness. *Nature, 434*(7029), 32–33.

Mainka, S. A., & Howard, G. W. (2010). Climate change and invasive species: Double jeopardy. *Integrative Zoology, 5*(2), 102–111.

Malhi, Y., Franklin, J., Seddon, N., Solan, M., Turner, M. G., Field, C. B., & Knowlton, N. (2020). Climate change and ecosystems: Threats, opportunities and solutions. *Philosophical Transactions of the Royal Society B: Biological Sciences, 375*, 20190104.

Manhães, A. P., Loyola, R., Mazzochini, G. G., Ganade, G., Oliveira-Filho, A. T., & Carvalho, A. R. (2018). Low-cost strategies for protecting ecosystem services and biodiversity. *Biological Conservation, 217*, 187–194.

Maroyi, A. (2022). Traditional uses of wild and tended plants in maintaining ecosystem services in agricultural landscapes of the Eastern Cape Province in South Africa. *Journal of Ethnobiology and Ethnomedicine, 18*(1), 17.

Millennium Ecosystem Assessment. (2005). *Ecosystems and human well-being*. Island Press.

Mir, T. A., Jan, M., Khare, R. K., & Bhat, M. H. (2021). Medicinal plant resources: Threat to its biodiversity and conservation strategies. In *Medicinal and aromatic plants: Healthcare and industrial applications* (pp. 717–739). Springer.

Mohri, H., Lahoti, S., Saito, O., Mahalingam, A., Gunatilleke, N., Hoang, V. T., Hitinayake, G., Takeuchi, K., & Herath, S. (2013). Assessment of ecosystem services in homegarden systems in Indonesia, Sri Lanka, and Vietnam. *Ecosystem Services, 5*, 124–136.

Mullu, D. (2016). A review on the effect of habitat fragmentation on ecosystem. *Journal of Natural Sciences Research, 6*(15), 1–15.

Muluneh, M. G. (2021). Impact of climate change on biodiversity and food security: A global perspective—A review article. *Agriculture & Food Security, 10*(1), 1–25.

Mumby, P. J., Chollett, I., Bozec, Y.-M., & Wolff, N. H. (2014). Ecological resilience, robustness and vulnerability: How do these concepts benefit ecosystem management? *Current Opinion in Environmental Sustainability, 7*, 22–27.

Munir, T., Hussain, M., & Naseem, S. (2016). Water pollution-a menace of freshwater biodiversity: A review. *Journal of Entomology and Zoology Studies, 4*(4), 578–580.

Ndayizeye, G., Imani, G., Nkengurutse, J., Irampagarikiye, R., Ndihokubwayo, N., Niyongabo, F., & Cuni-Sanchez, A. (2020). Ecosystem services from mountain forests: Local communities' views in Kibira National Park, Burundi. *Ecosystem Services, 45*, 101171.

Newman, D. J., & Cragg, G. M. (2020). Natural products as sources of new drugs over the nearly four decades from 01/1981 to 09/2019. *Journal of Natural Products, 83*(3), 770–803.

Ninan, K. N., & Kontoleon, A. (2016). Valuing forest ecosystem services and disservices–case study of a protected area in India. *Ecosystem Services, 20*, 1–14.

Oguh, C., Obiwulu, E., Umezinwa, O., Ameh, S., Ugwu, C., & Sheshi, I. (2021). Ecosystem and ecological services; need for biodiversity conservation-A critical review. *Asian Journal of Biology, 11*(4), 1–14.

Onoh, U. C., Ogunade, J., Owoeye, E., Awakessien, S., & Asomah, J. K. (2024). Impact of climate change on biodiversity and ecosystem services. *IIARD International Journal of Geography and Environmental Management, 10*, 77–93.

Palombo, M. R. (2021). Thinking about the biodiversity loss in this changing world. *Geosciences, 11*(9), 370.

Pecl, G. T., Araújo, M. B., Bell, J. D., Blanchard, J., Bonebrake, T. C., Chen, I. C., Clark, T. D., Colwell, R. K., Danielsen, F., & Evengård, B. (2017). Biodiversity redistribution under climate change: Impacts on ecosystems and human well-being. *Science, 355*(6332), eaai9214.

Peng, L., Feng, W., Yan, Z., Li, Y., Zhou, X., & Shimizu, S. (2021). Privacy preservation in permissionless blockchain: A survey. *Digital Communications and Networks, 7*(3), 295–307.

Pievani, T. (2014). The sixth mass extinction: Anthropocene and the human impact on biodiversity. *Rendiconti Lincei, 25*, 85–93.

Pinto, R., de Jonge, V. N., & Marques, J. C. (2014). Linking biodiversity indicators, ecosystem functioning, provision of services and human well-being in estuarine systems: Application of a conceptual framework. *Ecological Indicators, 36*, 644–655.

Quijas, S., & Balvanera, P. (2013). Biodiversity and ecosystem services. In *Encyclopedia of biodiversity* (Vol. 1, pp. 341–356). Academic Press.

Rao, V. R. (2018). Ecosystem functions, services and biodiversity: Past, present and future implicationsDOUBLEHYPHENII. *Asian Agri-History, 22*(4).

Rathoure, A. K., & Kumar, V. (2024). Biodiversity and ecosystem services: An interconnected web. In *Impact of societal development and infrastructure on biodiversity decline* (pp. 23–36). IGI Global.

Rawat, U. S., & Agarwal, N. K. (2015). Biodiversity: Concept, threats and conservation. *Environment Conservation Journal, 16*(3), 19–28.

Riera, R., Menci, C., Sanabria-Fernández, J. A., & Becerro, M. A. (2016). Do recreational activities affect coastal biodiversity? *Estuarine, Coastal and Shelf Science, 178*, 129–136.

Rockström, J., Gaffney, O., Rogelj, J., Meinshausen, M., Nakicenovic, N., & Schellnhuber, H. J. (2017). A roadmap for rapid decarbonization. *Science, 355*(6331), 1269–1271.

Sarrazin, F., & Lecomte, J. (2016). Evolution in the anthropocene. *Science, 351*(6276), 922–923.

Schickhoff, U., Bobrowski, M., Offen, I., & Mal, S. (2024). The biodiversity crisis in the anthropocene. In *Geography and the anthropocene* (pp. 79–111). Istanbul University Press.

Sekercioglu, C. H. (2010). Ecosystem functions and services. In *Conservation biology for all* (pp. 45–72). Oxford University Press.

Semeraro, T., Scarano, A., Buccolieri, R., Santino, A., & Aarrevaara, E. (2021). Planning of urban green spaces: An ecological perspective on human benefits. *Land, 10*(2), 105.

Shackleton, C. M. (2021). Ecosystem provisioning services in Global South cities. In *Urban ecology in the Global South* (pp. 203–226). Springer.

Shah, M., Khan, S. M., Manan, F., Hussain, J., Din, Z. U., Shah, I., Arshad, M., & Ahmad, I. (2025). Role of the plant nurseries in spread of invasive alien plant species in Pakistan's subtropical region; A threat for urban greening. *Urban Forestry & Urban Greening, 105*, 128688. https://doi.org/10.1016/j.ufug.2025.128688

Sharma, I., & Birman, S. (2024). Biodiversity loss, ecosystem services, and their role in promoting sustainable health. In *The climate-health-sustainability nexus: understanding the interconnected impact on populations and the environment* (pp. 163–188). Springer.

Shivanna, K. R. (2020). The sixth mass extinction crisis and its impact on biodiversity and human welfare. *Resonance, 25*(1), 93–109.

Sigmund, G., Ågerstrand, M., Antonelli, A., Backhaus, T., Brodin, T., Diamond, M. L., Erdelen, W. R., Evers, D. C., Hofmann, T., & Hueffer, T. (2023). Addressing chemical pollution in biodiversity research. *Global Change Biology, 29*(12), 3240–3255.

Smith, P., Cotrufo, M., Rumpel, C., Paustian, K., Kuikman, P., Elliott, J., McDowell, R., Griffiths, R., Asakawa, S., & Bustamante, M. (2015). Biogeochemical cycles and biodiversity as key drivers of ecosystem services provided by soils. *The Soil, 1*(2), 665–685.

Tamburini, G., Bommarco, R., Wanger, T. C., Kremen, C., Van Der Heijden, M. G., Liebman, M., & Hallin, S. (2020). Agricultural diversification promotes multiple ecosystem services without compromising yield. *Science Advances, 6*(45), eaba1715.

Tapale, B. K. (2024). *Environment conservation, challenges & threats in conservation of biodiversity*. Academic Guru Publishing House.

Terlizzi, A., Scuderi, D., Fraschetti, S., & Anderson, M. J. (2005). Quantifying effects of pollution on biodiversity: A case study of highly diverse molluscan assemblages in the Mediterranean. *Marine Biology, 148*(2), 293–305.

Thomaz, S. M. (2023). Ecosystem services provided by freshwater macrophytes. *Hydrobiologia, 850*(12), 2757–2777.

Thorn, J. P., Thornton, T. F., Helfgott, A., & Willis, K. J. (2020). Indigenous uses of wild and tended plant biodiversity maintain ecosystem services in agricultural landscapes of the Terai Plains of Nepal. *Journal of Ethnobiology and Ethnomedicine, 16*, 1–25.

Trew, B. T., & Maclean, I. M. D. (2021). Vulnerability of global biodiversity hotspots to climate change. *Global Ecology and Biogeography, 30*(4), 768–783.

Ullah, T., Khan, S. M., Abdullah, A., Sulaiman, N., Ullah, A., Khan, M. S., Zeb, S. A., & Pieroni, A. (2024). Bio-cultural diversity for food security: Traditional wild food plants and their folk cuisine in Lakki Marwat Northwestern Pakistan. *Diversity, 16*(11), 684. https://doi.org/10.3390/d16110684

Ullah, S., Khan, F. M. S., Khan, S. M., Manan, F., Shah, M., & Ullah, T. (2025). Ecological variation and geographic distribution of vegetation in Karak northwestern Pakistan. *Pakistan Journal of Botany, 57*(6). https://doi.org/10.30848/PJB2025-6(11)

Vauhkonen, J., & Packalen, T. (2018). Uncertainties related to climate change and forest management with implications on climate regulation in Finland. *Ecosystem Services, 33*, 213–224.

Vilà, M., & Hulme, P. E. (2017). *Impact of biological invasions on ecosystem services*. Springer.

Weiskopf, S. R., Rubenstein, M. A., Crozier, L. G., Gaichas, S., Griffis, R., Halofsky, J. E., Hyde, K. J. W., Morelli, T. L., Morisette, J. T., & Muñoz, R. C. (2020). Climate change effects on biodiversity, ecosystems, ecosystem services, and natural resource management in the United States. *Science of the Total Environment, 733*, 137782.

Wilby, A., Mitchell, C., Blumenthal, D., Daszak, P., Friedman, C. S., Jutro, P., Mazumder, A., Prieur-Richard, A.-H., Desprez-Loustau, M. L., & Sharma, M. (2009). Biodiversity, food provision, and human health. In *Biodiversity change and human health* (pp. 13–39). Island Press.

Yuan, R., Zhang, N., & Zhang, Q. (2024). The impact of habitat loss and fragmentation on biodiversity in global protected areas. *Science of the Total Environment, 931*, 173004.

Zeb, S. A., Abdullah, A., Ahmad, Z., Haq, Z. U., Ali, S., Zeb, T. A., Afza, R., Ahmad, I., Jatt, T., & Khan, S. M. (2026). Soil nutrients control the ecological characteristics of the riparian vegetation. *Pakistan Journal of Botany, 58*(6), 10.30848/PJB2026-6 10.30848/PJB2026-6(7).

Innovative Strategies for the Conservation of Endangered Species: Saving Lives Facing Critical Threat

Tejal Barkhade and Subair Kalathil

Abstract The present ecosystem is facing a great challenge in terms of biodiversity loss, wherein various species have already reached the juncture of extinction due to climate change, habitat loss, pollution, human activities, and poaching, along with several other anthropogenic pressures. The present chapter deals with emerging and possible approaches for the conservation of endangered species, integrating both conventional and advanced technologies along with community-level initiatives. Innovative technology usage will prevent the possibility of poaching and will conserve life on the way to extinction. Thus, habitat restoration, genetics-based rescue methods, artificial intelligence for population trend, monitoring, and prediction of threats are important areas of focus. The interventions of the government have a significant role in the protection and establishment of policies, regulations, and funding mechanisms commensurate with conservation goals. The present chapter also reflects community-based involvement conservation models such as participatory approaches, ecotourism, education, and awareness campaigns. The synergetic approach and collaborative contributions among scientists, policymakers, NGOs, and local communities are necessary for crafting measured and sustainable solutions. This integrated solution to saving endangered species needs urgent and coordinated action to preserve life on Earth.

Keywords Endangered · Conservation · Biodiversity · Threat · Approaches

T. Barkhade (✉)
Department of Life Sciences, Parul Institute of Applied Sciences, Parul University, Vadodara, Gujarat, India
e-mail: tejal.barkhade37399@paruluniversity.ac.in

S. Kalathil
Parul Institute of Law, Parul University, Vadodara, India
e-mail: subair.kalathil26145@paruluniversity.ac.in

S. Siddiqui et al. (eds.), *Frontiers in Sustainability*,
https://doi.org/10.1007/978-3-032-19740-5_3

1 Introduction

Biodiversity denotes the diversity of living organisms on Earth, including plants, animals, microorganisms, and the ecosystems they create (DeLong Jr, 1996). It has a fundamental role in preserving the balance of nature, encourages food production, regulates climate, and maintains clean water. Regardless of its importance, biodiversity is facing a serious risk around the globe due to various factors, including human actions and natural causes (Rawat & Agarwal, 2015). The existing loss of biodiversity is disquieting because numerous classes of species are endangered at a level much higher than the natural disappearance rate. The endangered species are a group of species that are at a very high risk of disappearance soon, due to habitat destruction, deforestation, overexploitation, pollution, climate change, and rapid urbanization (UNEP, 2020; Bralower & Millet, 2021). Scientists and scholars point out that this is the sixth mass extinction, and, contrasting earlier ones due to natural causes, the present one is largely caused by humans' activities (Pimm et al., 2014). Forests, oceans, and other ecosystems are being contaminated/degraded, resulting in the loss of habitat, food chains, and natural cycles (Adla et al., 2022). To offer a clear and organized outline of the chapter, Fig. 1 demonstrates a flowchart outlining the main thematic areas of the discussion. Starting with an introduction to

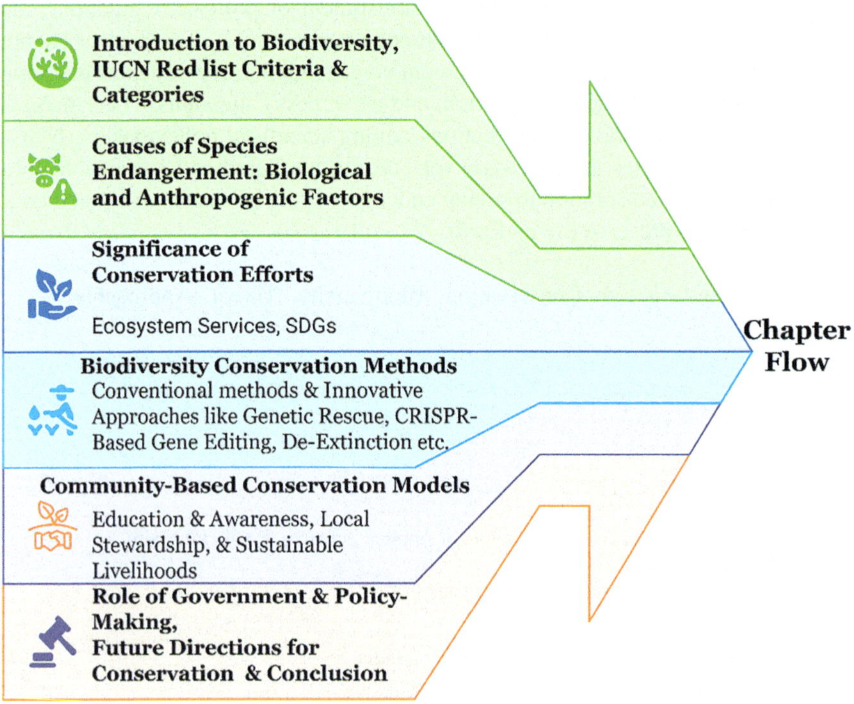

Fig. 1 Overview of the chapter structure and key thematic areas

biodiversity and International Union for Conservation of Nature (IUCN) Red List classifications, the chapter explores the biological and anthropogenic factors driving species endangerment. It then moves into the significance of conservation. A comprehensive discussion of conventional and innovative biodiversity conservation methods. Further, the chapter includes community-based models, underscoring the importance of local stewardship and education. Finally, it delves into the role of governmental policy-making and envisions future directions for sustainable conservation. This graphic flowchart allows readers to direct the chapter's progress in a consistent and attractive manner.

Given the severity of this biodiversity crisis, it is essential to understand how species are assessed in terms of their extinction risk. The worldwide known standard for evaluating the conservation status of species is given by the IUCN Red List of Threatened Species. This system uses a broad set of scientific criteria to categorize species based on their risk of extinction. Understanding these categories and criteria in the next subsection is essential for designing effective conservation strategies (IUCN, 2022).

1.1 IUCN Categories and Criteria for Species Classification

The IUCN Red List serves as the most authoritative global framework for assessing the conservation status of plant and animal species. It classifies species into nine categories: Extinct (EX), Extinct in the Wild (EW), Critically Endangered (CR), Endangered (EN), Vulnerable (VU), Near Threatened (NT), Least Concern (LC), Data Deficient (DD) and, Not Evaluated (NE) based on quantitative criteria that measure extinction risk, population trends, and threats (IUCN, 2022). These categories are shown in Fig. 2. This classification system helps conservation strategies by visibly recognizing species requiring urgent action based on extinction risk.

The IUCN Red List delivers a severe five main criteria (A-E) for classifying the risk of extinction for a species, shown in Table 1 (IUCN, 2022). A species is listed as Endangered when it meets any one of five scientifically distinct criteria, reflecting population trends, geographic distribution, and extinction probability. Each criterion is supported by thresholds and evidence based to confirm objectivity and comparability through taxa and regions. The use of these criteria allows conservationists, scientists, and policymakers to recognize species at substantial risk and prioritize conservation plans accordingly. By thoroughly evaluating species using these criteria, the IUCN Red List plays an important role in notifying global conservation hard work, tracking advancement toward biodiversity targets, and raising public awareness about the urgent need to guard endangered species on our planet.

Based upon this understanding of the increasing crisis, the next section states the current global biodiversity crisis, its underlying causes, and recent alarming outcomes.

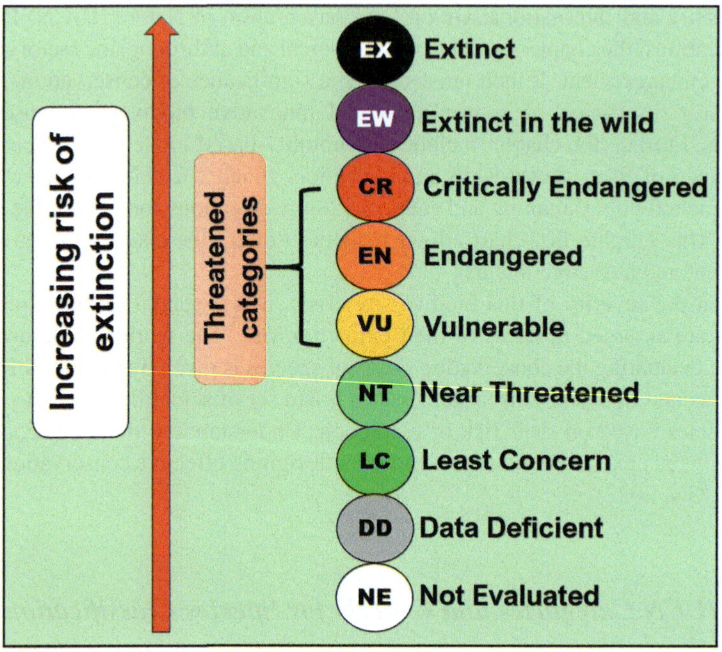

Fig. 2 IUCN Red List categories for species hierarchical classification based on extinction risk (IUCN, 2022)

Table 1 IUCN Red List Criteria (A–E) for Endangered Species Classification (IUCN, 2022)

Criterion	Title	Description
A	Population reduction	A drop of ≥50% over the past 10 years or three generations, whichever is longer. Causes may include habitat degradation, exploitation, invasive species, pollution, disease, or other threats.
B	Restricted geographic range	Extent of occurrence <5000 km^2 or area of occupancy <500 km^2, along with habitat fragmentation, declining habitat quality, or severe population fluctuations.
C	Small population with continuing decline	Fewer than 2500 mature individuals, with ongoing decline in population size, structure, or habitat quality.
D	Very small or restricted population	Total population of fewer than 250 mature individuals, or species occupies a highly restricted area vulnerable to chance events or human activities.
E	Quantitative analysis	A ≥ 20% probability of extinction in the wild within the next 20 years or five generations, based on formal quantitative risk analysis.

2 Causes of Species Endangerment: Biological and Anthropogenic Factors

Biological factors considerably contribute to species endangerment by controlling their ability to survive and recover. *(i) Low reproductive rates* refer to species that produce few offspring or have long gestation periods (e.g., rhinos, whales), making population recovery slow. *(ii) Restricted geographic range* means species limited to small or precise areas, making them more vulnerable to habitat disturbance. *(iii) Specialized habitat or diet requirements* are species dependent on unique conditions or food sources, limiting adaptability. *(iv) Genetic bottlenecks and inbreeding* involve a decrease in genetic variety in small populations, weakening their flexibility and increasing weakness to disease and environmental stress (Zhao et al., 2025).

The 2019 Intergovernmental Science-Policy Platform on Biodiversity and Ecosystem Services (IPBES) report revealed that over 1 million species face extinction, many within decades (IPBES, 2019). This crisis results from a complex interplay of biological vulnerabilities and predominantly anthropogenic pressures. IPBES identifies five direct drivers: habitat destruction, climate change, invasive species, overexploitation, and pollution, all largely human-induced.

(i) *Habitat destruction* remains the leading cause, driven by deforestation, urbanization, unregulated agriculture, mining, and infrastructure expansion (Bodo et al., 2021). In biodiversity hotspots like the Amazon and Western Ghats, large-scale forest clearance is rampant (Bhattacharya, 2025). This not only disturbs wildlife but also threatens endemic plant species such as *Cycas beddomei* and *Myristica malabarica*, both of which have highly localized ranges. A recent example includes the removal of 100 acres of forest in Hyderabad, India (The Hindu Bureau, 2025). Globally, over 420 million hectares of forest were lost between 1990 and 2020, particularly in tropical regions (FAO, 2020).

(ii) *Climate change* disrupts species' survival due to rising temperatures, erratic rainfall, and extreme weather (Muluneh, 2021). Coral reefs, highly sensitive to thermal stress, could face 70–90% loss if global temperatures rise above 1.5 °C (IPCC, 2018). Likewise, high-altitude plants like *Saussurea obvallata* (Brahma Kamal) and *Rheum nobile* in the Himalayas are at risk due to glacier melt, shifting climatic zones, and monsoon shifts (Smith, 1989).

(iii) *Invasive species,* introduced intentionally or accidentally, compete with natives for resources, causing native declines (CBD, 2009). The Nile perch in Lake Victoria led to the extinction of over 200 local fish species (Goldschmidt et al., 1993). In India, species like the African catfish and water hyacinth disrupt local ecosystems (Singh & Lakra, 2011). Also, invasive aquatic plants such as water hyacinth and Parthenium hysterophorus suppress native wetland flora, affecting biodiversity and local livelihoods (Rai & Singh, 2020).

(iv) *Overexploitation* through poaching, logging, and unsustainable fishing significantly reduces species populations. Examples include the African elephant and Indian pangolin, it also impacts medicinal plants such as *Taxus wallichiana* (Himalayan Yew), *Rauvolfia serpentina*, and *Nardostachys jatamansi*, which are overharvested for their pharmaceutical value. Overfishing has impacted species like Atlantic cod and affected dependent communities (Hinsley et al., 2023; WWF, 2022).

(v) *Pollution* from agriculture, industries, and households intensifies biodiversity loss. Pesticides like endosulfan have ruined aquatic flora and soil microbes (Kolawole & Iyiola, 2023), while the Kerala endosulfan tragedy led to long-term ecological and health destruction (Irshad & Joseph, 2015). Rivers like the Ganga and Yamuna are severely polluted, distressing submerged plant species and aquatic habitats (Sharma et al., 2024). Moreover, plastic and microplastic pollution impacts not only animals but also plant health and growth by polluting soils and water systems (Thushari & Senevirathna, 2020).

According to the WWF Living Planet Report (2022), global vertebrate populations have declined by 69% since 1970, with freshwater species facing an 83% reduction. In Latin America, the decline reaches 94%. Amphibians are highly threatened, with 41% at risk of extinction. The IUCN (2022) lists over 42,000 species as threatened. In India, over 90,000 animal and 45,000 plant species exist (NBAP, 2008), including endangered species like the Himalayan Brown Bear and Great Indian Bustard, facing threats from habitat loss and climate impacts.

Table 2 summarizes the quantitative results of the conservation status of taxonomic groups based on IUCN Red List assessments (IUCN, 2022). The data represented in the table are sourced from the IUCN's 2022 Red List summary statistics and were compiled by Carrington (2022) in a public biodiversity report published by *The Guardian*. To better understand the scale of this crisis, the findings show a strong intertaxonomic heterogeneity in the extinction susceptibility. Cycads arise as the most endangered group with over half of the assessed species (53%) estimated as threatened (CR + EN + VU), which further demonstrates their high priority of conservation attention. Other highly-endangered taxa include reef-building corals (35%), conifers (25%), which highlights the threatened status of foundation plant and marine life. Amphibians (28%) and mammals (23%) ranked higher in the proportion of threats than followed reptiles (19%) and birds (11%). Insects, gastropods, and crustaceans exhibit moderately high values of endangered species, but surprisingly the highest levels of Data Deficient species. By comparison, cephalopods are the group least under threat (<10% under threat). These results show that not only are more endangered groups that need to be given special protection (cycads, corals, amphibians), but data coverage in other groups that are poorly assessed (insects, gastropods, crustaceans) also needs to be enhanced.

Table 2 Quantitative analysis of conservation status across major taxonomic groups (IUCN, 2022; Carrington 2022)

Taxonomic groups	Total assessed	Threatened % (CR + EN + VU)	Least concern (LC) %	Data deficient (DD) %	Key observation
Cycads	359	53	25	2	Most threatened (>50%)
Reef-forming corals	950	35	40	7	High risk (~35%)
Conifers	695	25	45	15	High risk (~25%)
Amphibians	8459	28	45	13	Highly threatened (~28%)
Mammals	6577	23	50	15	High threatened (~23%)
Reptiles	11,690	19	50	21	Moderate (~19%)
Birds	11,162	11	55	27	Moderate-low (~11%)
Insects	1,053,578	16	30	42	Large DD (~42%)
Gastropods	86,465	12	40	40	Large DD (~40%)
Crustaceans	80,122	13	42	35	Large DD (~35%)
Sharks, rays, and chimaeras	1224	18	55	15	Moderate (~18%)
Dicots	246,366	12	40	38	Large DD (~38%)
Bony fishes	36,058	12	55	25	Low (~12%)
Cephalopods	838	7	45	41	Least threatened (<10%)

3 Significance of Conservation Efforts

Biodiversity plays a crucial role in maintaining the ecosystem's functional robustness. These ecosystems offer amenities that are fundamental to human welfare, such as fresh water, fertile soil, clean air, food, medicine, and climate regulation (MEA, 2005). The loss of biodiversity deteriorates the capability of ecosystems to execute these functions, resulting in severe consequences for all domains of all life forms, including humans. One of the significant reasons to preserve biodiversity is to sustain ecological balance. Each class of species plays a vital role in the food chain or the conservation of ecosystem processes (Sala et al., 2000). For instance, bees and other pollinators support the reproduction of plants that are the prime source of food for herbivores, and so on up the chain. If pollinators become extinct, it would distress the reproduction of plants, crop harvests, and food accessibility (Potts et al., 2010). Likewise, tigers and wolves help regulate herbivore populations, preventing overgrazing and shielding vegetation (Ripple et al., 2014).

Biodiversity provides ecosystem services that are clustered into four categories (*i*) provisioning (food, water, timber, medicinal plants), (*ii*) regulating (climate control, flood regulation, disease regulation), (*iii*) supporting (soil formation, nutrient cycling, pollination), and (*iv*) cultural services (recreational, spiritual, educational) (MEA, 2005). These services are often taken for granted but are essential for survival. For instance, wetlands act as natural water filters and help to control floods. Forests store carbon dioxide and help control the Earth's climate. Mangroves guard coastal areas from tornadoes and tidal waves. These services are gifts of nature to all living organisms, and their fortification is economically and environmentally important (Barbier et al., 2011).

Moreover, biodiversity-based sectors such as agriculture, forestry, fisheries, and tourism contribute a considerable share to the economy, particularly in developing countries (UNEP, 2011). Ecotourism is a rapidly growing industry in the Western Ghats, Sundarbans, and Kaziranga National Park in India, creating jobs and income for native people while endorsing wildlife protection. World Bank (2021) points out that nature-based tourism produces over $600 billion yearly globally. Vigorous ecosystems reduce the necessity for expensive artificial solutions for calamities, including flood control dams or water treatment plants (World Bank, 2021). Conservation is an asset in natural capital that cares for livelihoods and sustainable development.

In the medical field, numerous life-saving medicines are derived from plants and animals. For instance, the rosy periwinkle from Madagascar is used for leukemia and Hodgkin's lymphoma treatment. Likewise, compounds from snakes, frogs, and marine sponges have been identified for their possible use in treatment for heart disease, infections, and cancer (Chivian & Bernstein, 2008). In this perspective, the extinction of species is directly proportional to losing possible medicines for deadly illnesses.

The United Nations Sustainable Development Goals (SDGs), particularly Goals 13 (Climate Action), 14 (Life below Water), and 15 (Life on Land), emphasize the importance of conserving ecosystems and biodiversity (United Nations, 2015). The Convention on Biological Diversity (CBD) and the Aichi Biodiversity Targets urge countries to halt biodiversity loss (CBD, 2010). India, as a party to these agreements, has legal and moral obligations to take strong conservation measures. The conservation efforts are deeply linked to environmental health, economic development, ethical responsibility, and human survival. The loss of biodiversity threatens our food, water, health, safety, and quality of life. As the global population continues to grow and consume more natural resources, the pressure on ecosystems will increase. Conservation is no longer the concern of ecologists alone; it is now a central issue for policymakers, educators, businesses, and communities. An integrated, inclusive, and well-funded approach to conservation is essential to ensure a sustainable and just future for all (UNEP, 2021).

The chapter provides a comprehensive overview of the global biodiversity crisis, its underlying causes, and the urgent need for conservation. It examines the current state of biodiversity, key drivers of species endangerment, and the cascading effects of species loss on ecosystems. The discussion highlights the ecological, ethical, cultural, and socio-economic significance of conservation. It also outlines both conventional and innovative strategies, emphasizes community involvement, and underscores the role of policy and international collaboration in safeguarding biodiversity. In the subsequent sections, approaches, including conventional conservation methods, innovative conservation strategies, and community-based conservation models, are explored in detail to highlight their roles in safeguarding endangered species.

4 Conservation Methods

As given in Table 3, traditional methods emphasize habitat preservation, species recovery, public education, and legal enforcement. Tools like protected area networks, afforestation programs, and wildlife legislation have played a vital role in mitigating habitat destruction and population decline. Community involvement and environmental education are also essential in creating responsiveness and developing sustainable practices at the ground level. The above-mentioned approaches of conservation are equally important for working in harmony to make a solution for the multifaceted challenges confronted by biodiversity (Adom, 2018). In response to the increasing difficulty of conservation challenges, a series of innovative approaches have appeared, including cutting-edge technologies and interdisciplinary approaches. Table 4 outlines these new strategies, which include the use of artificial intelligence, remote sensing, genetic biobanking, and environmental DNA (eDNA) for more accurate and efficient biodiversity monitoring. These new approaches add-on traditional methods and embrace significant potential to develop

Table 3 Traditional approaches to biodiversity conservation

Approach	Description
Protected Areas and sanctuaries	Legally designated areas aimed at conserving wildlife and ecosystems by restricting human exploitation.
Biosphere reserves	Conservation of species outside natural habitats, like botanical gardens, zoos.
Conservation reserves	Protecting species in their natural habitats, such as national parks/ sanctuaries and reserves.
Captive breeding and reintroduction	Breeding of endangered species in captivity with the goal of releasing them into the wild to restore populations.
Environmental education	Raising public awareness about the importance of conservation.
Legal and policy frameworks	Statutory instruments at national and international levels to support biodiversity protection through regulations.
Habitat restoration	Rehabilitating ecosystems that have been degraded or fragmented.

Table 4 Innovative strategies for biodiversity conservation

Strategy	Description
Genetic rescue and assisted reproduction	Enhances genetic diversity using gene flow and reproductive technologies like IVF.
Cloning	Produces genetically identical individuals from preserved cells to revive endangered or extinct species.
CRISPR-based gene editing	Precisely modifies DNA to boost disease resistance and genetic fitness.
De-extinction	Uses cloning and genome editing to potentially restore extinct species.
Artificial Intelligence (AI) and Machine Learning (ML)	Enables real-time species tracking, predictive modeling, and automated poaching detection.
Remote sensing and GIS	Provides spatial data for monitoring land-use change, habitat loss, and species distribution.
Drone Surveillance & Internet of things (IoT)	Tracks wildlife and detects threats using drones, sensors, and real-time data transmission.

conservation results. Incorporating such advances into mainstream conservation policy is needed to address the continuing biodiversity crisis more efficiently (Mascia & Mills, 2018).

The aim is not to directly compare traditional and modern approaches, but to highlight how each method can synergistically contribute to the overarching goal of conserving endangered species. Detailed information on both approaches is provided in the following sections.

4.1 Conventional or Traditional Conservation Methods

4.1.1 Protected Areas and Wildlife Sanctuaries

Protected areas are the basis of global conservation approaches, serving as designated areas where natural resources and biodiversity are lawfully secured from over-exploitative human actions (UNEP-WCMC, 2020). The protected areas and wildlife sanctuaries play an important role in preserving the habitat of endangered species, preserving ecological processes, and providing opportunities for education, research, and sustainable tourism (Chape et al., 2005). National Parks are maintained by the governments explicitly to protect natural biodiversity. These parks contribute considerably to ecosystem services such as water purification, climate regulation, and carbon sequestration (MEA 2005). Any form of exploitation, such as logging, hunting, and/or mining, is prohibited in the premises of these parks, as these are protected regions as part of the conservation of wildlife and endangered species. The examples from India include Jim Corbett National Park protecting Bengal tigers, and similarly Kaziranga National Park that preserves the one-horned rhinoceros, and their habitats (Wildlife Institute of India, 2021).

Biosphere Reserves denote a larger conservation model designed by UNESCO under its Man and the Biosphere Programme (MAB) (UNESCO, 2017). These reserves aim to merge biodiversity conservation with the sustainable use of natural resources. A biosphere reserve is usually classified into core, buffer, and transition zones, permitting different levels of human activities. The Nilgiri Biosphere Reserve and Sundarbans Biosphere Reserve from India underscore the amalgamation of ecological conservation with community development programs and research. Conservation Reserves are protected areas neighboring national parks and sanctuaries are established to make connectivity corridors for wildlife movement (Kothari et al., 1998). The government, with the help of native people, has been handling these areas for safeguarding both ecological integrity and livelihood support. They contribute a crucial function in the reduction of habitat destruction and enhancing landscape conservation. These protected areas offer asylum for flora and fauna, serve as climate change mitigation zones, and support national and international biodiversity obligations (CBD, 2010). Nevertheless, their success depends on operational management, funding, administration of regulations, and inclusive governance that includes indigenous and native communities.

4.1.2 Captive Breeding and Reintroduction

Captive breeding and reintroduction are other important conservation approaches that have been used to save species on the edge of extinction (Snyder et al., 1996). This method involves breeding individual species in controlled environments, including zoos, wildlife centers, and/or specialized facilities, and later releasing them into their normal and/or restored habitats to reinforce wild populations. One of the remarkable accomplished stories is that of the extinct species Arabian oryx that was revived as part of the international captive breeding programs initiated in the 1960s (Spalton, 1993). The Operation Oryx project headed to the successful restoration of the oryx into protected areas in Oman, Saudi Arabia, and the United Arab Emirates (UAE), where the species now flourishes under continued surveillance and administration.

In the same way, the California condor largest North American land bird, faced extinction in the 1980s with only 27 birds remaining. Conservationists captured the remaining wild condors and established a captive breeding program. Currently, over 300 condors have been reinstated across California, Baja California, Arizona, Utah, and Mexico, and the majority of them are living in the forest (Walters et al., 2010). These examples emphasize the possibility of captive breeding and reinstatement with the amalgamation of scientific research, policy support, habitat restoration, and citizen engagement. Nevertheless, these programs require long-term commitment, resources, and adaptive administration to ensure reintroduced inhabitants become self-sustaining and resilient in the face of environmental deviations.

4.1.3 Legal and Policy Frameworks: National and International Mechanisms

The Legal and policy frameworks give an incredible biodiversity preservation obligation by providing structured, enforceable mechanisms to safeguard endangered species and their habitats. These frameworks function at both the international and national levels and include a combination of treaties, conventions, and legal statutes that shape conservation efforts globally and locally (CBD, 2010). At the international level, several legally binding treaties and conventions aim to encourage the conservation and sustainable use of biodiversity. Among the most significant are given in Table 5 (CITES Secretariat, 2022). India has passed several legal statutes and policies intended at protecting biodiversity and ecosystems given in Table 6 (Government of India, 1972) and (United Nations, 1992).

Table 5 International conventions and treaties for biodiversity conservation

Law/Policy	Description
Wildlife (Protection) Act, 1972	Provides protection to wildlife species and establishes protected areas such as national parks and sanctuaries.
Forest Conservation Act, 1980	Restricts diversion of forest land for non-forest purposes without central approval.
Biological Diversity Act, 2002	Implements the objectives of the Convention on Biological Diversity (CBD), focusing on conservation, sustainable use, and equitable benefit-sharing.
Environment Protection Act, 1986	Serves as an umbrella legislation for environmental protection, including biodiversity.
National Biodiversity Action Plan	Provides a strategic framework for integrating biodiversity concerns into national planning.

Table 6 National legal frameworks for biodiversity conservation in India

Treaty/Convention	Scope
Convention on Biological Diversity (CBD), 1992	Global framework for biodiversity conservation, sustainable use, and benefit-sharing. India is a signatory.
Convention on International Trade in Endangered Species of Wild Fauna and Flora (CITES), 1975: Convention on International Trade in Endangered Species	Regulates international trade in wild fauna and flora to prevent overexploitation
Ramsar Convention (1971)	Focuses on the conservation and wise use of wetlands of international importance.
UNFCCC (1992) United Nations Framework Convention on Climate Change	Addresses biodiversity indirectly through climate policies affecting ecosystems.
Convention on Migratory Species (CMS)	Aims to protect migratory species across their range states. India hosted CMS COP13 in 2020.

By evidently differentiating between national statutes and international conventions, this section presents a multi-layered legal construction for biodiversity conservation. Real operation of both levels of regulation, reinforced by international cooperation and national accountability, remains necessary to accomplishing long-term conservation results.

4.1.4 Community Forest Management and Indigenous Knowledge

The Community Forest Management (CFM) and native/indigenous/tribal knowledge systems are recognized as one of the effective conservation strategies across the globe (Agrawal & Gibson, 1999). This approach highlights the active role of native/local and indigenous/tribal communities as protectors of biodiversity, focusing on their indigenous ecological knowledge, sustainable practices and habits, and spiritual association with the land. Internationally, CFM has been practiced in numerous landscapes, including the tropical rainforests of the Amazon and Congo Basin, and the boreal forests of Scandinavia and Canada (Chhatre & Agrawal, 2009). In this method, native people are vigorously engaged in dealing with wild resources, shielding ecosystems, and nurturing biodiversity. In Nepal, the community forest user groups manage more than 2.3 million hectares of forest land, contributing to major enhancements in forest rejuvenation and carbon sequestration, along with giving livelihoods for native people and strengthening local governance (Gilmour, 2016). In Papua New Guinea, traditional landowners handle approximately 97% of the land, providing a successful balance between ecological conservation and cultural practice (Foale et al., 2011).

The Indigenous/tribal/native knowledge systems are deep-rooted in centuries of observation and engagement with nature. The Maasai people of East Africa have long preserved cropping patterns that support savanna biodiversity and wildlife movement. In Australia, Aboriginal fire management practices such as controlled burns are being reintegrated into modern conservation to control large wildfires and endorse a healthy ecosystem (Bliege Bird et al., 2008). Various scientific studies have consistently revealed that forests managed by indigenous/native/tribal peoples frequently have equal or higher biodiversity and lower deforestation rates compared to government-protected areas. This is due in part to their citizen scientific knowledge of local species, climatic patterns, and sustainable resource use. It is very important to make sure the secure land tenure, support indigenous/tribal rights, encourage inclusive decision-making, and incorporate traditional knowledge with modern conservation science for a complete understanding of the potential of CFM and tribal knowledge. Such collaborative frameworks provide whole and sustainable pathways for global biodiversity preservation, climate resilience, and community empowerment.

While conventional conservation methods such as protected areas, captive breeding, and community forest management have significantly contributed to biodiversity preservation, they are not without limitations. Many protected zones suffer from inadequate enforcement, limited funding, and isolation, which can restrict genetic flow and make populations vulnerable to environmental changes (Mace, 2014).

Captive breeding could face barriers including behavioral maladaptation, genetic blockages, and inadequate habitat appropriateness for reintroduction. Community-based efforts can face challenges due to socio-political conflicts, lack of capability, and/or inadequate scientific amalgamation. Consequently, these traditional strategies, though foundational, must now be complemented by cutting-edge technologies and advanced approaches. Emerging tools like artificial intelligence, drones, genetic engineering, and real-time monitoring systems can help to forecast the threats in advance, prevent poaching, reinstate habitats more accurately, and foresee the extinction risks with better accuracy. When combined with the grassroots knowledge and participation of local communities, these technologies can form a dynamic, multidimensional conservation framework capable of addressing the complex and evolving challenges of species extinction (Pimm et al., 2015).

4.2 Innovative Conservation Approaches

4.2.1 Genetic Rescue and Assisted Reproduction

The genetic rescue and assisted reproduction denote an innovative approach in conservation biology, proposing a new hope for stimulating endangered or extinct species. This approach influences developments in reproductive science, genetics, and biotechnology to take care of the challenges, including inbreeding, genetic bottlenecks, and population decline (Frankham, 2015). The genetic rescue involves introducing novel genetic material into a population to increase genetic diversity and reduce the risk of extinction. This can be attained through regulated gene flow between the isolated populations or the usage of procreative technologies, including artificial fertilization and in vitro fertilization (IVF) (Hohenlohe et al., 2021). The populations can develop more resistance to disease, environmental change, and other survival risks by enhancing the genetic variation. Three modern biotechnological approaches like cloning, Clustered Regularly Interspaced Short Palindromic Repeats (CRISPR) based gene editing, and de-extinction are mushrooming at the vanguard of conservation science, providing extraordinary potential to rescue endangered species, reinstate lost genetic diversity, and possibly revive extinct organisms (Pimm et al., 2014).

4.2.2 Cloning

Cloning is a technique of assisted reproduction, where genetically identical individuals are produced from conserved cells (Wilmut et al., 1997). In conservation biology, cloning functions as a prospective instrument for reviving endangered and/or extinct species by reproducing the individuals from conserved genetic materials, including tissue or frozen cells. The process uses somatic cell nuclear transfer (SCNT), where the nucleus of a donor cell is injected into an egg cell that has had its nucleus removed. This egg is encouraged to split and develop into an embryo that is implanted into a surrogate mother (Loi et al., 2007). Cloning helps preserve genetic diversity and can be valuable when only a few individuals of a species survive. Nevertheless, the cloned animals suffer from health issues, low survival rates, and a lack of behavioral adaptation to the wild. Furthermore, cloning does not address the root causes of extinction. The method of cloning is best used in conjunction with broader habitat and population conservation strategies.

A remarkable example of cloning is the black-footed ferret in the USA, where a clone named Elizabeth Ann was born in 2020 from preserved cells of a female ferret that lived in the 1980s (Revive & Restore, 2021). This effort reinstated valuable genetic diversity to the species' gene pool. Przewalski's horse was cloned in 2021, which is another success story of cloning. The National Dairy Research Institute (NDRI) in India efficaciously cloned a buffalo calf named Garima-II in 2013 using SCNT (NDRI, 2013). Even though the cloning method is not implemented yet for endangered species in particular, India's progress in cloning offers a robust foundation for future conservation cloning efforts. Endangered species like the Asiatic lion or Hangul deer to preserve their genetic diversity by using potential cloning methods to rescue them from extinction. However, it must be incorporated with ecological management and habitat restoration for lasting conservation impact.

4.2.3 CRISPR-Based Gene Editing

CRISPR is a revolutionary gene-editing technology that empowers scientists to create accurate alterations to an organism's deoxyribonucleic acid (DNA) (Doudna & Charpentier, 2014). CRISPR allows for accurate editing of DNA to rectify harmful mutations, introduce advantageous characteristics, or escalate genetic variation. In conservation, CRISPR is being explored to improve disease resistance in species like the Tasmanian devil that faces population deterioration due to a contagious cancer. CRISPR is also being researched to improve climate resilience in various species. The coral conservation, where CRISPR is experimented with to proliferation of thermal tolerance in corals facing bleaching due to climate change. It also holds potential in preventing hereditary disorders in small, isolated populations. This technique guides ribonucleic acid (RNA) to target particular sequences in the genome with a Cas9 enzyme that cuts the DNA at the exact location. Scientists can add, delete, or alter segments of DNA to attain the desired trait (Bay & Palumbi, 2015). In India, while CRISPR is mainly used in agriculture and medical research, its potential for wildlife conservation is being explored. The Indian Council of Agricultural Research (ICAR) and the Centre for Cellular and Molecular Biology (CCMB) in Hyderabad are doing the CRISPR-based research in livestock and aquatic species (ICAR, 2021; CCMB, 2022).

Through the increasing genome mapping of endangered Indian species like the great Indian bustard, CRISPR would play a crucial role in preventing genetic disorders or enhancing reproductive traits in the coming days. However, CRISPR is facing ethical challenges and ecological concerns, such as unintended mutations, genetic homogenization, or interference with natural evolutionary processes.

4.2.4 De-extinction

De-extinction represents the scientific procedure of bringing back extinct species using advanced biotechnological tools such as cloning, selective breeding, and genome editing. It includes recovering and renovating the DNA of extinct organisms, often from preserved tissue or fossilized remains and introducing the genetic material into the genome of a closely associated living species to produce a biological proxy or hybrid (Shapiro, 2015). The objective is not only just to recreate the species but to restore ecological functions lost with its extinction.

On April 7, 2025, the Texas-based biotechnology firm named Colossal Biosciences announced that it had successfully resurrected the dire wolf (*Canis dirus*), an ancient predator that went extinct more than 12,000 years ago. This achievement shows an important landmark in the field of de-extinction, where ancient DNA is reconstructed and incorporated with that of modern relatives to revive extinct species. While the dire wolf was genetically distinct from today's gray wolf, the Colossal's team leveraged advanced CRISPR gene-editing, synthetic biology, and surrogate breeding to restore an organism with key traits of the original species (Colossal Biosciences, 2025).

This transformative achievement has triggered a deliberation across the globe on the various dimensions of the de-extinction procedure. It illustrates the strength of synthetic genomics to possibly restore lost biodiversity and ecosystems. However, it attracts critical concerns about ecosystem amalgamation, genetic integrity, animal wellbeing, and the redirection of resources from endangered species conservation to speculative projects. The case of the dire wolf highlights that while de-extinction may captivate the imagination, its responsible application requires robust socio-legal frameworks, ecological risk assessments, and public deliberations. The concept of de-extinction, bringing back extinct species like the woolly mammoth or passenger pigeon through DNA reconstruction and surrogate breeding, is also being researched (Colossal Biosciences, 2025). It indicates the dual potential and risk of applying biotechnology in conservation. Even though genetic rescue and assisted reproduction embrace enormous promise and their success is based on inclusive planning, habitat conservation, and long-term monitoring. These innovations are seen as complementary tools that must be used in tandem with traditional conservation methods and community-based efforts.

In India, de-extinction is in the conceptual phase, but it garners interest due to the extinction of several regional species like the Indian cheetah. At the same time, India has reintroduced African cheetahs into Madhya Pradesh's Kuno National Park. De-extinction through genetic revival of the original Indian cheetah is theoretically possible in the future with sufficient DNA samples and technological advancements (BBC News, 2023). Moreover, the extinct subspecies of the pink-headed duck or Himalayan quail are sporadically discussed in the research community for potential revival. While the enthusiasm for de-extinction is rising, India presently prioritizes conserving endangered species before extinction, stressing habitat protection, anti-poaching, and community-based models over speculative resurrection.

4.2.5 Artificial Intelligence and Machine Learning

Artificial Intelligence (AI) and Machine Learning (ML) are rapidly transforming conservation biology by enabling data-driven, real-time decision-making in ways unachievable through traditional methods. These technologies are demonstrating priceless in the monitoring, analysis, and protection of endangered species and their habitats (Wich et al., 2015; Schneider et al., 2020). One of the most influential applications of AI is in predictive modeling for population trends. By processing large datasets including climate records, satellite imagery, migration patterns, and historical species counts, AI algorithms can detect patterns and forecast population dynamics. These models allow conservationists to predict how the species could respond to habitat loss, climate change, or disease outbreaks, permitting for active planning and targeted involvement. For example, AI models have been used to evaluate elephant movements in African savannas and forecast reproductive results for endangered birds in coastal habitats (Firth-Butterfield, 2023).

Another advanced tool is the use of AI-based camera traps and automated poaching detection systems. Traditional camera traps required manual review of thousands of images, which was a time-consuming and inefficient process. AI systematizes this task through deep learning to accurately identify species, track individual animals, and detect strange behaviors, and provides quicker responses and increases monitoring efficacy. In addition, AI-powered systems combined with acoustic sensors and drones will be helpful to identify gunshots, vehicle movement, or illegal human existence in secured areas, alarming authorities in real time to potential poaching cases (Norouzzadeh et al., 2018). These innovations not only increase the accuracy and speed of conservation actions but also reduce the human resources, making monitoring more scalable and cost-effective. The efficacious applications of AI require technical proficiency, high-quality data, and sufficient infrastructure. AI and ML increase the effectiveness of protecting endangered species and safeguarding ecological resilience in the dynamic world when used in combination with indigenous knowledge and community-based conservation.

4.2.6 Remote Sensing and Geographic Information Systems (GIS)

Remote Sensing and Geographic Information Systems (GIS) have become essential tools in conservation science, contributing detailed, real-time insights into Earth's vigorous ecosystems (Pettorelli et al., 2014). This powerful technology enables us to collect, analyze, and envisage geospatial data in earlier unimaginable. By integrating various layers of information from topography and vegetation cover to species distributions and human actions, GIS delivers a broad view of ecosystems and their changing nature. These technologies empower large-scale monitoring of biodiversity and environmental alteration, permitting conservationists to make informed, timely, and data-driven decisions for the safeguarding of endangered species and their environments (Cushman et al., 2016). Remote sensing involves the gathering of information on Earth's surface through satellites, aerial sensors, and/or drones with indirect contact. It offers high-resolution imagery and spectral data that can detect changes in vegetation cover, water bodies, forest degradation, and land-use patterns.

The technology can be meritoriously used to detect forest fires, deforestation, recognizing habitations of endangered animals, and assess the impacts of natural calamities and human intrusion on ecosystems. These systems allow for the formation of detailed habitat plans, enabling researchers to identify biodiversity centers, evaluate habitat appropriateness, and model species dispersal under various environmental situations. The GIS-based habitat modeling has been used to trace the nomadic patterns of species like snow leopards, elephants, and sea turtles, seconding the setting up of transboundary passages and protected areas (Pettorelli et al., 2014). One of the most impactful applications of GIS is location change detection, which involves comparing satellite images over time to identify tendencies such as deforestation, wetland reduction, and urban extension. This facility is crucial for well-timed alarm systems and lasting conservation planning, allowing immediate response to risks before catastrophe occurs. The conservationists will be able to monitor enormous and unreachable areas with accuracy, enabling effective resource distribution and prioritizing of conservation efforts by integrating remote sensing and GIS. These technologies not only improve our understanding of species ecology and environmental changes but also support proof-based policy framing, habitat rebuilding, and climate resilience tactics.

4.2.7 Drone Surveillance and Internet of Things (IoT) for Real-Time Monitoring

Drone surveillance and the Internet of Things (IoT) are revolutionizing real-time monitoring in conservation, offering precise, scalable, and cost-effective solutions for protecting endangered species and their habitats (Zhang et al., 2016). These technologies are principally impactful in anti-poaching operations and environmental condition tracking, where rapid detection and response are needed. Drones or Unmanned Aerial Vehicles (UAVs), fitted out with high-resolution cameras, thermal imaging, and the Global Positioning System (GPS) technology, allow conservationists to monitor remote landscapes with less human intrusion. In anti-poaching efforts, drones offer aerial surveillance over protected areas, empowering real-time tracing of both wildlife and unlawful human activities. Thermal cameras are capable of spotting body heat, which can lead to the detection of poachers even in the dark or under thick canopy cover. These systems communicate live footage to ground teams, thus considerably decreasing action time and improving the probability of defensive intervention (Gade & Moeslund, 2014).

The Internet of Things (IoT) devices, like environmental monitoring units, camera traps, and acoustic sensors that are scientifically collecting data on animals and birds' movements, sound activity, temperature, humidity, and water quality. This data is transferred instantaneously to the cloud database, where it is thoroughly scrutinized for anomalies, including the escalation in gunshot sounds or sudden reductions in water levels, enabling the timely detection of potential risks (Sultana & Wahid, 2019).

For example, in Kenya's Maasai Mara National Reserve, a combination of drones and IoT-enabled collars on rhinos and elephants has led to a considerable decline in poaching events by permitting rangers to monitor animal movements and reply quickly to unlawful interferences (Wasser et al., 2015). India has introduced the most advanced technologies to monitor and safeguard the Bengal tiger from the risk of extinction. The National Tiger Conservation Authority (NTCA) of India has implemented the monitoring system for Tigers-Intensive Protection and Ecological Status (M-STrIPES), a GPS-based application that allows forest officers and/or guards to collect and analyze the data on locations of tigers, deaths, and ecological observations throughout patrols (NTCA, 2018).The drone surveillance, along with IoT, makes a comprehensive conservation intelligence system, which not only enhancements the protection of endangered species but also supports ecosystem health monitoring and research. When combined with traditional field patrols and community engagement, these technologies deliver an authoritative toolset for progressing active, real-time conservation approaches.

A geographically diverse perspective on the application of advanced conservation strategies is presented in the global map, Fig. 3. This map shows country-wise some examples where innovative technologies such as cloning, de-extinction,

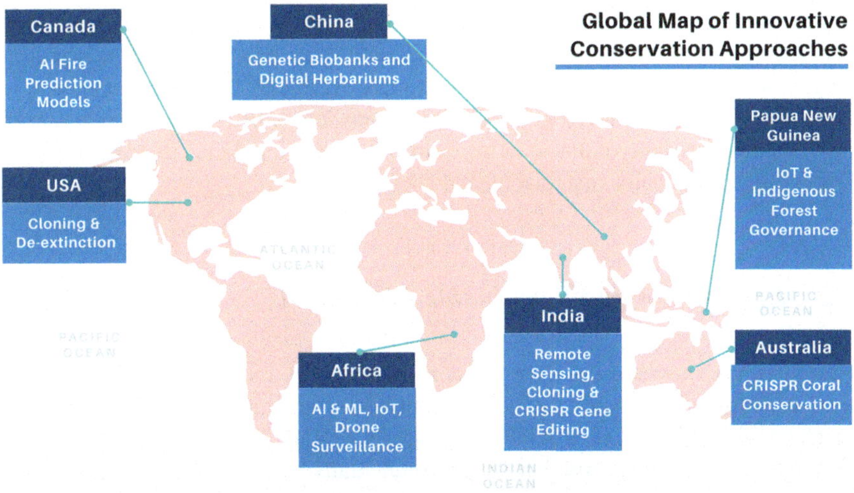

Fig. 3 Global distribution of innovative conservation strategies for endangered species with country-specific examples

CRISPR-based gene editing, artificial intelligence, drone surveillance, and Internet of Things (IoT) have been successfully employed for the conservation of endangered species.

5 Community-Based Conservation Models

5.1 Participatory Rural Appraisal (PRA) and Local Stewardship

Participatory Rural Appraisal (PRA) is considered a vivacious technique that vigorously encourages native people in planning, executing, and monitoring conservation initiatives (Chambers, 1994). It highlights knowledge exchange, mutual respect, and empowerment, empowering various local stakeholders to identify their conservation issues and identify locally suitable solutions. PRA takes indigenous ecological knowledge, traditional land management practices, and social structures that often go unnoticed in top-down models through tools such as mapping, seasonal calendars, and resource ranking. In India, the Dongria Kondh tribe used PRA techniques to document their spiritual and ecological relationship with the Niyamgiri Hills, helping stop devastating mining projects (Bhagwat et al., 2015). Likewise, in Nepal, community-managed forests covering over 2 million hectares have revealed improved biodiversity and forest cover due to a PRA-driven governance approach (Gilmour, 2016). Internationally, Papua New Guinea has embraced clan-based stewardship where traditional landowners/feudal use participatory tools to manage forests, contributing to both conservation and carbon credit schemes (WCS, 2024). The First Nations communities of Canada have fruitfully used participatory methods to co-manage marine areas, integrating indigenous knowledge with modern science (Alexander et al., 2019). These cases demonstrate that conservation is prolific when communities are considered not as mere recipients, but as co-managers of natural resources, acknowledging their local knowledge. PRA is not simply nurturing local responsibility but also helps form conservation policies that are culturally respectful, ecologically sustainable, and socially just.

5.2 Ecotourism and Sustainable Livelihoods

Ecotourism is known as the model of conservation that associates biodiversity protection with financial gains for local communities, generating sustainable employment while educating environmental awareness (Kiss, 2004). It operates on the principle of low-impact tourism in environmentally sensitive areas, ensuring that local communities directly benefit from the preservation of natural resources. Ecotourism, such as guided wildlife visits, safari, homestays, field trips, and craft sales, creates revenue which is invested in local conservation practices and development, which motivates communities to protect the ecosystem (Kiss, 2004). For instance, the tribal communities from Periyar Tiger Reserve in India participate in forest patrolling, eco-guided treks, and cultural exposition programs. In Kenya and Tanzania, the Maasai people achieve conservancies neighboring national parks, where ecolodges generate income used for community improvement and the safety of wildlife (Perera, 2024). These models have significantly decreased poaching incidents and improved the economic flexibility of local populations.

In Costa Rica, community-run ecolodges in regions like Monteverde and Osa Peninsula give wildlife tours and conservation experiences, and funds generated from this are used in education, reforestation, and marine protection (Campbell, 2002). In the Amazon Basin, native communities are handling ecotourism to support sustainable hunting, forest 24-h care, and traditional skills industries. Ecotourism has assisted in aligning conservation with advanced goals, confirming that environmental protection is economically achievable. However, its success depends on equal benefit-sharing, community leadership, and environmental planning. When accomplished reliably, ecotourism is a potential tool for conservation, cultural preservation, and rural empowerment.

5.3 Education and Awareness Campaigns

Encouraging a conservation mindset across generations and geographical areas needs education and awareness initiatives (Jacobson et al., 2015). Integrating environmental education into school curricula supports young learners to understand biodiversity, ecosystems, and the role of humans in nature. Through creative learning approaches like nature walks, eco-clubs, and wildlife-themed classroom activities, students gain ecological literacy and a sense of responsibility for conservation at a young age. For example, National Green Corps in India endorses programs such as the eco-clubs in schools, while in South Africa, conservational and ecological education is part of the Life Orientation curriculum (Ministry of Environment, Forest and Climate Change, 2001). These steps prepare students with practical conservation knowledge and principles. On the other side, media outreach and citizen participation play transformative roles as well. Millions of people throughout the world have been motivated by global environmental initiatives such as BBC's Planet Earth, the United Nations Environment Programme's (UNEP) Wild for Life campaign, and the World Wildlife Fund's (WWF) Earth Hour. In Latin America, radio programs in native languages raise conservation awareness in remote regions for the indigenous community, while in North America and Europe, social media engages urban youth. Youth involvement is growing rapidly through global networks such as Fridays for Future and Roots & Shoots (UNEP, 2019). These platforms provide young people with a voice in climate and conservation policy dialogs. Education and outreach are foundations of long-term conservation achievement, connecting people to nature and turning awareness into action worldwide.

5.4 *Women and Youth in Conservation Leadership*

Women and youth are often underutilized forces in conservation leadership. Empowering these groups not only endorses gender and generational equity but also takes various standpoints, novelties, and public trust into conservation practice. Women, particularly in indigenous and rural communities, are often the major stewards of land, water, and biodiversity, possessing deep ecological knowledge carried through generations (NWF, 2020). The all-female anti-poaching organization Akashinga in Zimbabwe has received international appreciation for blending social empowerment, women's rights, with conservation of biodiversity (Mkono et al., 2023). In Nepal, women constitute above 30% of community forest user group frontrunners that encourage sustainable and robust forest management (FAO, 2015). Likewise, through conservation cooperatives, indigenous women in Mexico are revitalizing agroforestry techniques and traditional seed systems. Global conservation narratives are also being reshaped by youth movements. Millions are being mobilized to demand more robust environmental policies through initiatives like Youth4Nature, EcoTok, and Greta Thunberg's Fridays for Future (UNEP, 2019). Pushing money into inclusive leadership assures that conservation is not just a responsibility of science, but also that society has a greater role to make a socially and environmentally robust future. It is important to understand that when women and youth are encouraged and empowered through governmental support, conservation gets momentum at its maximum, leads innovation, and long-term sustainability for fighting not just ecological issues, but also social equity and generational continuity.

6 Role of Government and Policy-Making

Governments across the globe have the greater responsibility for protecting endangered species and guaranteeing sustainable ecosystems. Their obligations include forming legal and policy frameworks, executing policies, allotting funding, coordinating with international organizations and bodies, and activating resources to shield biodiversity loss. A dynamic and inclusive governance is needed to overcome risks like unlawful wildlife trade, human-wildlife conflict, habitat devastation, and more on climate change in an ever-changing environmental landscape (CBD Secretariat, 2020).

One of the foremost responsibilities of governments is to enact and enforce wildlife protection laws. National legislation, like the U.S. Endangered Species Act (1973), the UK's Wildlife and Countryside Act (1981), and India's Wildlife Protection Act (1972), functions as the legal pillar for the protection of species. These laws offer the label of endangered species, guidelines for hunting and trade, the construction of sanctuaries and national parks, and a tribunal for wildlife crimes. Furthermore, policies can equip local authorities and forest departments more empowered with the legal support needed to execute conservation at the local level

(Rosen & Smith, 2010). The multilateral environmental agreements, such as the CITES, the Convention on Biological Diversity (CBD), and the Ramsar Convention, necessitate dynamic involvement and compliance from various state governments across the globe. Countries can diminish the transboundary risks, keep track of wildlife trade, and share data and resources with other countries that can helpfully understand the scenario (CITES Secretariat, 2022; CBD, 2010).

In contemporary times, the policy-making process has advanced to include community involvement and decentralized governance by identifying that protection and/or conservation cannot flourish at its peak without citizen engagement. At this juncture, governments have introduced frameworks and guidelines such as Community Conservancies in Africa, Joint Forest Management in India, and Co-Management Agreements in Canada. These models offer rights of local and indigenous communities to take care of and benefit from biodiversity, thus endorsing equity and sustainability (Armitage et al., 2020). Moreover, governments influence conservation through land-use planning, environmental impact assessments (EIA), and climate change adaptation policies. Integrating biodiversity conservation into urban development, agriculture, mining, and infrastructure policies helps reduce unintended threats to species and habitats. Policies that activate green infrastructure, buffer zones, and ecological corridors allow species to travel and familiarize themselves in response to climate change.

Transfer of Technology (ToT) and innovation are encouraged through government policy. Initiatives associated with AI-based monitoring systems, satellite tracking of endangered species, and genomic research are progressively part of national conservation strategies and/or schemes (Ullah et al., 2025). Public–Private Partnerships (PPP) models foster innovation along with guaranteeing accountability. Eventually, the triumph of conservation policy rests on political will, public awareness, cross-sector partnership, and robust execution of machineries (Groves et al., 1995). Governments must work constantly to frame, adapt, and enact policies to address new emerging risks, invest in education and capacity-building programs, and guarantee transparent and inclusive governance structures. From the point of view of guardians of national biodiversity, governments have the resources and the authority to shoulder the responsibility of conserving endangered species.

7 Synergetic Collaboration for Conservation of Endangered Species

Conserving endangered species cannot be achieved through isolated efforts, and it necessitates an extremely cooperative and progressive methodology that brings together various sectors, embraces innovation, understands and addresses systemic gaps, and anticipates possible future risks. A synergetic framework that ties stakeholders altogether internationally, shares knowledge, and forms resilient strategies and/or policies that will explain the conservation in the future (Carlson et al., 2021).

7.1 *Multi-Stakeholder Engagement*

Conservation success hinges on meaningful engagement among a diverse group of stakeholders, scientists, NGOs, policymakers, local communities, and the private sector. Scientists come up with innovative research and data, which provide provisions for conservation activities. NGOs plan and execute programs at the grassroots levels and regularly work as dynamic associations between policy and natives, simultaneously with public administration. Governments dispense the legislative and institutional frameworks that guarantee conservation activities are scalable and enforceable. In the postmodern period, the private players have flourished as a key partner, contributing their share through sustainable corporate/business practices, corporate social responsibility (CSR) initiatives, and funding mechanisms. If you want to make true engagement, it is necessary to include native/indigenous peoples, who contribute immense traditional knowledge and long-standing stewardship of natural resources (Jolibert & Wesselink, 2012).

7.2 *Data Sharing and Open Science for Conservation*

The conservation landscape is being transformed by the rise of open science and global data sharing. Platforms like the Global Biodiversity Information Facility (GBIF), and Spatial Monitoring and Reporting Tool (SMART), and the IUCN Red List assist conservationists in getting access and providing instantaneous data on species distribution, risks, and trends (IUCN, 2022; SMART, 2025). Open-access biodiversity data enriches the efficacy of research, assists in making robust decisions, and encourages transparency. Citizen science projects similar to iNaturalist and eBird are growing that assisting the monitoring efforts, public engagement simultaneously creating enormous datasets (Chandler et al., 2017). Alliance through open science guarantees that conservation is not hindered by data silos but instead benefits from collective knowledge and insight.

7.3 *Framework for a Global Conservation Coalition*

It should be noted that the international community must come together under the banner of a conservation framework for combating the acceleration of the biodiversity crisis effectively. A Global Conservation Coalition, similar to the Paris Agreement for climate, would endorse calculable objectives, align policies, guarantee resource equity, and offer political activism for biodiversity enhancement (UNEP, 2023). The present initiatives, like the Post-2020 Global Biodiversity framework, provide strong primary strategies and approaches, but at the same time, a socially and environmentally robust association must move forward by incorporating citizen voices, local knowledge, climate resilience, and inclusive development into global biodiversity governance.

8 Future Directions for Conservation of Endangered Species

The future of conservation of endangered species will not be simply based on expanding contemporary available practices but on reimagining them in an ever-changing world. Moving away from the traditional celebrated methods, the future prerequisites inclusive, interdisciplinary, innovative, and dynamic strategies and programs. As the acceleration of the biodiversity crisis because to unscientific human activities that lead to climate change and ecological imbalances, conservation efforts must progress enthusiastically. The below given future direction will offer a strategic roadmap for advancing the conservation of endangered species and providing a sustainable ecosystem.

8.1 Synthetic Habitats and Climate-Adaptive Ecosystems

In the coming decades, the traditional habitations would collapse due to the shift of climate zones, which leads to restoring nature into an outdated idea. Future conservationists will need to create synthetic habitats, prudently designed ecosystems that combine native and non-native species to make steady, strong biological communities. Coral reefs made partially of heat-resistant hybrids, or grasslands planted with species altered to drought and floods, would emerge. Conservation will become not simply a protective act but an ingenious, ecological expertise, building new designs for life itself (Solé et al., 2024).

8.2 Bio-Protection Shields for Cities

The future cities will aggressively need to use the living bio-shields, enormous networks of trees, vertical gardens, restored riverbanks, and wetlands, to survive the potential climate risks. Conservation will not just happen somewhere in remote forests, but it will invade the cityscapes too, for the survival of species along with human beings. Green walls that host native birds, moss-covered cooling towers, amphibian-friendly canals, and pollinator skyscrapers will become biodiversity headquarters by design. Saving endangered species will start on your street, not in a distant jungle (Aronson et al., 2017).

8.3 Genetic Biobanks in every Country

Preserving genetic diversity will no longer be the domain of a few frozen vaults in rich countries. Every nation, even the smallest, will operate national genetic biobanks for endangered species. Local communities, universities, and indigenous groups will participate in decentralized cryobanking, preserving DNA samples, seeds, tissues, and gametes. A future extinction will not mean total loss: every ecosystem will have a biological "memory" stored safely, offering a second chance for restoration and resilience (Austin et al., 2003).

8.4 Building Climate-Resilient Conservation Models

One of the most important shifts in conservation in the future will be learning how to work with the changing climate. The future conservation models must focus on crafting protected areas that anticipate future climate situations rather than past ones. This could help the species to migrate to adaptable habitats or design landscapes that can manage extreme climate situations. Building ecosystems that are flexible and adaptive will be essential to safeguarding biodiversity in a rapidly changing world (Antoszewski et al., 2022).

8.5 Urban Biodiversity Corridors

In the coming years, conservationists will need to consider cities as potential allies for sustainable biodiversity. Future activities should keep focus on crafting green corridors that comprise networks of parks, gardens, rooftop gardens, and tree-lined streets, which permit animals to travel securely through urban areas. City planning must incorporate biodiversity objectives that guarantee urban development supports and respects natural ecosystems. It is possible that the human being should avoid the idea of anthropocentrism and embrace egocentrism (Peng et al., 2017).

8.6 Mainstreaming Indigenous and Local Knowledge Systems

Indigenous peoples and local communities have always been the protectors of biodiversity, and their contributions are ignored by the scientific community. Nowadays, it's highly important to acknowledge their informal knowledge of indigenous people and the local knowledge of local people to address future risks. In future conservation activities, it is crucial to vigorously incorporate indigenous leadership into all levels of decision-making and acknowledge their tremendous contributions to the protection of biodiversity. Respect for indigenous rights to land, territories, and resources is not simply a moral or constitutional obligation; it signifies one of the most operational strategies for guaranteeing the protection and resilience of ecosystems (Bennett et al., 2017).

9 Conclusion

The conservation of endangered species indicates a multifaceted problem that requires an inclusive and integrated response from the entire world. As this chapter has discussed, the biodiversity crisis is driven by a combination of anthropogenic pressures, habitat destruction, climate change, invasive species, overexploitation, and pollution, all of which threaten the ecological foundations upon which life depends. Addressing these risks necessitates a well-planned amalgamation of traditional conservation methods, advanced technological innovations, community-based models, robust legal frameworks, and democratic governance.

This chapter deals with a varied collection of traditional conservation strategies, including protected areas, captive breeding programs, and wildlife protection laws, which have laid the foundation for biodiversity conservation worldwide. However, the acceleration of environmental deprivation requires the embracing of advanced tools, including genetic rescue technologies, artificial intelligence, remote sensing, and Internet of Things monitoring systems. These novelties provide extraordinary openings for precision, early intervention, and large-scale ecosystem management. The chapter has pointed out that the integration of indigenous and local knowledge systems into modern scientific conservation practices is equally important for a robust, sustainable, and futuristic ecosystem. Indigenous peoples and local communities have long been protectors of biodiversity through traditional practices centered in cultural, spiritual, and ecological wisdom that was gathered from their lived and co-existence experiences. Mainstreaming their knowledge, leadership, and participatory governance models is crucial for framing conservation frameworks that are socially equitable, ecologically sound, and globally significant.

Future directions for the conservation of endangered species, as mentioned in this chapter in detail, must be forward-looking, interdisciplinary, and adaptive. Schemes/programs/policies/strategies like crafting urban biodiversity corridors, constructing synthetic habitats, constructing climate-resilient ecosystems, and improving genetic diversity conservation through biobanking will be fundamental in responding to existing and emerging global challenges. The success of these methods will be based on the capability to synchronize scientific advancements with ethical considerations and the requirements of society. Conserving endangered species needs a revolutionary alteration from isolated, fragmented efforts to integrated, inclusive, and innovation-driven frameworks. It needs sustained commitment, cross-sectoral partnerships, political will, democratic and inclusive governance, and appreciation of the essential value of all life forms. If suitably associated, the approaches and directions deliberated in this chapter provide an operational path toward retrogressive biodiversity damage, reinstating ecosystem functionality, and guaranteeing a sustainable future for the natural world.

References

Adla, K., Dejan, K., Neira, D., & Dragana, Š. (2022). Degradation of ecosystems and loss of ecosystem services. In *One health* (pp. 281–327). Academic Press.

Adom, D. (2018). Traditional biodiversity conservation strategy as a complement to the existing scientific biodiversity conservation models in Ghana. *Environment and Natural Resources Research, 8*(3), 1–24.

Agrawal, A., & Gibson, C. C. (1999). Enchantment and disenchantment: The role of community in natural resource conservation. *World Development, 27*(4), 629–649.

Alexander, S. M., Provencher, J. F., Henri, D. A., Taylor, J. J., Lloren, J. I., Nanayakkara, L., & Cooke, S. J. (2019). Bridging indigenous and science-based knowledge in coastal and marine research, monitoring, and management in Canada. *Environmental Evidence, 8*(1), 1–24.

Antoszewski, P., Krzyżaniak, M., & Świerk, D. (2022). The future of a climate-resilient and climate-neutral city in the temperate climate zone. *International Journal of Environmental Research and Public Health, 19*(7), 4365.

Armitage, D., Mbatha, P., Muhl, E. K., Rice, W., & Sowman, M. (2020). Governance principles for community-centered conservation in the post-2020 global biodiversity framework. *Conservation Science and Practice, 2*(2), e160.

Aronson, M. F. J., Lepczyk, C. A., Evans, K. L., Goddard, M. A., Lerman, S. B., & MacIvor, J. S. (2017). Biodiversity in the city: Key challenges for urban green space management. *Frontiers in Ecology and the Environment, 15*(4), 189–196.

Austin, M. A., Harding, S., & McElroy, C. (2003). Genebanks: A comparison of eight proposed international genetic databases. *Public Health Genomics, 6*(1), 37–45.

Barbier, E. B., Hacker, S. D., Kennedy, C., Koch, E. W., Stier, A. C., & Silliman, B. R. (2011). The value of estuarine and coastal ecosystem services. *Ecological Monographs, 81*(2), 169–193.

Bay, R. A., & Palumbi, S. R. (2015). Rapid acclimation ability mediated by transcriptome changes in reef-building corals. *Genome Biology and Evolution, 7*(6), 1602–1612.

BBC News. (2023, August 3). *India Cheetah Project: Fears over big cat deaths in Kuno National Park*. BBC. https://www.bbc.com/news/world-asia-india-66392483

Bennett, N. J., Roth, R., Klain, S. C., Chan, K. M., Clark, D. A., Cullman, G., et al. (2017). Mainstreaming the social sciences in conservation. *Conservation Biology, 31*(1), 56–66.

Bhagwat, S. A., Kushalappa, C. G., Williams, P. H., & Brown, N. D. (2015). A battle lost? Report on mining activities in the Niyamgiri Hills, India. *Conservation and Society, 13*(1), 29–39.

Bhattacharya, P. (2025). Forest biodiversity conservation. In *Textbook of forest science* (pp. 123–142). Springer Nature Singapore. https://doi.org/10.1007/978-981-97-8289-5_7

Bliege Bird, R., Bird, D. W., Codding, B. F., Parker, C. H., & Jones, J. H. (2008). The "fire stick farming" hypothesis: Australian aboriginal foraging strategies, biodiversity, and anthropogenic fire mosaics. *Proceedings of the National Academy of Sciences, 105*(39), 14796–14801.

Bodo, T., Gimah, B. G., & Seomoni, K. J. (2021). Deforestation and habitat loss: Human causes, consequences, and possible solutions. *Journal of Geographical Research, 4*(2), 22.

Bralower, T., & Millet, A. (2021). Endangered species and ecosystems. In *Communities in crisis: Student voices on climate change*.

Campbell, L. M. (2002). Conservation narratives and the "received wisdom" of ecotourism: Case studies from Costa Rica. *International Journal of Sustainable Development, 5*(3), 300–325. https://doi.org/10.1504/IJSD.2002.003755

Carlson, R. R., Evans, L. J., Foo, S. A., Grady, B. W., Li, J., Seeley, M., & Asner, G. P. (2021). Synergistic benefits of conserving land-sea ecosystems. *Global Ecology and Conservation, 28*, e01684.
Carrington, D. (2022, December 6). *The biodiversity crisis in numbers—A visual guide*. The Guardian. Retrieved from https://www.theguardian.com/environment/2022/dec/06/the-biodiversity-crisis-in-numbers-a-visual-guide-aoe
Centre for Cellular and Molecular Biology (CCMB). (2022). *CRISPR-based genome editing research in livestock and aquatic species*. Council of Scientific and Industrial Research. https://www.ccmb.res.in
Chambers, R. (1994). Participatory rural appraisal (PRA): Challenges, potentials and paradigm. *World Development, 22*(10), 1437–1454.
Chandler, M., See, L., Copas, K., Bonde, A. M., López, B. C., Danielsen, F., & Turak, E. (2017). Contribution of citizen science towards international biodiversity monitoring. *Biological Conservation, 213*, 280–294.
Chape, S., Harrison, J., Spalding, M., & Lysenko, I. (2005). Measuring the extent and effectiveness of protected areas as an indicator for meeting global biodiversity targets. *Philosophical Transactions of the Royal Society B: Biological Sciences, 360*(1454), 443–455.
Chhatre, A., & Agrawal, A. (2009). Trade-offs and synergies between carbon storage and livelihood benefits from forest commons. *Proceedings of the National Academy of Sciences, 106*(42), 17667–17670.
Chivian, E., & Bernstein, A. (Eds.). (2008). *Sustaining life: How human health depends on biodiversity*. Oxford University Press.
CITES Secretariat. (2022). *What is CITES?* https://cites.org/eng/disc/what.php
Colossal Biosciences. (2025). *Colossal resurrects the dire wolf: A new era of de-extinction begins*. Retrieved from https://www.colossal.com/
Convention on Biological Diversity. (2009). *Invasive alien species: Introduction and spread*. Secretariat of the Convention on Biological Diversity. https://www.cbd.int/invasive/WhatareIAS.shtml
Convention on Biological Diversity (CBD) Secretariat. (2020). *Global biodiversity outlook 5*. Retrieved April 28, 2025, from https://www.cbd.int/gbo5
Cushman, S. A., Chase, M., & Griffin, C. (2016). Mapping landscape resistance to identify corridors and barriers for elephant movement in southern Africa. *Landscape Ecology, 25*(3), 315–328.
DeLong, D. C., Jr. (1996). Defining biodiversity. *Wildlife Society Bulletin*, 738–749.
Doudna, J. A., & Charpentier, E. (2014). The new frontier of genome engineering with CRISPR-Cas9. *Science, 346*(6213), 1258096.
Firth-Butterfield, K. (2023, March 3). *Here's how AI is helping Africa's endangered elephants*. World Economic Forum. https://www.weforum.org/stories/2023/03/africa-endangered-forest-elephants-ai/
Foale, S., Cohen, P., Januchowski-Hartley, S., Wenger, A., & Macintyre, M. (2011). Tenure and taboos: Origins and implications for fisheries in the Pacific. *Fish and Fisheries, 12*(4), 357–369.
Food and Agriculture Organization of the United Nations (FAO). (2015). *Understanding women's participation in forestry in Nepal* https://www.fao.org/fileadmin/templates/rap/files/meetings/2015/150210_Nepal_brief.pdf
Food and Agriculture Organization of the United Nations (FAO). (2020). *Global forest resources assessment 2020: Main report*. FAO. https://doi.org/10.4060/ca9825en
Frankham, R. (2015). Genetic rescue of small inbred populations: Meta-analysis reveals large and consistent benefits of gene flow. *Molecular Ecology, 24*(11), 2610–2618.
Gade, R., & Moeslund, T. B. (2014). Thermal cameras and applications: A survey. *Machine Vision and Applications, 25*, 245–262.
Gilmour, D. (2016). *Forty years of community-based forestry: A review of its extent and effectiveness*. http://www.fao.org/3/a-i5415e.pdf

Goldschmidt, T., Witte, F., & Wanink, J. (1993). Cascading effects of the introduced Nile perch on the detritivorous/phytoplanktivorous species in the sublittoral areas of Lake Victoria. *Conservation Biology, 7*(3), 686–700.

Government of India. (1972). *Wildlife Protection Act, 1972*. https://www.india.gov.in/wildlife-protection-act-1972-3

Groves, C. R., Klein, M. L., & Breden, T. F. (1995). Natural heritage programs: Public-private partnerships for biodiversity conservation. *Wildlife Society Bulletin*, 784–790.

Hinsley, A., Willis, J., Dent, A. R., Oyanedel, R., Kubo, T., & Challender, D. W. (2023). Trading species to extinction: Evidence of extinction linked to the wildlife trade. *Cambridge Prisms: Extinction, 1*, 10.1017/ext.2023.7.

Hohenlohe, P. A., Funk, W. C., & Rajora, O. P. (2021). Population genomics for wildlife conservation and management. *Molecular Ecology, 30*(1), 62–82. https://doi.org/10.1111/mec.15720

Indian Council of Agricultural Research (ICAR). (2021). *Annual report 2021–22*. ICAR. https://icar.org.in/annual-report

Intergovernmental Panel on Climate Change. (2018). *Global warming of 1.5°C: An IPCC special report on the impacts of global warming of 1.5°C above pre-industrial levels and related global greenhouse gas emission pathways*. https://www.ipcc.ch/sr15/

International Union for Conservation of Nature (IUCN). (2022). *The IUCN red list of threatened species—Summary statistics*. Retrieved from. https://www.iucnredlist.org

IPBES. (2019). *Summary for policymakers of the global assessment report on biodiversity and ecosystem services* (S. Díaz, J. Settele, E. S. Brondízio, H. T. Ngo, M. Guèze, J. Agard, C. Zayas, Eds.). Intergovernmental Science-Policy Platform on Biodiversity and Ecosystem Services. https://files.ipbes.net/ipbes-web-prod publicfiles/inline/files/ipbes_global_assessment_report_summary_for_policymakers.pdf

Irshad, S. M., & Joseph, J. (2015). An invisible disaster: Endosulfan tragedy of Kerala. *Economic and Political Weekly*, 61–65.

Jacobson, S. K., McDuff, M. D., & Monroe, M. C. (2015). *Conservation education and outreach techniques* (2nd ed.). Oxford University Press. https://global.oup.com/academic/product/conservation-education-and-outreach-techniques-9780198716693?cc=in&lang=en&

Jolibert, C., & Wesselink, A. (2012). Research impacts and impact on research in biodiversity conservation: The influence of stakeholder engagement. *Environmental Science & Policy, 22*, 100–111.

Kiss, A. (2004). Is community-based ecotourism a good use of biodiversity conservation funds? *Trends in Ecology & Evolution, 19*(5), 232–237.

Kolawole, A. S., & Iyiola, A. O. (2023). Environmental pollution: Threats, impact on biodiversity, and protection strategies. In *Sustainable utilization and conservation of Africa's biological resources and environment* (pp. 377–409). Springer Nature.

Kothari, A., Pathak, N., & Anuradha, R. V. (1998). *Conservation with communities: Towards a policy and practice of equity, sustainable use, and livelihood security*. Indian Institute of Public Administration. https://www.cbnrm.net/pdf/kothari_a_002.pdf

Loi, P., Galli, C., & Ptak, G. (2007). Cloning of endangered mammalian species: Any progress? *Trends in Biotechnology, 25*(5), 195–200.

Mace, G. M. (2014). Whose conservation? *Science, 345*(6204), 1558–1560. https://doi.org/10.1126/science.1254704

Mascia, M. B., & Mills, M. (2018). When conservation goes viral: The diffusion of innovative biodiversity conservation policies and practices. *Conservation Letters, 11*(3), e12442.

Millennium Ecosystem Assessment (MEA). (2005). *Ecosystems and human Well-being: Biodiversity synthesis*. World Resources Institute. https://www.millenniumassessment.org/en/Synthesis.html

Ministry of Environment, Forest and Climate Change. (2001). *National Green Corps (NGC) Eco Clubs. Government of India*. Retrieved April 28, 2025, from https://eepmoefcc.nic.in/

Mkono, M., Rastegar, R., & Ruhanen, L. (2023). Empowering women to protect wildlife in former hunting tourism zones: A political ecology of Akashinga, Zimbabwe. *Journal of Sustainable Tourism, 31*(5), 1090–1106.

Muluneh, M. G. (2021). Impact of climate change on biodiversity and food security: A global perspective—A review article. *Agriculture & Food Security, 10*(1), 1–25.

National Biodiversity Action Plan (NBAP). (2008). *Ministry of Environment and Forests, Government of India.* https://www.moef.gov.in

National Dairy Research Institute (NDRI). (2013). *Achievements in animal biotechnology.* https://ndri.res.in/

National Tiger Conservation Authority (NTCA). (2018). *Monitoring System for Tigers–Intensive Protection and Ecological Status (M-STrIPES).* Retrieved from https://ntca.gov.in/our-work-m-stripes/

National Wildlife Federation (NWF). (2020). *Women in conservation leadership.* National Wildlife Federation. https://www.nwf.org/Our-Work/People/Women-in-Conservation-Leadership

Norouzzadeh, M. S., Nguyen, A., Kosmala, M., Swanson, A., Palmer, M. S., Packer, C., & Clune, J. (2018). Automatically identifying, counting, and describing wild animals in camera-trap images with deep learning. *Proceedings of the National Academy of Sciences, 115*(25), E5716–E5725.

Peng, J., Zhao, H., & Liu, Y. (2017). Urban ecological corridors construction: A review. *Acta Ecologica Sinica, 37*(1), 23–30.

Perera, B. (2024). *Ecotourism and its contributions to biodiversity conservation and community development.* IntechOpen. https://www.intechopen.com/chapters/1211297

Pettorelli, N., Laurance, W. F., O'Brien, T. G., Wegmann, M., Nagendra, H., & Turner, W. (2014). Satellite remote sensing for applied ecologists: Opportunities and challenges. *Journal of Applied Ecology, 51*(4), 839–848.

Pimm, S. L., Jenkins, C. N., Abell, R., Brooks, T. M., Gittleman, J. L., Joppa, L. N., & Sexton, J. O. (2014). The biodiversity of species and their rates of extinction, distribution, and protection. *Science, 344*(6187), 1246752.

Pimm, S. L., Alibhai, S., Bergl, R., Dehgan, A., Giri, C., Jewell, Z., Joppa, L., Kays, R., & Loarie, S. (2015). Emerging technologies to conserve biodiversity. *Trends in Ecology & Evolution, 30*(11), 685–696.

Potts, S. G., Biesmeijer, J. C., Kremen, C., Neumann, P., Schweiger, O., & Kunin, W. E. (2010). Global pollinator declines: Trends, impacts and drivers. *Trends in Ecology & Evolution, 25*(6), 345–353.

Rai, P. K., & Singh, J. S. (2020). Invasive alien plant species: Their impact on environment, ecosystem services and human health. *Ecological Indicators, 111*, 106020.

Rawat, U. S., & Agarwal, N. K. (2015). Biodiversity: Concept, threats and conservation. *Environment Conservation Journal, 16*(3), 19–28.

Revive & Restore. (2021). *Elizabeth Ann, the first cloned black-footed ferret.* Retrieved from https://reviverestore.org/

Ripple, W. J., Estes, J. A., Beschta, R. L., Wilmers, C. C., Ritchie, E. G., Hebblewhite, M., & Wirsing, A. J. (2014). Status and ecological effects of the world's largest carnivores. *Science, 343*(6167), 1241484.

Rosen, G. E., & Smith, K. F. (2010). Summarizing the evidence on the international trade in illegal wildlife. *EcoHealth, 7*, 24–32. https://doi.org/10.1007/s10393-010-0317-y

Sala, O. E., Stuart Chapin, F. I. I. I., Armesto, J. J., Berlow, E., Bloomfield, J., Dirzo, R., & Wall, D. H. (2000). Global biodiversity scenarios for the year 2100. *Science, 287*(5459), 1770–1774.

Schneider, S., Taylor, G. W., & Kremer, S. C. (2020). Deep learning object detection methods for ecological camera trap data. *Methods in Ecology and Evolution, 11*(9), 1122–1134.

Secretariat of the Convention on Biological Diversity (CBD). (2010). *Strategic Plan for Biodiversity 2011–2020 and the Aichi Biodiversity Targets.* Retrieved April 27, 2025, from https://www.cbd.int/doc/strategic-plan/2011-2020/aichi-targets-en.pdf

Shapiro, B. (2015). *How to clone a mammoth: The science of de-extinction*. Princeton University Press. https://press.princeton.edu/books/hardcover/9780691157054/how-to-clone-a-mammoth?utm_source=chatgpt.com
Sharma, M., Rawat, S., Kumar, D., Awasthi, A., Sarkar, A., Sidola, A., & Kotecha, K. (2024). The state of the Yamuna River: A detailed review of water quality assessment across the entire course in India. *Applied Water Science, 14*(8), 175.
Singh, A. K., & Lakra, W. S. (2011). Risk and benefit assessment of alien fish species of the aquaculture and aquarium trade in India. *Reviews in Aquaculture, 3*(1), 3–18.
SMART. (2025). *About us*. SMART Conservation Tools. https://smartconservationtools.org/en-us/About/About-Us
Smith, J. M. (1989). The causes of extinction. *Philosophical Transactions of the Royal Society of London. Series B, Biological Sciences, 325*(1228), 241–252.
Snyder, N. F., Derrickson, S. R., Beissinger, S. R., Wiley, J. W., Smith, T. B., Toone, W. D., & Miller, B. (1996). Limitations of captive breeding in endangered species recovery. *Conservation Biology, 10*(2), 338–348.
Solé, R., Maull, V., Amor, D. R., Mauri, J. P., & Núria, C. P. (2024). Synthetic ecosystems: From the test tube to the biosphere. *ACS Synthetic Biology, 13*(12), 3812–3826.
Spalton, A. (1993). A brief history of the reintroduction of the Arabian oryx *Oryx leucoryx* into Oman 1972–1992. *International Zoo Yearbook, 32*(1), 81–90.
Sultana, T., & Wahid, K. A. (2019). IoT-guard: Event-driven fog-based video surveillance system for real-time security management. *IEEE Access, 7*, 134881–134894.
The Hindu Bureau. (2025, April 4). *Supreme Court stays tree felling in Hyderabad's Kancha Gachibowli*. The Hindu. https://www.thehindu.com/news/cities/Hyderabad/supreme-court-stays-tree-felling-in-hyderabads-kancha-gachibowli/article69407115.ece.
Thushari, G. G. N., & Senevirathna, J. D. M. (2020). Plastic pollution in the marine environment. *Heliyon, 6*(8).
Ullah, F., Saqib, S., & Xiong, Y. C. (2025). Integrating artificial intelligence in biodiversity conservation: Bridging classical and modern approaches. *Biodiversity and Conservation, 34*(1), 45–65.
UNEP-WCMC. (2020). *Protected planet report 2020*. United Nations Environment Programme World Conservation Monitoring Centre. https://www.unep-wcmc.org/en/protected-planet
UNESCO. (2017). *Biosphere Reserves—Learning Sites for Sustainable Development*. https://en.unesco.org/biosphere
United Nations. (1992). *Convention on biological diversity*. https://www.cbd.int/doc/legal/cbd-en.pdf
United Nations. (2015). *The 17 goals. United Nations sustainable development goals*. Retrieved April 27, 2025, from https://sdgs.un.org/goals
United Nations Environment Programme (UNEP). (2011). *Towards a green economy: Pathways to sustainable development and poverty eradication*. UNEP. https://www.unep.org/resources/report/towards-green-economy
United Nations Environment Programme (UNEP). (2019). *Fridays for Future Movement: Champions of the Earth laureate*. https://www.unep.org/championsofearth/laureates/2019/fridays-future-movement
United Nations Environment Programme (UNEP) (2021). *Making peace with nature: A scientific blueprint to tackle the climate, biodiversity, and pollution emergencies*. United Nations. Retrieved April 27, 2025, from https://www.unep.org/resources/making-peace-nature
United Nations Environment Programme (UNEP). (2023). *Aligning climate action and biodiversity conservation to support the Paris Agreement*. https://www.unep.org/news-and-stories/speech/aligning-climate-action-and-biodiversity-conservation-support-paris
Walters, J. R., Derrickson, S. R., Fry, D. M., Haig, S. M., Marzluff, J. M., & Wunderle, J. M. (2010). Status of the California Condor (Gymnogyps californianus) and efforts to achieve its recovery. *The Auk, 127*(4), 969–1001. https://digitalcommons.unl.edu/cgi/viewcontent.cgi?article=1674&context=usgsstaffpub

Wasser, S. K., Brown, L., Mailand, C., Mondol, S., Clark, W., Laurie, C., & Weir, B. S. (2015). Genetic assignment of large seizures of elephant ivory reveals Africa's major poaching hotspots. *Science, 349*(6243), 84–87.

Wich, S., Dellatore, D., Houghton, M., Ardi, R., & Koh, L. P. (2015). A preliminary assessment of using conservation drones for Sumatran orang-utan (Pongo abelii) distribution and density. *Journal of Unmanned Vehicle Systems, 4*(1), 45–52.

Wildlife Conservation Society (WCS). (2024, August). *In Papua New Guinea, Inaugl Tribe members commit to legally protect more than 12,000 ha (46 square miles) of high biodiversity forest.* WCS Newsroom. https://newsroom.wcs.org/News-Releases/articleType/ArticleView/articleId/20469/In-Papua-New-Guinea-Inaugl-Tribe-Members-Commit-to-Legally-Protect-More-Than-12000-ha-46-square-miles-of-High-Biodiversity-Forest.aspx

Wildlife Institute of India. (2021). *Annual report 2020–2021.* Wildlife Institute of India. Retrieved from https://wii.gov.in/

Wilmut, I., Schnieke, A. E., McWhir, J., Kind, A. J., & Campbell, K. H. (1997). Viable offspring derived from fetal and adult mammalian cells. *Nature, 385*(6619), 810–813.

World Bank. (2021). *The economic case for nature: A global earth-economy model to assess development policy pathways.* https://www.worldbank.org/en/news/feature/2021/06/03/the-economic-case-for-nature

World Wildlife Fund (WWF). (2022). *Living planet report 2022—Building a nature-positive society.* https://www.worldwildlife.org/publications/living-planet-report-2022

Zhang, J., Hu, J., Lian, J., Fan, Z., Ouyang, X., & Ye, W. (2016). Seeing the forest from drones: Testing the potential of lightweight drones as a tool for long-term forest monitoring. *Biological Conservation, 198*, 60–69.

Zhao, Y., Liu, J., Wang, F., Zhang, Y., & Ma, Z. (2025). Targeted conservation actions are required to save species from extinction. *Biological Conservation, 310*, 111340.

Understanding Climate Change in the Himalaya: Causes, Impacts, and Projections

Rupesh Dhyani, Mayank Shekhar, and Amalava Bhattacharyya

Abstract The Himalaya, known as the "Third Pole" and the "Water Tower of Asia," is undergoing rapid and unprecedented climatic transformation. As one of the most ecologically and culturally significant mountain regions on Earth, the Himalayan region experiences disproportionate impacts from global climate change due to elevation-dependent warming, regional atmospheric pollution, and fragile geophysical systems. This comprehensive synthesis examines the primary drivers, observed impacts, and projected future scenarios of climate change across the Himalayan arc. It highlights cryosphere degradation, shifts in monsoon dynamics, increasing frequency of extreme events, and cascading effects on biodiversity, water security, agriculture, and Indigenous livelihoods. Drawing on paleoclimate records, remote sensing data, and local observations, this chapter emphasizes the urgent need for coordinated climate action. It advocates for integrating Indigenous and Local Knowledge with scientific research, strengthening community-based adaptation and fostering transboundary cooperation among Himalayan nations. With over 1.5 billion people dependent on its rivers and ecosystems, the fate of the Himalaya holds global significance. This chapter calls for immediate mitigation of greenhouse gas emissions, more equitable climate governance and sustained investment in resilience strategies to safeguard the regions environmental and socio-cultural future.

Keywords Climate variability · Global warming · Paleoclimate · Cryosphere · Indigenous communities

---

R. Dhyani (✉)
Department of Geography, Justus-Liebig-University Giessen, Giessen, Germany

M. Shekhar · A. Bhattacharyya
Birbal Sahni Institute of Palaeosciences, Lucknow, India

S. Siddiqui et al. (eds.), *Frontiers in Sustainability*,
https://doi.org/10.1007/978-3-032-19740-5_4

1 Introduction

1.1 Himalaya: The Fragile Crown of Asia

The Himalaya, often referred to as the "Roof of the World" or the "Third Pole," constitutes the world's highest and youngest mountain range, stretching over 2400 km across five countries, India, Nepal, Bhutan, Pakistan, and China (Tibet Autonomous Region) (Gansser, 1974; Wani et al., 2023; Wester et al., 2019). Rising over 8000 m at its tallest peaks and shaped by the ongoing collision of the Indian and Eurasian tectonic plates, the Himalaya is a dynamic and geologically active region (Gansser, 1974). It is home to 10 of the world's 14 highest peaks, including Mount Everest, and houses more than 54,000 glaciers and vast seasonal snowfields, which together comprise the largest ice mass outside the polar regions (Kindlmann, 2011).

But beyond its sheer physical magnitude, the Himalaya holds extraordinary ecological, cultural, and hydrological significance. It is recognized as a Global Biodiversity Hotspot by Conservation International (Myers et al., 2000), supporting over 10,000 plant species, 300 mammal species, and numerous endemic birds, amphibians, and reptiles (Chettri & Tamang, 2024). Beyond this biological wealth, the region is not just a biological treasure trove, it is also home to millions of Indigenous peoples and highland communities with diverse cultural traditions, languages, and religious beliefs (Wester et al., 2019).

Equally crucial is the Himalaya's role as the "Water Tower of Asia," a term reflecting its centrality to hydrological systems across the continent (Islam et al., 2024). Ten of Asia's major rivers including the Indus, Ganges, Brahmaputra, Yangtze, Mekong, Yellow River, Salween, and others originate in or are fed by Himalayan glaciers and snowmelt (Immerzeel, 2010; Pritchard, 2019). These rivers provide drinking water, irrigation, and hydroelectric power to over 1.5 billion people downstream, underpinning agricultural productivity and economic development in densely populated nations such as India, Pakistan, Bangladesh, and China (Wester et al., 2019). Seasonal glacial melt provides an important buffer against monsoon variability, maintaining water flow during the dry months an ecosystem service is increasingly threatened by accelerating climate change (Krishnan et al., 2019; Wester et al., 2019).

Recent studies confirm that the Himalaya is warming at twice the global average, a phenomenon driven by elevation-dependent warming (EDW), which amplifies temperature increases at higher altitudes (Panday & Thibeault, 2015; You et al., 2017). This warming is compounded by local and regional drivers such as black carbon (soot) deposition from biomass burning and fossil fuels, which darkens snow and ice surfaces and accelerates melting (Gertler et al., 2016; Li et al., 2016). Recent research shows that Himalayan glaciers have experienced a significant acceleration in ice loss since the Little Ice Age, with current melt rates far exceeding historical averages due to factors like climate warming, debris cover, and proglacial lake expansion (Lee et al., 2021; Shekhar et al., 2017).

The climate sensitivity of the Himalayan region is further shaped by its complex interactions between the cryosphere, hydrosphere, and atmosphere. Small changes in temperature or precipitation can induce disproportionately large effects, such as glacial lake outburst floods (GLOFs), avalanches, and sudden shifts in river discharge (Khadka et al., 2025; Sattar et al., 2022; Shah, 2025; Vishwakarma et al., 2022). The 2013 Kedarnath disaster in Uttarakhand and the 2021 Chamoli glacier burst are stark reminders of how climate-amplified hazards can lead to catastrophic impacts in mountainous terrain (Allen et al., 2016).

At the same time, these transformations are deeply entwined with local livelihoods and knowledge systems. Highland communities, many of whom depend on subsistence agriculture, transhumance, or ecotourism, are often the first to observe and respond to climatic changes (Goodrich et al., 2017). Yet they are also among the most vulnerable to disruption due to limited infrastructure, economic marginalization, and restricted access to state resources (Yadav & Singh, 2024). Traditional ecological knowledge, including practices related to seasonal forecasting, water sharing, and biodiversity stewardship, holds crucial adaptive potential, yet it remains underrepresented in national and international policy frameworks (Biswas et al., 2025). This gap highlights a critical need for integrating local knowledge with scientific assessment to develop effective adaptation strategies.

In sum, the Himalaya is not just a regional concern, it is a critical barometer of planetary change. Its glaciers, ecosystems, and communities offer unique insights into the early warning signals of a warming world. Understanding climate change in this fragile crown of Asia is key to anticipating and mitigating some of the most profound social and ecological transformations of the twenty-first century.

2 Causes of Climate Change in the Himalaya

The accelerating pace of climate change in the Himalaya is the result of a complex interplay between global climate forcing, regional pollution sources, and local geophysical feedback mechanisms. While the foundational driver is the anthropogenic rise in greenhouse gas (GHG) concentrations (IPCC, 2023), the region's unique elevation, atmospheric conditions, and land use pressures amplify its vulnerability, resulting in a rate and intensity of warming that far exceeds global averages (Wester et al., 2019).

2.1 Insights from the Past: Himalayan Paleoclimate Records

To comprehend the scale and speed of current climate changes in the Himalaya, it is essential to understand the region's paleoclimatic history. Paleoenvironmental data from lake sediments, ice cores, tree rings, and speleothems (cave formations) offer a long-term perspective on past climatic variability and the natural range of temperature and precipitation regimes in the region (Dhyani et al., 2025).

Studies from lake sediments in Ladakh and the Tibetan Plateau have revealed alternating periods of wetter and drier phases over the Holocene, indicating a dynamic monsoon system influenced by solar forcing, volcanic activity, and global teleconnections like ENSO (Gasse et al., 1991; Herzschuh, 2014). Ice cores from the Himalaya and the Karakoram, including the Dasuopu and Guliya cores, preserve high-resolution records of temperature, dust, and atmospheric composition dating back thousands of years (Thompson et al., 2000). These records show that while past warming events did occur, the rate and synchronicity of contemporary warming are unprecedented in the last several millennia.

Tree-ring data from species such as Himalayan cedar (*Cedrus deodara*) and fir (*Abies pindrow*) provide precise reconstructions of temperature and precipitation fluctuations over the past millennium (Bhattacharyya et al., 2023). These proxies confirm that recent decades have been exceptionally warm compared to the Medieval Warm Period and the Little Ice Age (Mann & Jones, 2003). Moreover, paleoclimate records support the idea that even modest natural changes in climate historically caused significant shifts in glacier extent, vegetation cover, and hydrological patterns, underscoring the potential magnitude of anthropogenically driven changes (Dhyani et al., 2025).

Understanding these past climates provides a crucial baseline to distinguish anthropogenic effects from natural variability and helps refine climate models to improve future projections (Bhattacharyya et al., 2023). Integrating paleoclimate insights into adaptation and mitigation planning can enhance resilience by identifying thresholds, feedbacks, and historical analogs relevant to present and future risks (Braconnot et al., 2012).

2.2 *Global Greenhouse Gas Emissions: The Primary Forcing*

At the core of the climate crisis in the Himalaya lies the global increase (GHGs), primarily carbon dioxide (CO_2), methane (CH_4), and nitrous oxide (N_2O) resulting from fossil fuel combustion, large-scale deforestation, and intensive agriculture (IPCC, 2021). These gases trap longwave radiation in the atmosphere, leading to a warming effect that is global in scale but regionally variable in impact. In the Himalaya, this manifests as **non-uniform and intensified warming**, especially in snow and ice-dominated zones that are sensitive to even small shifts in temperature (Pepin et al., 2015). Multiple studies have shown that mean surface temperatures in many subregions of the Himalaya have increased by approximately 0.3–0.7 °C per decade over the last few decades, significantly higher than the global average warming rate (Shrestha et al., 1999; Xu et al., 2009).

This accelerated regional warming disrupts glacial balance, modifies precipitation regimes, and intensifies monsoon variability. Even minor temperature increases can push glaciers past critical melt thresholds, destabilizing high-altitude hydrological systems that feed major Asian rivers (Immerzeel, 2010). Consequently, global GHG emissions serve as the background climate forcing, but the Himalayan cryosphere and hydrological systems response is disproportionately severe.

2.3 *Regional Amplifiers of Warming*

While global emissions set the stage for climate change, several regional drivers further accelerate warming in the Himalaya (Wester et al., 2019). These include atmospheric pollutants such as black carbon and land use transformations that reduce ecological resilience and alter local energy balances.

One of the most potent of these regional drivers is **black carbon (BC)**, a short-lived but highly effective warming agent formed through the incomplete combustion of biomass and fossil fuels. Key sources include residential cookstoves, diesel engines, open burning of agricultural residues, and industrial activity in the Indo-Gangetic Plain. Transported by atmospheric circulation into the Himalaya, black carbon is deposited on snow and ice surfaces, where it reduces albedo and significantly accelerates surface melting by enhancing solar absorption (Flanner et al., 2009; Ramanathan & Carmichael, 2008). Additionally, BC in the atmosphere contributes to localized heating, influencing temperature inversions and snowmelt timing. Modeling studies by Kopacz et al. (2011) and Yasunari et al. (2010) have estimated that black carbon may be responsible for up to 30% of glacier melt in some areas, particularly in the western and central Himalaya, where post-monsoon atmospheric cleansing is limited.

Alongside atmospheric pollutants, **land use change and unregulated development** act as persistent stressors. Deforestation for agriculture, road construction, urban expansion, and hydropower infrastructure reduces carbon sequestration capacity, disrupts microclimates, and increases surface heat absorption (Singh & Thadani, 2015). These activities often occur without adequate ecological assessments, especially in fragile alpine zones. Furthermore, tourism-driven development introduces localized "heat islands" and increases pollution and waste generation. In areas like Himachal Pradesh and Sikkim, poorly managed urbanization and land degradation have exacerbated landslide frequency, altered watershed hydrology, and weakened the buffering capacity of natural ecosystems (Tiwari et al., 2018).

2.4 *Unique Physical and Climatic Dynamics*

What makes the Himalayan climate particularly sensitive to global warming is the presence of **unique atmospheric and geophysical processes** that amplify climate signals (Wester et al., 2019). These include elevation-dependent warming and shifts in large-scale atmospheric circulation patterns, both of which intensify regional climate responses.

Elevation-dependent warming (EDW) refers to the observed phenomenon where high-altitude regions are warming faster than the surrounding lowlands. In the Himalaya, regions above 5000 m are experiencing temperature increases that are nearly double the global average (Pepin et al., 2015). This accelerated warming is driven by several feedback mechanisms. In particular, snow-albedo feedback plays a critical role: as snow cover melts earlier in the season, the exposed darker surfaces or ice surfaces absorb more heat, creating a self-reinforcing cycle of warming and melt. Additionally, increases in atmospheric water vapor at high elevations, where humidity is typically low, amplify the greenhouse effect (Pepin et al., 2022). Changes in cloud cover and the vertical structure of the atmosphere also affect how solar radiation is distributed, leading to further warming of the upper troposphere (Palazzi et al., 2017). These processes collectively intensify the retreat of glaciers, threaten permafrost stability, and alter alpine ecosystems. The Hindu Kush Himalaya Assessment warns that under current warming trajectories, one-third of the region's glaciers could disappear by the end of the century, even if global warming is limited to 1.5 °C (IPCC, 2019; Wester et al., 2019).

Equally important are the **shifts in atmospheric circulation patterns**, especially the South Asian Summer Monsoon and the Westerlies. The monsoon, which brings critical rainfall to the region, is becoming more **erratic** in its timing, intensity, and spatial distribution. Observations show a trend toward delayed onset, shorter duration, and more extreme rainfall events changes which are increasing flood risks and reducing agricultural predictability (Turner & Annamalai, 2012). In the western Himalaya, winter precipitation is heavily influenced by **western disturbances**, which are weakening and shifting due to altered global circulation patterns (Ridley et al., 2013). This leads to a reduction in snow accumulation critical for glacier health. Moreover, there is growing evidence that the frequency of extreme weather events such as heatwaves, cloudbursts, and out-of-season snow fall is increasing across the region, indicating heightened atmospheric instability (Dimri et al., 2019).

Taken together, these global, regional, and local factors reveal that the Himalaya is not merely warming; but undergoing a profound climatic reorganization. The system is becoming more **chaotic, unpredictable, and prone to sudden thresholds or tipping points**, with cascading implications for water security, biodiversity, and human livelihoods across South and Central Asia (Wester et al., 2019).

3 Observed and Measured Impacts

The effects of climate change are no longer theoretical in the Himalaya; they are visible, measurable, and accelerating (Wester et al., 2019). Over the past two decades, empirical research and satellite data have provided compelling evidence of wide-ranging impacts across the cryosphere, hydrosphere, biosphere, and human systems. The Himalaya is undergoing a profound ecological and social transformation, driven by rising temperatures, shifting precipitation patterns, and increasing climate extremes (Wester et al., 2019).

3.1 Cryosphere Degradation

The most immediate and visible impacts of climate change in the Himalaya are occurring in the **cryosphere**, particularly in the form of glacial retreat (Wang et al., 2024), snowpack reduction (Krishnan et al., 2019), and permafrost instability (Gruber, 2012).

3.2 Glacier Retreat

Glaciers across the Himalaya are retreating at unprecedented rates, with significant implications for downstream hydrology and disaster risk. Remote sensing and field-based studies consistently report negative mass balances in most glacier systems (Bolch et al., 2012; Jasrotia et al., 2023). The **Hindu Kush Himalaya Assessment** (Wester et al., 2019) found that glaciers in the region lost nearly 30–50% of their ice mass in the twentieth century, with the rate of loss doubling since the turn of the millennium. This retreat manifests in terminus withdrawal, surface lowering, and thinning of ice bodies. The eastern Himalaya, in particular, has exhibited **greater glacial mass loss** due to warmer, wetter conditions compared to the western range (Maurer et al., 2019).

3.3 Snowpack Reduction

Rising temperatures are also diminishing **seasonal snow cover**, particularly at lower and mid-elevations, the snowline is shifting upwards, and snow accumulation periods are becoming shorter (Krishnan et al., 2019). Studies show that **earlier onset of snowmelt** disrupts the natural timing of meltwater release that traditionally supported spring agriculture and early season river flows (Barnett et al., 2005). The result is a less predictable water cycle, with both ecological and socioeconomic consequences.

4 Hydrological Shifts and Extreme Events

Climate change is fundamentally altering the hydrological systems of the Himalaya. From shifting monsoon patterns (Krishnan et al., 2019) to intensified flood risks (Wester et al., 2019), water availability is becoming more variable and less reliable.

4.1 Changing Monsoon Patterns

The **South Asian Monsoon**, which provides the bulk of annual rainfall, is becoming increasingly erratic. While overall monsoon rainfall trends are spatially and temporally complex, evidence suggests **greater variability**, with short bursts of **extreme rainfall events** leading to landslides and flash floods, interspersed with prolonged dry spells (Turner & Annamalai, 2012). Studies indicate a **westward shift** and **weakening of the monsoon circulation**, resulting in reduced precipitation in some catchments and excess in others (Roxy et al., 2017).

4.2 Glacial Lake Outburst Floods (GLOFs)

As glaciers retreat, meltwater accumulates behind fragile moraine dams, forming unstable **glacial lakes.** The number and volume of these lakes have increased dramatically in recent decades (Wester et al., 2019). A breach of these lakes, often triggered by avalanches, seismic activity, or heavy rainfall, can result in catastrophic **GLOFs** that destroy downstream infrastructure and threaten lives. ICIMOD (2011) documented over 200 potentially dangerous glacial lakes in Nepal, Bhutan, and Tibet. Events such as the **2021 Chamoli disaster** in Uttarakhand have raised urgent concerns about the risks posed by rapidly changing glacial dynamics (Shugar et al., 2021).

4.3 Altered River Flow Regimes

Changes in the timing and volume of glacier and snowmelt are altering **seasonal river discharge patterns**. Many rivers now show **earlier peak flows** in spring, followed by reduced flows in the summer dry season, an inversion of the traditional Himalayan hydrological rhythm. This threatens irrigation cycles, drinking water supplies, and hydropower generation across multiple countries that depend on Himalayan runoff (Immerzeel, 2010; Pritchard, 2019).

5 Ecosystem Disruption

Himalayan ecosystems, which are finely tuned to climatic gradients, are undergoing rapid restructuring. Changes in temperature, precipitation, and snow cover are affecting biodiversity, ecosystem services, and agricultural productivity (Anjum et al., 2023; Dhyani et al., 2025; Wester et al., 2019).

5.1 Biodiversity Loss and Range Shifts

Species across taxonomic groups are shifting their ranges **upslope** in response to warming temperatures. However, species already restricted to alpine or subnival zones have no available habitat to migrate into and face heightened extinction risks (Xu et al., 2009). Changes in **phenology,** such as earlier flowering or breeding, are disrupting ecological interactions like plant-pollinator relationships, with cascading effects on food webs. Moreover, invasive species, better adapted to disturbance and higher temperatures, are gaining competitive advantages in altered ecosystems (Telwala et al., 2013).

5.2 Water Resource Stress

The reliability of **freshwater supplies** is being compromised by changes in snowmelt timing, monsoon variability, and glacial retreat (Immerzeel, 2010). Water stress is not confined to highland communities but extends to downstream cities and agricultural zones in the Indo-Gangetic and Brahmaputra plains. Studies project declining **dry season water availability** in major river basins such as the Ganges, which supports over 500 million people (Wester et al., 2019).

5.3 Agricultural Vulnerability

Himalayan agriculture, largely rain-fed and subsistence-oriented, is **acutely sensitive** to climatic variability. Climate-induced variability—including erratic rainfall, shifting growing seasons, increased crop disease incidence, and declining soil moisture—is undermining traditional cropping systems and intensifying threats to food security (Gentle & Maraseni, 2012). In some areas, farmers are abandoning previously viable crops such as apples and barley due to rising temperatures and water scarcity, shifting toward more resilient or lower-demand alternatives.

6 Socio-Cultural Impacts

The impacts of climate change in the Himalaya extend beyond physical systems, deeply affecting the social fabric, livelihoods, and cultural identity of mountain communities (Wester et al., 2019). Climate change acts as a threat multiplier, exacerbating existing vulnerabilities and triggering complex chains of effects that range from outmigration and the erosion of Indigenous knowledge (Gerlitz et al., 2017) to mental health stressors (Ayeb-Karlsson et al., 2020) and the weakening of traditional governance structures.

6.1 Threats to Indigenous Communities

Highland communities with long-standing **cultural and spiritual ties** to the land are facing displacement due to GLOFs, landslides, and persistent water shortages (Ives, 2004). Pastoralists and transhumant groups are losing access to traditional grazing routes as pastures degrade and become seasonally inaccessible (Sharma et al., 2019). Infrastructure loss and declining agricultural yields are driving **outmigration**, disrupting family and village structures, and eroding cultural cohesion (Rastogi, 2025).

6.2 Erosion of Traditional Knowledge Systems

Climate change is outpacing the **adaptive capacity embedded in Indigenous and local knowledge systems,** which for generations have guided land use, weather forecasting, resource management, and spiritual practices(Bhardwaj & Pande, 2025). These systems, once grounded in predictable environmental cycles, are now losing efficacy and authority. As a result, there is a growing disconnection between generations and a loss of resilience rooted in community-based adaptation (Byg & Salick, 2009; Tiwari et al., 2020).

7 Projections and Future Scenarios: Navigating Uncertainty

The future of the Himalayan climate is marked by a mixture of **high-confidence trends and deep uncertainties**, especially in relation to regional and seasonal precipitation (Dixit et al., 2023). However, climate models, satellite data, and regional downscaling efforts now offer a growing body of evidence that the Himalaya will continue to experience **accelerated and potentially destabilizing changes** throughout the twenty-first century. These transformations will profoundly affect ecosystems, water systems, agriculture, and geopolitical stability across South and Central Asia (Wester et al., 2019).

7.1 *Temperature and Precipitation Trends*

Projections from global climate models (GCMs) and regional climate models (RCMs) under various Representative Concentration Pathways (RCPs) and Shared Socioeconomic Pathways (SSPs) suggest significant warming across the entire Himalayan arc. Under high-emission scenarios such as RCP8.5 or SSP5-8.5, mean annual temperatures (Fig. 1) are projected to rise by 3 °C to 5 °C or more by the end of the century (Krishnan et al., 2019). Warming is expected to be nonlinear and

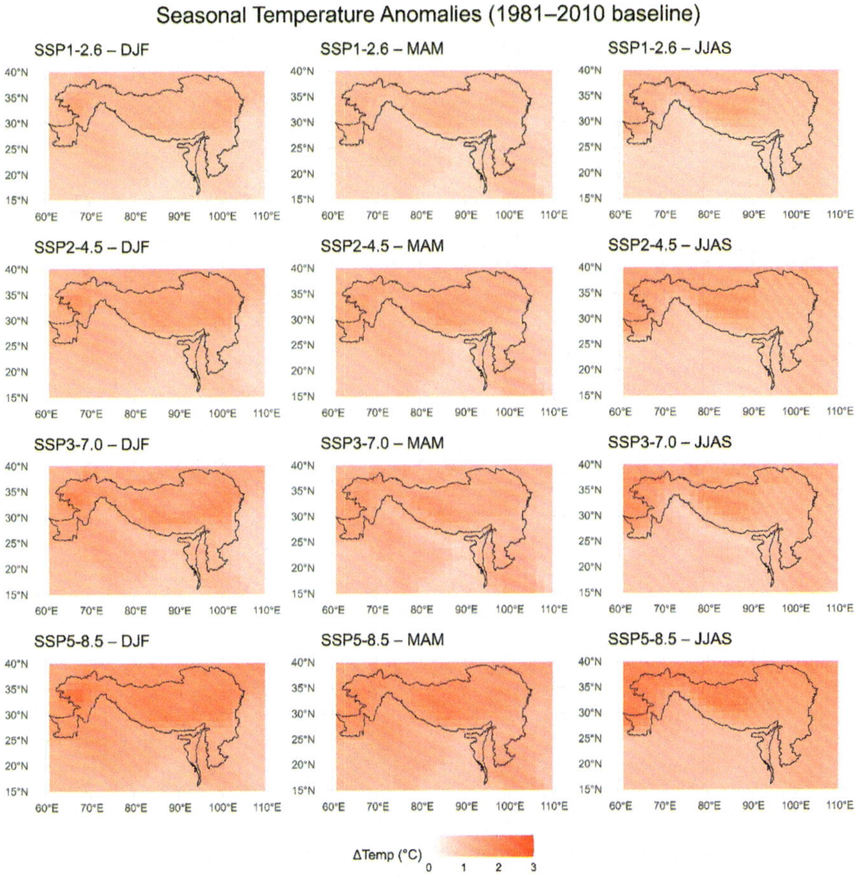

Fig. 1 Projected Seasonal Temperature Anomalies (°C) Relative to 1981–2010 Baseline Across HKH. This figure shows projected seasonal surface air temperature anomalies for HKH under four different Shared Socioeconomic Pathways (SSPs)—SSP1-2.6, SSP2-4.5, SSP3-7.0, and SSP5-8.5—across three seasons: DJF (December–February), MAM (March–May), and JJAS (June–September). The maps indicate a clear increase in temperature anomalies with higher emission scenarios, with SSP5-8.5 exhibiting the most pronounced warming, especially during DJF and MAM

elevation-dependent, with upper-altitude regions warming faster than lowland zones, exacerbating cryosphere melt and permafrost degradation (Pepin et al., 2015).

In contrast, precipitation projections remain more uncertain, particularly in terms of spatial and seasonal distribution (Fig. 2). Most models suggest an overall increase in total annual precipitation, but with greater interannual variability and more frequent intense rainfall events (Palazzi et al., 2017). For instance, some projections indicate an increase in winter precipitation in the western Himalayaprimarily due to stronger westerlies, while other regions may experience drying trends during crucial growing seasons (Bhardwaj & Pande, 2025). This makes the Himalayan climate

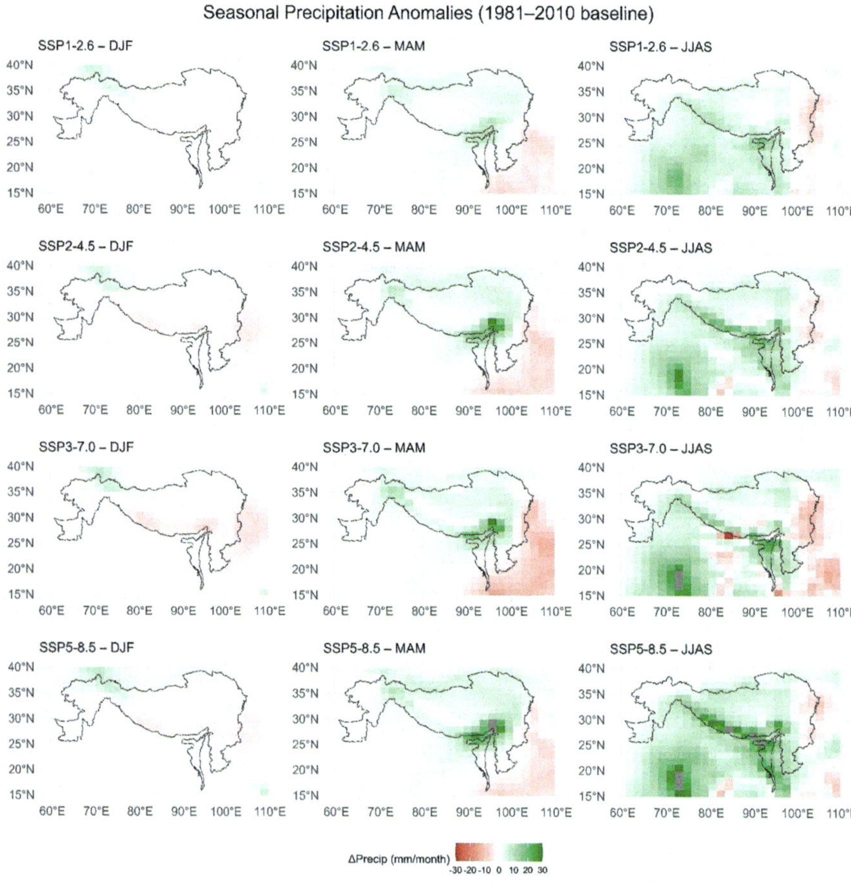

Fig. 2 Projected Seasonal Precipitation Anomalies (mm/month) Relative to 1981–2010 Baseline Across HKH. This figure illustrates projected changes in seasonal precipitation across HKH under four Shared Socioeconomic Pathways (SSPs)—SSP1-2.6, SSP2-4.5, SSP3-7.0, and SSP5-8.5—for three seasons: DJF (December–February), MAM (March–May), and JJAS (June–September). Green areas indicate increased precipitation, while red areas represent declines, relative to the 1981–2010 climatological baseline

especially difficult to forecast with precision and adds complexity to water management and agricultural planning (Negi & Maikhuri, 2013).

7.2 Cryosphere Futures

Perhaps the most alarming and robust projections relate to the future of the **Himalayan cryosphere**. Modeling studies based on current emission trajectories suggest that the region could lose between **30% and 70% of its glacier volume** by 2100, even under moderate mitigation scenarios like RCP4.5 (Kraaijenbrink et al., 2017; Marzeion et al., 2020). Under high-emission pathways, many **smaller glaciers in the eastern Himalaya and Hindu Kush** may disappear entirely (Wester et al., 2019).This will be accompanied by a continued **decline in snow cover duration**, reduced snowfall-to-rainfall ratios, and an upward shift in the snowline, fundamentally altering seasonal water storage dynamics (Kraaijenbrink et al., 2017). These changes undermine the "natural reservoir" function of snow and glaciers, which historically released meltwater gradually during the dry season (Yang & Bai, 2024).The concept of **"peak water"** is critical here. As glaciers shrink, annual runoff may initially increase due to enhanced melting. However, once glacier volume declines past a certain threshold, **annual discharge will begin to drop**, leading to **long-term reductions in river flow**, particularly during dry seasons when glacier melt is most crucial (Huss & Hock, 2018). This shift has already been observed in some sub-basins and is projected to worsen throughout the century.

7.3 Water Security Risks

The implications for **water security** in the greater Himalayan region are profound. The Himalaya feeds ten major river systems, including the **Indus, Ganges, and Brahmaputra**, which support more than **1.5 billion people** (Shrestha et al., 2015). As glacier buffers diminish and precipitation becomes less predictable, the **seasonality and reliability of water flow** will be disrupted (Edwards et al., 2010).

In the **Indus Basin**, which is highly glacier-dependent and already water-stressed, flow reductions of up to **35% by 2100** are possible under business-as-usual scenarios, threatening **agriculture, food production, and hydropower** across Pakistan and northwest India (Immerzeel, 2010; Lutz et al., 2014). The **Ganges Basin**, with its reliance on monsoonal rains and snowmelt, may experience both **floods during peak monsoon** and **dry season shortages**, endangering urban water supplies, irrigation, and deltaic ecosystems in Bangladesh.

Moreover, these rivers cross **national boundaries**, raising concerns about **transboundary water conflicts**. Without cooperative frameworks and integrated water governance, the growing water stress could exacerbate **regional tensions and**

geopolitical instability, particularly in contexts already marked by fragile diplomatic relations (Scott et al., 2019).

7.4 Tipping Points and Cascading Effects

While gradual climatic trends pose serious long-term challenges, perhaps the greatest risks to the Himalayan region arise from abrupt, nonlinear changes the so-called tipping points, which, once crossed, could result in irreversible and potentially catastrophic transformations (Khadka et al., 2025). The Himalaya is increasingly understood as a nexus of multiple, interacting tipping elements, each with the capacity to destabilize ecological and human systems far beyond the region itself. One of the most concerning is monsoon destabilization. Climate models indicate that ongoing warming, particularly in the Indian Ocean, could disrupt the thermal gradients and ocean-atmosphere feedbacks that sustain the South Asian monsoon (Krishnan et al., 2019). In extreme scenarios, this may weaken or even collapse monsoonal circulation in some parts of South Asia, with profound consequences for rainfall-dependent agriculture, freshwater availability, and regional food security (Levermann et al., 2009).

Another critical threshold relates to cryosphere collapse. As glacier systems retreat beyond their accumulation zones and lose mass at accelerating rates, the capacity of these "water towers" to buffer seasonal hydrological cycles diminishes (Immerzeel, 2010). In many sub-basins, this may eventually result in long-term hydrological collapse, especially during the dry seasons when glacier melt has historically sustained river flow (Islam et al., 2024). Even under moderate mitigation scenarios, several smaller and mid-sized glaciers in the central and eastern Himalaya are projected to become functionally extinct within decades (Huss & Hock, 2018; Kraaijenbrink et al., 2017).

Simultaneously, the Himalaya is at risk of ecosystem regime shifts, where forests, alpine meadows, and wetlands may transition to entirely different vegetation types under prolonged drought, increased fire frequency, or pest outbreaks. Such diebacks can sharply reduce carbon sequestration, destabilize local climates, and lead to the loss of endemic biodiversity (Negi & Maikhuri, 2013). These transitions are especially likely in ecotonal areas already stressed by human encroachment, invasive species, and shifting climatic envelopes (Telwala et al., 2013; Xu et al., 2009).

Importantly, these tipping points are not isolated events. The Himalaya operates as a coupled socio-ecological system, where a breakdown in one component, such as the cryosphere, can trigger cascading effects in hydrology, food systems, ecosystems, and social stability (Bhatta et al., 2015). For example, the sudden collapse of a glacier-fed lake could lead to catastrophic flooding, destroying farmland and displacing entire communities (Manandhar et al., 2011). This displacement may strain already vulnerable urban areas, disrupt labor markets, and heighten competition over limited resources, increasing the likelihood of resource-based conflicts and inter-community tensions (Scott et al., 2019).

In worst-case scenarios, climate-induced shocks may catalyze mass displacement, undermine regional food and water security, and create "hotspots of vulnerability" where climate impacts and conflict risks compound each other (Ojha et al., 2016). Given the Himalaya's role as a regional water tower and global biodiversity corridor, these cascading effects could extend far beyond national boundaries, influencing transboundary cooperation, international aid, and even regional geopolitical dynamics (Bandyopadhyay & Gyawali, 1994). The stakes of crossing these thresholds are therefore not only local or national, they are of planetary significance.

8 The Indispensable Role of Local Knowledge and Adaptation

Effectively addressing the climate crisis in the Himalaya requires going beyond scientific analysis and technological interventions. While climate models, remote sensing, and hydrological data are essential for understanding broad trends, they are not sufficient on their own to develop grounded, effective, and culturally appropriate adaptation strategies. The lived experiences, knowledge systems, and adaptive capacities of Himalayan communities particularly Indigenous peoples and traditional land users are indispensable in navigating the region's mounting environmental challenges (Sharma et al., 2019).

8.1 Community-Led Responses: Local Solutions in Action

Across the Himalayan arc, communities are already demonstrating **innovative and locally tailored responses** to climate impacts. These include both revival of traditional practices and novel adaptations born from necessity (Negi et al., 2022). In **Nepal**, for example, mountain villages have implemented **community-based early warning systems** to monitor rising water levels in glacial lakes and river systems, using low-cost sensors and communication networks to trigger timely evacuations (ICIMOD, 2011). Similarly, in **Ladakh**, the centuries-old practice of constructing **zings** small stone-lined water harvesting tanks is being revived and adapted to modern water needs, while in **Himachal Pradesh**, traditional **Kuhl irrigation systems** have been restored through participatory watershed management (Bandyopadhyay & Gyawali, 1994).

Elsewhere, communities are shifting toward **drought-resistant crops**, **integrating agroforestry**, and adopting **diversified livelihoods** such as ecotourism and value-added niche products like medicinal herbs and artisanal goods (Bhatta et al., 2015). Community forest management groups such as **Van Panchayats** in

Uttarakhand and **Community Forest User Groups (CFUGs)** in Nepal have emerged as critical institutions for managing natural resources, regulating grazing, and coordinating watershed restoration (Agrawal & Ostrom, 2001; Kheder & Zugravu, 2012).

These examples illustrate that **resilience is already being built from the ground up**, even in the absence of formal state support. However, the broader challenge lies in **recognizing, scaling, and institutionalizing** these practices within larger policy frameworks (Ojha et al., 2016).

8.2 Indigenous and Local Knowledge (ILK): A Vital Resource

Indigenous and local knowledge (ILK) systems represent centuries of empirical observation and adaptation to environmental variability (Sherpa, 2014). They encompass not only technical practices such as terrace farming, seed saving, or seasonal migration patterns but also worldviews that link ecological health with social cohesion, spiritual life, and moral obligations to land and water (Negi & Maikhuri, 2013). In the Himalaya, ILK provides **long-term climatic baselines**, often transmitted through oral histories, seasonal indicators, and ecological narratives (Weaver et al., 2022). Communities track subtle changes in **plant phenology, animal behavior, snowfall patterns**, and river flow to make critical decisions about planting, harvesting, and livestock movement (Manandhar et al., 2011). Such insights offer **context and nuance** often missing from satellite data or climate models.

Yet, ILK is increasingly at risk. Rapid environmental changes, generational shifts, formal education systems, and market-driven livelihood transitions are contributing to its erosion (Weaver et al., 2022). Moreover, development agencies and climate planners often overlook or marginalize local knowledge in favor of externally imposed "solutions," undermining both the **legitimacy and effectiveness** of adaptation efforts (Nyong et al., 2007). Recognizing ILK as a **complementary knowledge system**, rather than a substitute for science, is critical. Co-produced approaches that combine ILK with scientific data such as participatory climate monitoring or hybrid forecasting toolscan yield more holistic and actionable insights (Roué et al., 2022).

9 Policy Integration and Recommendations

For climate adaptation in the Himalaya to be effective, equitable, and sustainable, **policy frameworks must move beyond top-down approaches** and meaningfully engage local communities as key actors in both knowledge production and implementation (Subedi et al., 2023). One of the most promising strategies is the **co-production of knowledge**, where community members, researchers, and policymakers collaboratively design and implement adaptation strategies. This approach repositions local stakeholders not as passive recipients of external expertise but as active contributors to research design, data collection, interpretation, and problem-solving. Such collaboration enhances the **relevance, legitimacy, and uptake** of adaptation interventions.

An equally important priority is the **mainstreaming of Indigenous and Local Knowledge (ILK)** into formal climate policies (Nyong et al., 2007). Validated ILK like traditional seasonal forecasting indicators or customary water-sharing agreements offers critical insights into local climate variability and resource management (Joshi et al., 2023). When incorporated into **national adaptation plans (NAPs)**, **disaster risk reduction (DRR) strategies**, or **ecosystem-based management policies**, ILK enhances cultural appropriateness, trust, and resilience (Bhardwaj & Pande, 2025). For example, recognizing customary water rights in basin-scale planning can promote equitable access while reinforcing traditional stewardship systems (Kumar, 2010).

To institutionalize these approaches, governments must also **strengthen community-level governance structures**, such as **Van Panchayats in India**, **Community Forest User Groups (CFUGs) in Nepal**, and traditional village councils (Sherpa, 2014). These local institutions are often more agile, context-sensitive, and trusted than centralized agencies, particularly in remote or marginalized areas (Rastogi, 2025). Empowering them with decision-making authority and technical support can greatly enhance adaptive capacity (Ojha et al., 2016).

Moreover, there is a pressing need to **prioritize Locally-Led Adaptation (LLA)** in climate finance mechanisms (Coger et al., 2022). Funding flows should be restructured to **directly support community initiatives**, ensuring that adaptation resources are flexible, long-term, and responsive to evolving local priorities (Rahman et al., 2023). Donors and governments must design mechanisms for **transparent and accountable community finance**, recognizing that sustained adaptation depends on the autonomy and ingenuity of those most affected (Bhanye, 2025).

Finally, adaptation efforts must be grounded in **transboundary collaboration**, as climate impacts do not adhere to political borders. The Himalaya spans multiple countries, each facing shared challenges such as glacial retreat, water insecurity, and increased disaster risk. Science-policy platforms like the **Hindu Kush Himalaya Science-Policy Forum**, coordinated by ICIMOD, should be supported and expanded to promote **collaborative research, data sharing, joint early warning systems**, and **coordinated policy action** across the region. Such cooperation is essential for building **regional resilience** in the face of an increasingly unstable and interconnected climate system (Wester et al., 2019).

10 Conclusion: An Urgent Call for Action and Understanding

The Himalaya stands at the precipice of profound and potentially irreversible transformation, driven by the intensifying impacts of global climate change. The evidence is clear and compelling: glaciers are retreating at unprecedented rates, monsoon patterns are becoming increasingly erratic, ecosystems are under growing stress, and mountain communities face mounting vulnerabilities. These are not isolated phenomena, but early indicators of deeper systemic shifts that could profoundly affect water security, food systems, biodiversity, and human well-being across much of Asia. Projections underscore a future marked by escalating riskswhere the degradation of cryospheric systems may jeopardize the very rivers that sustain over a billion lives.

Addressing this crisis requires a level of urgency, coordination, and international cooperation that has so far been lacking. At the global level, rapid and sustained reductions in greenhouse gas emissions are essential. Without a sustained decline in emissions, even the most robust regional or local adaptation strategies will be overwhelmed by the scale of change. Regionally, the Himalayan nations must strengthen transboundary collaboration through shared monitoring systems, open data exchange, coordinated disaster response mechanisms, and equitable frameworks for water governance. Climate impacts do not respect national boundaries, and only through cooperation can the region manage shared risks effectively.

At the national and local scales, urgent investments must be made in robust, context-sensitive adaptation strategies that integrate both the latest scientific insights and the deep, place-based knowledge of Himalayan communities. Empowering these communities to shape and lead their own resilience strategies is not only just it is necessary. They are on the front lines of change and possess generations of experience in managing environmental variability and resource stewardship.

Nevertheless, significant knowledge gaps persist. There is an urgent need for expanded and enhanced monitoring infrastructure, including a denser network of high-altitude weather stations, improved glacier and snowpack tracking, and more comprehensive hydrological gaging. Climate models must be refined to better capture the region's complex topography and climate dynamics—especially regarding precipitation, where uncertainties remain high. Moreover, adaptation planning will benefit immensely from integrated research approaches that connect physical sciences with social, economic, and cultural dimensions. Equally critical is the promotion of open access to data across national borders to foster transparency, collaboration, and evidence-based policymaking.

The fate of the Himalaya is not just a regional concern, it is a global imperative. What happens in this fragile mountain system will shape the future of water availability, ecological balance, and human security across Asia and beyond. Protecting Himalayan environment requires scientific rigor, technological innovation, inclusive governance, and strong political commitment. Respect for Indigenous and local knowledge systems, investment in community-led adaptation, and meaningful global climate action are essential. The time for delay has passed; the moment for decisive and collaborative action is now.

References

Agrawal, A., & Ostrom, E. (2001). Collective action, property rights, and decentralization in resource use in India and Nepal. *Politics and Society, 29*(4), 485–514.

Allen, S. K., P. Rastner, M. Arora, C. Huggel, y M. Stoffel. 2016. «Lake outburst and debris flow disaster at Kedarnath, June 2013: Hydrometeorological triggering and topographic predisposition». Landslides 13 (6): 1479–1491. https://doi.org/10.1007/s10346-015-0584-3.

Anjum, N., Ridwan, Q., Sharma, M., Hanief, M., Pant, S., Wani, Z. A., & Bhat, J. A. (2023). Changing climatic scenarios: Impacts, vulnerabilities, and perception with special reference to the Indian Himalayan region. In *Climate change in the Himalayas* (pp. 201–215). Academic Press.

Ayeb-Karlsson, S., Kniveton, D., & Cannon, T. (2020). Trapped in the prison of the mind: Notions of climate-induced (im)mobility decision-making and wellbeing from an urban informal settlement in Bangladesh. *Palgrave Communications, 6*(1), 62. https://doi.org/10.1057/s41599-020-0443-2

Bandyopadhyay, J., & Gyawali, D. (1994). Himalayan water resources: Ecological and political aspects of management. *Mountain Research and Development, 14*, 1–24.

Barnett, T. P., Adam, J. C., & Lettenmaier, D. P. (2005). Potential impacts of a warming climate on water availability in snow-dominated regions. *Nature, 438*(7066), 303–309.

Bhanye, J. (2025). A review study on community-based flood adaptation in informal settlements in the Global South. *Discover Sustainability, 6*(1), 595.

Bhardwaj, M., & Pande, S. (2025). Traditional water systems in the lower Himalayan region. In *Handbook of nature-based drought solutions* (pp. 517–537). Elsevier.

Bhatta, L. D., van Oort, B. E. H., Stork, N. E., & Baral, H. (2015). Ecosystem services and livelihoods in a changing climate: Understanding local adaptations in the Upper Koshi, Nepal. *International Journal of Biodiversity Science, Ecosystem Services & Management, 11*(2), 145–155.

Bhattacharyya, A., Dhyani, R., Joshi, R., Shekhar, M., Kuniyal, J. C., Ranhotra, P. S., & Singh, S. P. (2023). Is survival of Himalayan Cedar (Cedrus deodara) threatened? An evaluation based on predicted scenarios of its growth trend under future climate change. *Science of the Total Environment, 882*, 163630. https://doi.org/10.1016/j.scitotenv.2023.163630

Biswas, T., Subramanian, S., Niyogi, R., Mylliemngap, W., Mitra, S., Das, D. K., & Sarkar, M. S. (2025). Conservation challenges and adaptation strategies for Indian Himalayan biodiversity in a changing climate. In *Climate change impact on Himalayan biodiversity* (pp. 179–224). Springer Nature.

Bolch, T., Kulkarni, A., Kääb, A., Huggel, C., Paul, F., Cogley, J. G., Frey, H., Kargel, J. S., Fujita, K., Scheel, M., Bajracharya, S., & Stoffel, M. (2012). The state and fate of Himalayan glaciers. *Science, 336*(6079), 310–314. https://doi.org/10.1126/science.1215828

Braconnot, P., Harrison, S. P., Kageyama, M., Bartlein, P. J., Masson-Delmotte, V., Abe-Ouchi, A., Otto-Bliesner, B., & Zhao, Y. (2012). Evaluation of climate models using palaeoclimatic data. *Nature Climate Change, 2*(6), 417–424. https://doi.org/10.1038/nclimate1456

Byg, A., & Salick, J. (2009). Local perspectives on a global phenomenon—Climate change in eastern Tibetan villages. *Global Environmental Change, 19*(2), 156–166.

Chettri, B., & Tamang, R. (2024). Status and trends of Herpetofaunal diversity in the Himalaya biodiversity hotspot. In *Biodiversity hotspot of the Himalaya* (pp. 215–244). Apple Academic Press.

Coger, T., Dinshaw, A., Tye, S., Kratzer, B., Aung, M. T., Cunningham, E., et al. (2022). *Locally led adaptation: From principles to practice*. Washington, DC.

Dhyani, R., Casas-Gómez, P., Schneider, L., Sánchez-Salguero, R., Thakur, S., Shekhar, M., Joshi, R., Kuniyal, J. C., Bhattacharyya, A., Singh, S. P., & Linares, J. C. (2025). Contrasting growth responses in Himalayan trees to future climate. *Science of the Total Environment, 986*, 179700.

Dimri, A. P., Bookhagen, B., Stoffel, M., & Yasunari, T. (Eds.). (2019). *Himalayan weather and climate and their impact on the environment*. Springer Nature.

Dixit, A., Sahany, S., Mishra, S. K., & Mesquita, M. D. (2023). Seasonal dependent suitability of physical parameterizations to simulate precipitation over the Himalayan headwater. *Scientific Reports, 13*(1), 4756.

Edwards, S. E., Lowe, C., Stanbrough, L., Walker, B., Kent, R., Oglesby, R., & Morton, K. (2010). *The waters of the third pole: Sources of threat*. Sources of Survival.

Flanner, M. G., Zender, C. S., Hess, P. G., Mahowald, N. M., Painter, T. H., Ramanathan, V., & Rasch, P. J. (2009). Springtime warming and reduced snow cover from carbonaceous particles. *Atmospheric Chemistry and Physics, 9*(7), 2481–2497. https://doi.org/10.5194/acp-9-2481-2009

Gansser, A. (1974). Himalayas. In A. M. Spencer (Ed.), *Mesozoic-Cenozoic Orogenic Belts: Data for Orogenic Studies* (pp. 267–278). Geological Society of London. https://doi.org/10.1144/GSL.SP.2005.004.01.15

Gasse, F., Arnold, M., Fontes, J., et al. (1991). A 13,000-year climate record from western Tibet. *Nature, 353*, 742–745. https://doi.org/10.1038/353742a0

Gentle, P., & Maraseni, T. N. (2012). Climate change, poverty and livelihoods: Adaptation practices by rural mountain communities in Nepal. *Environmental Science & Policy, 21*, 24–34. https://doi.org/10.1016/j.envsci.2012.03.007

Gerlitz, J.-Y., Macchi, M., Brooks, N., Pandey, R., Banerjee, S., & Jha, S. K. (2017). The role of indigenous knowledge in crafting adaptation strategies in the Hindu Kush Himalaya. In P. Wester, A. Mishra, A. Mukherji, & A. B. Shrestha (Eds.), *The Hindu Kush Himalaya assessment* (pp. 345–366). Springer.

Gertler, Charles G., Siva Praveen Puppala, Arnico Panday, Dorothea Stumm, y Joseph Shea. 2016. «Black carbon and the Himalayan cryosphere: A review». Atmospheric Environment, South Asian Aerosols And Anthropogenic Emissions: Regional And Global Climate Implications, vol. 125 (enero): 404–417. https://doi.org/10.1016/j.atmosenv.2015.08.078.

Goodrich, C. G., Manjari Mehta, M. M., & Suman Bisht, S. B. (2017). *Status of gender, vulnerabilities and adaptation to climate change in the Hindu Kush Himalaya: Impacts and implications for livelihoods, and sustainable mountain development*.

Gruber, S. (2012). Derivation and analysis of a high-resolution estimate of global permafrost zonation. *The Cryosphere, 6*(1), 221–233.

Herzschuh, U. (2014). Palaeo-moisture evolution in monsoonal Central Asia during the last 50,000 years. *Quaternary Science Reviews, 103*, 116–137.

Huss, M., & Hock, R. (2018). Global-scale hydrological response to future glacier mass loss. *Nature Climate Change, 8*(2), 135–140. https://doi.org/10.1038/s41558-017-0049-x

ICIMOD. (2011). *Glacial lakes and glacial lake outburst floods in Nepal*. International Centre for Integrated Mountain Development.

Immerzeel, W. W., van Beek, L. P. H., & Bierkens, M. F. P. (2010). Climate change will affect the Asian water towers. *Science, 328*(5984), 1382–1385. https://doi.org/10.1126/science.1183188

IPCC. (2019). *IPCC special report on the ocean and cryosphere in a changing climate* [H.-O. Pörtner, D.C. Roberts, V. Masson-Delmotte, P. Zhai, M. Tignor, E. Poloczanska, K. Mintenbeck, A. Alegría, M. Nicolai, A. Okem, J. Petzold, B. Rama, N.M. Weyer (Eds.)]. Cambridge University Press. (See Chapter 2: High Mountain Areas).

IPCC. (2021). *Climate change 2021: The physical science basis. Contribution of working group I to the sixth assessment report of the intergovernmental panel on climate change* [Masson-Delmotte, V., et al. (Eds.)]. Cambridge University Press. Chapter 5 covers global carbon and other biogeochemical cycles.

IPCC. (2023). *Climate change 2023: Synthesis report. Contribution of working groups I, II and III to the sixth assessment report of the intergovernmental panel on climate change* [H. Lee and J. Romero (Eds.)]. IPCC.

Islam, N., Vennemann, T., Büntgen, U., Cherubini, P., & Lane, S. N. (2024). Tree-ring hydrological research in the Himalaya: State of the art and future directions. *Progress in Physical Geography: Earth and Environment, 48*(3), 454–489.

Ives, J. (2004). *Himalayan perceptions: Environmental change and the well-being of mountain peoples*. Routledge.

Jasrotia, A. S., Ahmad, S., Thakur, P. K., Ridwan, Q., Wani, Z. A., Alamri, S. A. M., et al. (2023). Long-term geospatial observations of the Drang Drung and Pensilungpa Glaciers, North Western Himalaya, India, from 1976 to 2020. *Sustainability, 15*(20), 15067.

Joshi, M., Luitel, K., Barfal, S. S., Kuniyal, J. C., & Pande, K. (2023). Significance of indigenous knowledge systems in water conservation, management: A study from Sikkim himalaya. In *Traditional ecological knowledge of resource management in Asia* (pp. 159–174). Springer.

Khadka, N., Liu, W., Shrestha, M., et al. (2025). Multidisciplinary perspectives in understanding Himalayan glacial lakes in a climate challenged world. *Information Geography, 1*(1), 100002. https://doi.org/10.1016/j.infgeo.2025.100002

Kheder, S. B., & Zugravu, N. (2012). Environmental regulation and French firms location abroad: An economic geography model in an international comparative study. *Ecological Economics, 77*, 48–61.

Kindlmann, P. (2011). *Himalayan biodiversity in the changing world*. Springer Science & Business Media.

Kopacz, M., Mauzerall, D. L., Wang, J., Leibensperger, E. M., Henze, D. K., & Singh, K. (2011). Origin and radiative forcing of black carbon transported to the Himalayas and Tibetan plateau. *Atmospheric Chemistry and Physics, 11*(6), 2837–2852. https://doi.org/10.5194/acp-11-2837-2011

Kraaijenbrink, P. D., Bierkens, M. F., Lutz, A. F., & Immerzeel, W. W. (2017). Impact of a global temperature rise of 1.5 degrees Celsius on Asia's glaciers. *Nature, 549*(7671), 257–260.

Krishnan, R., Shrestha, A. B., Ren, G., Rajbhandari, R., Saeed, S., Sanjay, J., et al. (2019). Unravelling climate change in the Hindu Kush Himalaya: Rapid warming in the mountains and increasing extremes. In P. Wester, A. Mishra, A. Mukherji, & A. B. Shrestha (Eds.), *The Hindu Kush Himalaya assessment: Mountains, climate change, sustainability and people* (pp. 57–97). Springer. https://doi.org/10.1007/978-3-319-92288-1_3

Kumar, S. (2010). Community based Management of Traditional Water Resources in Western Himalayan region: A case study from the pilot locations in India, South East Asia. In *Sustainability in food and water: An Asian perspective* (pp. 147–158). Springer Netherlands.

Lee, E., Carrivick, J. L., Quincey, D. J., Cook, S. J., James, W. H. M., & Brown, L. E. (2021). Accelerated mass loss of Himalayan glaciers since the little ice age. *Scientific Reports, 11*(1), 24284. https://doi.org/10.1038/s41598-021-03805-8

Levermann, A., Schewe, J., Petoukhov, V., & Held, H. (2009). Basic mechanism for abrupt monsoon transitions. *Proceedings of the National Academy of Sciences, 106*(49), 20572–20577.

Li, C., Bosch, C., Kang, S., et al. (2016). Sources of Black Carbon to the Himalayan–Tibetan Plateau Glaciers. *Nature Communications, 7*(1), 12574. https://doi.org/10.1038/ncomms12574

Lutz, A. F., Immerzeel, W. W., Shrestha, A. B., & Bierkens, M. F. P. (2014). Consistent increase in High Asia's runoff due to increasing glacier melt and precipitation. *Nature Climate Change, 4*(7), 587–592. https://doi.org/10.1038/nclimate2237

Manandhar, S., Vogt, D. S., Perret, S. R., & Kazama, F. (2011). Adapting cropping systems to climate change in Nepal: A cross-regional study of farmers' perception and practices. *Regional Environmental Change, 11*(2), 335–348.

Mann, M. E., & Jones, P. D. (2003). Global surface temperatures over the past two millennia. *Geophysical Research Letters, 30*(15). https://doi.org/10.1029/2003GL017814

Marzeion, B., Hock, R., Anderson, B., Bliss, A., Champollion, N., Fujita, K., et al. (2020). Partitioning the uncertainty of ensemble projections of global glacier mass change. *Earth's Future, 8*(7), e2019EF001470. https://doi.org/10.1029/2019EF001470

Maurer, J. M., Schaefer, J. M., Rupper, S., & Corley, A. J. S. A. (2019). Acceleration of ice loss across the Himalayas over the past 40 years. *Science Advances, 5*(6), eaav7266.

Myers, Norman, Russell A. Mittermeier, Cristina G. Mittermeier, Gustavo A. B. da Fonseca, y Jennifer Kent. 2000. «Biodiversity hotspots for conservation priorities». Nature 403 (6772): 853–858. https://doi.org/10.1038/35002501.

Negi, V. S., & Maikhuri, R. K. (2013). Socio-ecological and religious perspective of agrobiodiversity conservation: Issues, concern and priority for sustainable agriculture, Central Himalaya. *Journal of Agricultural and Environmental Ethics, 26*(2), 491–512.

Negi, V. S., Tiwari, D. C., Singh, L., Thakur, S., & Bhatt, I. D. (2022). Review and synthesis of climate change studies in the Himalayan region. *Environment, Development and Sustainability, 24*(9), 10471–10502.

Nyong, A., Adesina, F., & Osman Elasha, B. (2007). The value of indigenous knowledge in climate change mitigation and adaptation strategies in the African Sahel. *Mitigation and Adaptation Strategies for Global Change, 12*(5), 787–797.

Ojha, H. R., Ghimire, S., Pain, A., Nightingale, A., Khatri, D. B., & Dhungana, H. (2016). Policy without politics: Technocratic control of climate change adaptation policy making in Nepal. *Climate Policy, 16*(4), 415–433.

Palazzi, E., Filippi, L., & von Hardenberg, J. (2017). Insights into elevation-dependent warming in the Tibetan Plateau-Himalayas from CMIP5 model simulations. *Climate Dynamics, 48*, 3991–4008. https://doi.org/10.1007/s00382-016-3316-z

Panday, P. K., Thibeault, J., & Frey, K. E. (2015). Changing temperature and precipitation extremes in the Hindu Kush-Himalayan region: An analysis of CMIP3 and CMIP5 simulations and projections. *International Journal of Climatology, 35*(10), 3058–3077. https://doi.org/10.1002/joc.4192

Pepin, N., Bradley, R. S., Diaz, H. F., Baraer, M., Caceres, E. B., Forsythe, N., et al. (2015). Elevation-dependent warming in mountain regions of the world. *Nature Climate Change, 5*(5), 424–430. https://doi.org/10.1038/nclimate2563

Pepin, N. C., Arnone, E., Gobiet, A., Haslinger, K., Kotlarski, S., Notarnicola, C., et al. (2022). Climate changes and their elevational patterns in the mountains of the world. *Reviews of Geophysics, 60*, e2020RG000730. https://doi.org/10.1029/2020RG000730

Pritchard, H. D. (2019). Asia's shrinking glaciers protect large populations from drought stress. *Nature, 569*(7758), 649–654.

Rahman, M. F., Falzon, D., Robinson, S.-a., Kuhl, L., Westoby, R., Omukuti, J., Schipper, E. L. F., et al. (2023). Locally led adaptation: Promise, pitfalls, and possibilities. *Ambio, 52*(10), 1543–1557.

Ramanathan, V., & Carmichael, G. (2008). Global and regional climate changes due to black carbon. *Nature Geoscience, 1*(4), 221–227. https://doi.org/10.1038/ngeo156

Rastogi, A. (2025). *Economic dynamics of Uttarakhand: Pillars of growth and sustainable development*. Perfect Writer Publishing.

Ridley, J., Wiltshire, A., & Mathison, C. (2013). More frequent occurrence of westerly disturbances in Karakoram up to 2100. *Science of the Total Environment, 468*, S31–S35. https://doi.org/10.1016/j.scitotenv.2013.03.074

Roué, M., Nakashima, D., & Krupnik, I. (Eds.). (2022). *Resilience through knowledge co-production: Indigenous knowledge, science, and global environmental change*. Cambridge University Press.

Roxy, M. K., Ghosh, S., Pathak, A., Athulya, R., Mujumdar, M., Murtugudde, R., Terray, P., & Rajeevan, M. (2017). A threefold rise in widespread extreme rain events over central India. *Nature Communications, 8*(1), 708. https://doi.org/10.1038/s41467-017-00744-9

Sattar, A., Haritashya, U. K., Kargel, J. S., & Karki, A. (2022). Transition of a small Himalayan glacier Lake outburst flood to a Giant Transborder flood and debris flow. *Scientific Reports, 12*(1), 12421. https://doi.org/10.1038/s41598-022-16337-6

Scott, C. A., Zhang, F., Mukherji, A., Immerzeel, W., & Wester, P. (2019). Water in the Hindu Kush Himalaya. In P. Wester, A. Mishra, A. Mukherji, & A. B. Shrestha (Eds.), *The Hindu Kush Himalaya assessment: Mountains, climate change, sustainability and people* (pp. 257–299). Springer. https://doi.org/10.1007/978-3-319-92288-1_8

Shah, S., & Ishtiaque, A. (2025). Adaptation to Glacial Lake Outburst Floods (GLOFs) in the Hindukush-Himalaya: A review. *Climate, 13*(3), 3. https://doi.org/10.3390/cli13030060

Sharma, E., Molden, D., Rahman, A., Khatiwada, Y. R., Zhang, L., Singh, S. P., et al. (2019). Introduction to the hindu kush himalaya assessment. In *The Hindu Kush Himalaya assessment: Mountains, climate change, sustainability and people* (pp. 1–16). Springer.

Shekhar, M., Bhardwaj, A., Singh, S., et al. (2017). Himalayan glaciers experienced significant mass loss during later phases of little ice age. *Scientific Reports, 7*(1), 10305. https://doi.org/10.1038/s41598-017-09212-2

Sherpa, P. (2014). Climate change, perceptions, and social heterogeneity in Pharak, Mount Everest region of Nepal. *Human Organization, 73*(2), 153–161.

Shrestha, A. B., Wake, C. P., Mayewski, P. A., & Dibb, J. E. (1999). Maximum temperature trends in the Himalaya and its vicinity: An analysis based on temperature records from Nepal for the period 1971–94. *Journal of Climate, 12*(9), 2775–2786. https://doi.org/10.1175/1520-0442(1999)012<2775:MTTITH>2.0.CO;2

Shrestha, A. B., Agrawal, N. K., Alfthan, B., Bajracharya, S. R., Maréchal, J., & Oort, B. V. (2015). *The Himalayan climate and water atlas: Impact of climate change on water resources in five of Asia's major river basins*. ICIMOD.

Shugar, D. H., Jacquemart, M., Shean, D., Bhushan, S., Upadhyay, K., Sattar, A., et al. (2021). A massive rock and ice avalanche caused the 2021 disaster at Chamoli, Indian Himalaya. *Science, 373*(6552), 300–306. https://doi.org/10.1126/science.abh4455

Singh, S. P., & Thadani, R. (2015). Complexities and controversies in Himalayan research: A call for collaboration and rigor for better data. *Mountain Research and Development, 35*(4), 401–409.

Subedi, A., Raut, N., & Gurung, S. (2023). How Himalayan communities are changing cultivation practices in the context of climate change. *Regional Sustainability, 4*(4), 378–389.

Telwala, Y., Brook, B. W., Manish, K., & Pandit, M. K. (2013). Climate-induced elevational range shifts and increase in plant species richness in a Himalayan biodiversity epicentre. *PLoS One, 8*(2), e57103. https://doi.org/10.1371/journal.pone.0057103

Thompson, L. G., Yao, T., Mosley-Thompson, E., Davis, M. E., Henderson, K. A., & Lin, P. N. (2000). A high-resolution millennial record of the South Asian monsoon from Himalayan ice cores. *Science, 289*(5486), 1916–1919. https://doi.org/10.1126/science.289.5486.1916

Tiwari, P. C., Tiwari, A., & Joshi, B. (2018). Urban growth in Himalaya: Understanding the process and options for sustainable development. *Journal of Urban and Regional Studies on Contemporary India, 4*(2), 15–27.

Tiwari, P. C., Aryal, K., & Joshi, B. (2020). Climate change and indigenous knowledge: Perceptions and adaptations of the Thakali community in the Annapurna Himalaya, Nepal. *Environmental Development, 36*, 100574. https://doi.org/10.1016/j.envdev.2020.100574

Turner, A. G., & Annamalai, H. (2012). Climate change and the South Asian summer monsoon. *Nature Climate Change, 2*(8), 587–595. https://doi.org/10.1038/nclimate1495

Vishwakarma, B. D., Ramsankaran, R., Azam, M. F., et al. (2022). Challenges in understanding the variability of the cryosphere in the Himalaya and its impact on regional water resources. *Frontiers in Water, 4*. https://doi.org/10.3389/frwa.2022.909246

Wang, L., Yang, S., Chen, K., Liu, S., Jin, X., & Xie, Y. (2024). A long-duration glacier change analysis for the Urumqi River valley, a representative region of Central Asia. *Remote Sensing, 16*(9), 1489.

Wani, Z. A., Fatima, T., Verma, O., Sharma, V., & Pant, S. (2023). Impacts of climate change on plants with special reference to the Himalayan region. In *Climate change adaptation, risk management and sustainable practices in the Himalaya* (pp. 237–251). Springer.

Weaver, S., Bakhmetyeva, T., & Rinn, D. W. (2022). Climate witness: Oral environmental history and community-based research—A case study from Trans-Himalayan India. *The Oral History Review, 49*(1), 56–76.

Wester, P., Mishra, A., Mukherji, A., & Shrestha, A. B. (2019). *The Hindu Kush Himalaya assessment: Mountains, climate change, sustainability and people*. Springer Nature. https://doi.org/10.1007/978-3-319-92288-1

Xu, J., Grumbine, R. E., Shrestha, A., Eriksson, M., Yang, X., Wang, Y., & Wilkes, A. (2009). The melting Himalayas: Cascading effects of climate change on water, biodiversity, and livelihoods. *Conservation Biology, 23*(3), 520–530. https://doi.org/10.1111/j.1523-1739.2009.01237.x

Yadav, A., & Singh, A. (2024). The Himalayas in the anthropocene. In *The himalayas in the Anthropocene: Environment and development* (pp. 1–31). Springer Nature. https://doi.org/10.1007/978-3-031-50101-2

Yang, Z. L., & Bai, P. (2024). Response of runoff and its components to climate change in the Manas River of the Tian Shan Mountains. *Advances in Climate Change Research, 15*(1), 62–74.

Yasunari, T. J., Bonasoni, P., Laj, P., Fujita, K., Vuillermoz, E., Marinoni, A., et al. (2010). Estimated impact of black carbon deposition during pre-monsoon season from Nepal Climate Observatory—Pyramid data and snow albedo changes over Himalayan glaciers. *Atmospheric Chemistry and Physics, 10*(14), 6603–6615. https://doi.org/10.5194/acp-10-6603-2010

You, Q.-L., Ren, G.-Y., Zhang, Y.-Q., et al. (2017). An overview of studies of observed climate change in the Hindu Kush Himalayan (HKH) region. *Advances in Climate Change Research, 8*(3), 141–147. https://doi.org/10.1016/j.accre.2017.04.001. Including special issue on climate change in the Hindu Kush Himalaya.

Climate Change as a Major Threat to Biodiversity

Navneeti Chamoli, Bharat Hamal, Jitender Kumar, Somvir Singh, Rajeev Kumar, Priyanka Sharma, and Arti Sharma

Abstract Biodiversity, encompassing genetic, species, and ecosystem diversity, forms the foundation of life on Earth and sustains essential ecological functions. It is essential for sustaining ecological stability, ensuring food security, regulating climate, and conserving natural resources. Biodiversity is confronting an unparalleled crisis due to human-induced pressures, with climate change identified as one of the most significant dangers. Increasing global temperatures, unpredictable precipitation patterns, severe weather events, and habitat degradation are causing alterations in species distributions, disturbing ecological interactions, and heightening extinction threats. India, recognized as a megadiverse country with four global biodiversity hotspots namely the Himalayas, the Indo-Burma region, the Western Ghats, and the Sundaland is highly vulnerable to climate-induced biodiversity loss. The retreat of Himalayan glaciers threatens alpine flora and fauna, while rising sea levels endanger the Sundarbans mangroves and coastal ecosystems. Changes in monsoon patterns disrupt

N. Chamoli
Department of Seed Science & Technology, HNB Garhwal University, Srinagar, Uttarakhand, India

B. Hamal
Department of Forestry, HNB Garhwal University, Srinagar, Uttarakhand, India

J. Kumar
Department of Plant Sciences, Central University of Himachal Pradesh, Kangra, Himachal Pradesh, India

S. Singh
Department of Biosciences, University Institute of Biotechnology, Chandigarh University, Mohali, Punjab, India

R. Kumar
School of Bioengineering and Biosciences, Lovely Professional University, Phagwara, Punjab, India

P. Sharma
Department of Bio Sciences, Career Point University, Hamirpur, Himachal Pradesh, India

A. Sharma (✉)
Faculty of Agriculture, Abhilashi University, Mandi, Himachal Pradesh, India

S. Siddiqui et al. (eds.), *Frontiers in Sustainability*,
https://doi.org/10.1007/978-3-032-19740-5_5

freshwater ecosystems and agricultural biodiversity, impacting food security and rural livelihoods. Also, invasive species, habitat fragmentation, and altered phenological cycles exacerbate the threats posed by climate change. To mitigate these challenges, urgent conservation efforts are needed, including ecosystem-based adaptation strategies, climate-resilient habitat restoration, and policy-driven interventions. Incorporating biodiversity conservation within climate action frameworks, advocating for sustainable land use practices, and including local populations in conservation efforts will be essential for enhancing resilience to climate change. An extensive and flexible strategy is crucial to protect India's diverse biodiversity and guarantee the enduring viability of its ecosystems amid a swiftly evolving climate. This chapter underscores the necessity of a multifaceted strategy to protect biodiversity and guarantee ecological sustainability among the intensifying effects of climate change.

Keywords Biodiversity · Ecosystem · Resilience · Threat · Acidity

1 Introduction

Biodiversity is the sum of all living things, including microbes, plants, animals, their habitats, and genetic variation both within and between species (CBD, 2010). It is essential to human existence in order to sustain life on Earth and support ecological processes. Biodiversity plays a major role in food security, economic livelihoods, public health, and cultural significance in addition to providing essential ecosystem services like pollination, water purification, and climate regulation (IPBES, 2019). Biodiversity is threatened by a variety of environmental and human-caused stressors. These include the spread of invasive species, pollution, overuse of natural resources, habitat destruction, and climate change (IPBES, 2019). Rapid population growth, economic expansion, technological advancement, and sociopolitical changes are the main causes of these risks. Despite being a natural process, human activity is primarily to blame for the continuous decline in biodiversity, which has led to previously unheard-of losses in species and ecosystems worldwide (UNEP, 2021).

Anthropogenic activities and climate change are causing biodiversity to be under more stress than it has ever been under in Earth's recent evolutionary history (IPCC, 2022). A major global issue, climate change has far-reaching effects on both the environment and human societies (Steffen et al., 2004). It concerns long-term alterations in all climate variables, mostly brought about by human activity (Kromp-Kolb & Formayer, 2005; Latif, 2007). Food production, water availability, and overall environmental stability are all impacted by these changes, which pose serious risks to ecosystems and human cultures (Millennium Ecosystem Assessment, 2005; IPCC, 2021). Global warming, rising sea levels, and extreme weather events are the results of these activities' significant increase in the atmospheric concentrations of CO_2, N_2O, and CH_4. All aspects of life are greatly impacted by climate change, but the loss of biodiversity is one of its most serious and worrisome consequences. The survival of species and the stability of ecological processes are at risk due to the unprecedented disruption of the complex ecosystem balance that has been established over millions of years (Steffen et al., 2018).

Biodiversity and climate change are interconnected. Increasing global temperatures are altering climatic zones, compelling animals to relocate to new habitats in pursuit of more favorable conditions. Although certain species may adapt, numerous others find it challenging to manage swift environmental changes, resulting in population reductions and, in some instances, extinction (Bellard et al., 2012). Coral reefs are particularly susceptible to temperature fluctuations, and rising ocean temperatures have led to extensive coral bleaching, jeopardizing marine biodiversity. According to Hughes et al. (2017), changing precipitation patterns and extended droughts are transforming freshwater ecosystems, diminishing water availability for aquatic animals and disrupting food webs. The swift melting of ice in polar regions is threatening the animals like polar bears and penguins, which rely on stable ice-covered habitats (Stirling & Derocher, 2012). Also, rising global temperatures are shifting climatic zones, forcing species to migrate to new habitats in search of favorable conditions. While some species may adapt, many others struggle to cope with rapid environmental changes, leading to Population reductions and, in certain instances, extinction ((Bellard et al., 2012). Coral reefs are particularly susceptible to temperature fluctuations, and elevated ocean temperatures have led to extensive coral bleaching, jeopardizing marine biodiversity (Hughes et al., 2017). Likewise, changing precipitation patterns and extended droughts are transforming freshwater ecosystems, diminishing water availability for aquatic animals and disrupting food webs. The swift melting of ice in polar regions jeopardizes species like polar bears and penguins, which rely on stable ice-covered habitats for their survival (Stirling & Derocher, 2012).

By changing species interactions and natural life cycles (Fig. 1), climate change also has an impact on biodiversity. Seasonal climate cues are essential for the migration, reproduction, and hibernation of both plants and animals (Malhi et al., 2009). These biological rhythms are upset by changes in precipitation or temperature, which results in inconsistencies in food availability and reproductive success. The survival and reproductive rates of migratory birds may be impacted by inadequate food sources if they arrive at their breeding grounds too early or too late (Atkinson et al., 2004). These days, severe weather events like hurricanes, floods, and wildfires are frequent and destructive, destroying habitats and forcing species to relocate. These changes weaken ecosystems, making them more vulnerable to additional stressors like pollution and invasive species (Duarte et al., 2013; Reynolds et al., 2007). Ecosystems' ability to control the climate deteriorates with time. In addition to destroying habitats, deforestation releases large amounts of stored carbon dioxide into thc atmosphere, accelerating global warming (Potts et al., 2010) Similarly, the loss of marine biodiversity, particularly seagrasses and phytoplankton, reduces the ocean's capacity to store carbon, which exacerbates climate change (Abdisa et al., 2017; Ahmed et al., 2017). This vicious cycle emphasizes how crucial it is to address biodiversity preservation and climate change mitigation at the same time. Human societies that rely on healthy ecosystems for sustenance, water, and climate

Fig. 1 Climate change is a worldwide occurrence predominantly caused by human actions that elevate greenhouse gas (GHG) emissions, resulting in global warming. The combustion of fossil fuels for energy generation, industrial processes, deforestation, and extensive agriculture substantially contributes to the accumulation of carbon dioxide (CO_2), methane (CH_4), and nitrous oxide (N_2O) in the atmosphere. These gases retain heat, disturbing Earth's energy equilibrium and resulting in a gradual increase in global temperatures. The ramifications of these alterations are significant, impacting ecosystems, biodiversity, and human communities globally

adaptation will become unstable if immediate action is not taken to slow down the rapid decline of biodiversity (CBD, 1992; Gadgil et al., 1993; Adger, 2000; Corbeels et al., 2014; UNFCCC, 2015).

2 Primary Drivers of Climate Change

While changes in land use, such as urbanization and industrial agriculture, disturb natural ecosystems and increase emissions, deforestation exacerbates climate change by reducing the number of trees that absorb CO_2. Furthermore, severe weather events, rising sea levels, and disruptions to global ecosystems are all consequences of global warming, which is made worse by industrial activities and an excessive reliance on non-renewable energy sources (CBD, 2003). The transportation industry also makes a substantial contribution to greenhouse gas emissions worldwide. Domestic transport accounts for a sizable amount of emissions in the EU, with road transport being the main source. Due to increased economic activity and mobility demands, emissions from passenger and freight transportation have continuously increased over time. Growing transport volumes are primarily responsible for the rise in emissions related to transportation. The number of vehicles on the road has tripled in the last three decades (EEA, 2006).

3 Impacts of Climate Change

3.1 Greenhouse Gas Emissions and Global Warming

The primary cause of climate change is the rise in GHG emissions (Fig. 2). Carbon emissions from burning coal, oil, and natural gas increased significantly during the industrial revolution (AbdElgawad et al., 2015; Chater et al., 2015). Deforestation exacerbates the issue by reducing the Earth's capacity to absorb CO_2. Methane emissions from livestock and the widespread use of nitrogen-based fertilizers that release N_2O are two ways that agriculture contributes (Frungillo et al., 2014; Jhun et al., 2015; Gupta & Igamberdiev, 2016).

3.2 Changing Weather Patterns

Global weather patterns have changed dramatically as a result of climate change. Variations in temperature and precipitation upset ecosystems, resulting in unseasonable rainfall, extended droughts, and unpredictable monsoon cycles. These changes have an effect on agriculture, reducing food security and agricultural yield. In some

Fig. 2 Impact of greenhouse gases on earth

areas, desertification renders land unusable for farming, while in others, increased flooding leads to soil erosion and degradation. Temperature variations also affect plant phenology, disrupting food web dynamics and pollination (Burney & Ramanathan, 2014; Rana et al., 2025).

3.3 Ocean Acidification and Rising Sea Levels

Ocean acidification is caused by the oceans absorbing too much CO_2 from the atmosphere, endangering marine life. As temperatures rise and acidity increases, coral bleaching affects coral reefs, which are vital marine ecosystems (Allemand & Osborn, 2019). Fish populations and fisheries decline as a result, affecting livelihoods that depend on marine resources and affecting marine food chains. One major worry is the rising sea levels brought on by the thawing of polar ice caps and glaciers. Communities are being uprooted and biodiversity in coastal and estuarine ecosystems is being lost as a result of the erosion of coastal habitats and the threat of island nations being submerged.

Ocean acidification is the process by which the ocean's pH level keeps going down because it absorbs CO_2 from the air. Oceanic pH has gone down by about 0.1 units since the Industrial Revolution, which is the same as a 30% rise in hydrogen ion concentration (Doney et al., 2009). This chemical change makes carbonate ions less available, which is bad for organisms that need them to build their shells, like molluscs, echinoderms, and some planktonic species. The physiological effects go beyond coral bleaching. For example, mussels (Mytilus edulis) have trouble forming shells (Gazeau et al., 2007), sea urchin larvae (*Strongylocentrotus purpuratus*) have developmental problems (Dupont, 2008), and fish have trouble regulating acid–base levels, which can affect their senses and metabolism (Ishimatsu et al., 2008).

The effects on ecosystems as a whole are also very important, not just on individual species. Researchers have found that oyster reefs break down when the water is acidic, which makes it harder for fish and invertebrates to find places to live (Waldbusser et al., 2011). Planktonic calcifiers like Foraminifera and Coccolithophores are important for the ocean's carbon cycle. They have lower calcification rates, which could change the way trophic levels work and the way carbon is stored (Fabry, 1990). In polar ecosystems, acidification and melting sea ice add extra stress, which is especially hard on species that are adapted to cold and grow slowly (Comeau et al., 2009).

Long-term mesocosm experiments in- and natural CO_2 vent studies in the Mediterranean Sea show that chronic acidification changes the types of species present, with a decrease in calcifying taxa and an increase in jellyfish and macroalgae dominance (Hall-Spencer et al., 2008). These kinds of changes show that acidified oceans could have less biodiversity and ecosystems that work differently.

3.4 Extreme Weather Events

Hurricanes, cyclones, wildfires, and intense storms are occurring with greater ferocity, devastating ecosystems and human settlements alike. Prolonged heatwaves contribute to forest fires, destroying large swathes of vegetation and wildlife (IPCC, 2014; Mittler & Blumwald, 2010). Flooding and landslides disrupt natural landscapes and human infrastructure, leading to economic losses and displacement of populations. The increasing unpredictability of weather events poses a challenge for disaster preparedness and mitigation efforts (Scheffers et al., 2016).

4 Climate Change: A Threat to Biodiversity

According to IPCC (Intergovernmental Panel on Climate Change, 2007b), the atmospheric concentration of CO_2 has increased from approximately 280–379 ppm in 2005. Methane (CH_4) levels have also surged from about 715–1774 ppb in 2005. Key sources of methane include agriculture (especially livestock digestion and rice paddies), landfills, and the extraction of fossil fuels. Similarly, nitrous oxide (N_2O) concentrations have increased from 270 ppb before industrialization to 319 ppb in 2005, primarily due to excessive use of nitrogen-based fertilizers.

These greenhouse gases contribute to an overall radiative forcing of +2.3 Wm^{-2}, an unprecedented rate of change over the past 10,000 years. The IPCC (2007a) asserts that global warming is now undeniable, as evidenced by increasing air and ocean temperatures, shrinking ice cover, and rising sea levels. Human-induced emissions since the mid-twentieth century have been the dominant driver of temperature increases, overwhelming natural climate variability. Recent data from NASA (2022) suggests that by 2024, Earth's temperature had risen by approximately 1.47 °C (2.65 °F) above pre-industrial levels. Similarly, the World Meteorological Organization (WMO) states that between 2011 and 2020, global temperatures were roughly 1.1 °C higher than the 1850–1900 baselines.

This warming trend has resulted in profound environmental changes. The world's oceans have absorbed substantial amounts of heat, with warming observed at depths exceeding 300 m. This has led to coral bleaching, disruptions in marine ecosystems, and the intensification of tropical storms. The Arctic has warmed at approximately twice the global rate over the past century, leading to a significant decline in sea ice, which has been shrinking by approximately 2.7% per decade. The melting of glaciers and ice sheets has significantly contributed to an average sea level rise of about 1.8 mm/year, endangering coastal communities and low-lying island nations. Beyond temperature increases, climate change has had far-reaching effects on global weather patterns. Precipitation regimes have shifted, with some regions experiencing prolonged droughts while others face intense and frequent rainfall events. Since the 1970s, severe and extended droughts have affected many regions, particularly the tropics and subtropics. These dry conditions have fuelled large-scale wildfires (Ranasinghe et al., 2022).

Studies suggest that storm strength of tropical cyclones also intensified since the 1970s, with stronger wind speeds and increased precipitation in north Atlantic region. While data from other regions remains inconclusive, there is growing evidence that storm activity at global level is increasing in severity. Temperature extremes have become more common, with a decline in cold days and nights and a rise in the frequency of heatwaves. These changes threaten biodiversity by altering ecosystems, reducing species' adaptive capacities, and disrupting food and water availability. Climate models project that future warming will largely depend on emission trajectories. If emissions remain unchecked beyond 2060, temperature increases could range from 4.9 to 6.1 °C and if greenhouse gas concentrations stabilize at early twenty-first century levels, the projected warming rate will be around

0.1 °C per decade. Under high-emission scenarios, this rate could double, leading to devastating environmental consequences. Future tropical cyclones are expected to intensify, exhibiting stronger wind speeds and increased precipitation. Even minor increases in sea level will amplify the risk of storm surges, intensifying coastal flooding and threatening marine and coastal ecosystems. A study by Frei et al. (2006) analyzed projected changes in extreme precipitation events across Europe using regional climate models. Their findings indicate that what were once-in-a-century rainfall events could occur every 20–40 years in some regions. Snow cover is expected to decline significantly, while Arctic and Antarctic Sea ice will continue to shrink. Heatwaves, often accompanied by heavy rainfall and flooding, will become more frequent and severe (IPCC, 2007a).

5 Methodologies for Assigning Biodiversity Alterations to Climate Change

Assigning biodiversity loss directly to climate change necessitate comprehensive, multifaceted scientific strategies. Long-term ecological monitoring is a potent instrument that yields temporal datasets, elucidating patterns in species abundance, distribution, and phenology spanning decades. The Global Biodiversity Information Facility (GBIF) and other national biodiversity monitoring networks have recorded shifts in species distributions toward polar regions or elevated terrains, aligning with regional warming trends (Chen et al., 2011; Parmesan & Yohe, 2003). These observations are frequently corroborated by species distribution models (SDMs), which utilize environmental variables to forecast alterations in suitable habitat under prospective climatic scenarios. Species Distribution Models (SDMs) can forecast future range alterations and extinction probabilities based on IPCC climatic scenarios (Thuiller et al., 2005).

Climate modification experiments, such warming chambers, rainfall exclusion, and CO_2 enrichment, enable researchers to isolate climate variables and evaluate their causative effects on ecosystem structure and species interactions (Rustad et al., 2001). The International Tundra Experiment (ITEX) has proved essential in elucidating the effects of warming on Arctic plant ecosystems. A prevalent method is meta-analysis, which synthesizes extensive datasets from several researches to discern similar patterns of biodiversity response to climatic variables. Integrative attribution frameworks that amalgamate observational data, mechanistic models, and statistical attribution methodologies are currently employed to differentiate the impacts of climate from other factors, including land use change and invasive species (Hansen & Cramer, 2015). These interdisciplinary methodologies are essential for enhancing causal inference and guiding conservation measures amidst the rapid progression of climate change.

While many climate impacts develop gradually, certain scenarios involve abrupt "shock events" that could drastically transform ecosystems (Tol et al., 2006; Van der

Sluijs & Turkenburg, 2006; Vellinga & Wood, 2002). Climate "tipping points"—critical thresholds beyond which rapid and irreversible changes occur—represent a significant threat to biodiversity (Jaeger, 2005). The following sections provide a detailed discussion of all the consequences of climate change.

5.1 Habitat Loss and Fragmentation

Wildfircs, droughts, and hurricanes, lead to the destruction and degradation of natural habitats (Williams et al., 2003). Rising sea levels also threaten coastal and wetland ecosystems, reducing the available space for many species (Parmesan, 2006). Human-induced deforestation and land conversion for agriculture and urbanization exacerbate the problem, fragmenting once-continuous habitats (Miles et al., 2004). This fragmentation isolates species populations, reducing genetic diversity and increasing their vulnerability to environmental changes, leading to decline in population declines and extinctions (Brooks et al., 2002; Fahrig, 2003).

5.2 Changes in Species Distribution and Migration

Many species must change their geographic range in order to survive when climatic conditions change. Marine species are migrating farther into the oceans and toward the poles, while terrestrial species are moving to higher latitudes or altitudes in search of cooler habitats. Localized extinctions result from species' inability to migrate or adapt quickly enough. Such changes also alter predator-prey dynamics and resource competition by upsetting established ecosystems and species interactions. Human-wildlife conflicts can escalate when species migrate and clash with human populations (Williams et al., 2023).

5.3 Disruptions in Reproductive Cycles and Life History Traits

In order to control their reproductive cycles and life history characteristics, many organisms depend on environmental cues like temperature, photoperiod, and precipitation. These cues have changed as a result of climate change, leading to mismatches between species and their environments. For instance, pollinators that rely on the nectar of plants may be impacted if they bloom earlier than usual. In a similar vein, birds may reduce their reproductive success by laying eggs earlier or later than when their food sources are at their best. These disturbances have the potential to ripple through ecosystems, impacting whole food chains and decreasing biodiversity in general (Chaparro-Pedraza & de Roos, 2019; Franks & Weis, 2009).

5.4 *Decline in Pollinators and Food Web Alterations*

Pollinators (bees, butterflies, and birds) play an important role in maintaining biodiversity by facilitating plant reproduction. However, climate change has led to significant declines in pollinator populations due to habitat destruction, temperature fluctuations, and changes in patterns of flowering (Vasiliev & Greenwood, 2021). A decline in pollinators disrupts food webs by reducing plant reproduction, which in turn affects herbivores and higher trophic levels. Altered temperature and precipitation patterns also impact primary producers such as phytoplankton, which form the foundation of many aquatic food webs, leading to cascading effects throughout ecosystems. Being ectothermic in nature, insect pollinators are sensitive to temperature and precipitation changes (Barnett & Facey, 2016; Burkle & Alarcon, 2011; Colinet et al., 2015), which make them particularly vulnerable. Climate change may also disrupt plant–pollinator interactions due to phenological asynchrony (Kudo & Ida, 2013). IPBES, 2019 has reported significant declines in pollinator populations (Fox, 2013; González-Varo et al., 2013; Goulson et al., 2015; Hallmann et al., 2017; Powney et al., 2019; Seibold et al., 2019; Weiner et al., 2014).

By enabling species from warmer regions to spread out their range, warming may have a positive impact on biodiversity in high-latitude regions, according to energy theory (Fuchs et al., 2015; Gerard et al., 2010; Goldewijk, 2001). Huntley et al., 2004; Hawkins et al., 2003). There have been conflicting findings from empirical research on climate change and pollinator biodiversity in the Northern Hemisphere, including reports of declines (Woods et al., 2008), mixed effects (Burns et al., 2016), and increases in species richness (Franzen & Ockinger, 2012). Variations in study design, scale, or confounding variables could be the cause of these discrepancies (Ewers & Didham, 2006). Despite the urgency, comprehensive climate change mitigation measures are frequently absent from national pollinator conservation strategies in the USA and Europe (Dicks et al., 2015; Underwood et al., 2017; Vasiliev & Greenwood, 2020). Conservation efforts may be weakened if the effects of climate change are misunderstood. Using a hierarchical framework (Noss, 1990), this review attempts to critically evaluate the effects of climate change on pollinator biodiversity, assessing changes at the genetic, species, and community levels. According to Gladstone-Gallagher et al. (2019), homogenisation at any level has the potential to compromise metacommunity resilience and result in the long-term loss of biodiversity.

5.5 *Increased Susceptibility to Diseases*

Climate change has also heightened species' vulnerability to diseases by creating favorable conditions for pathogens and their vectors. Rising temperatures and changing humidity levels have expanded the range of many infectious diseases, affecting both wildlife and human populations (Sgro et al., 2011) For instance, amphibians have experienced devastating declines due to the spread of fungal infections exacerbated by changing climatic conditions. Similarly, warmer temperatures have contributed to the expansion of insect-borne diseases, affecting mammals, birds, and humans (Patz et al., 2005). As species face increased physiological stress due to climate change, their immune responses weaken, making them more susceptible to infections and reducing overall population viability (Semenza & Paz, 2021). Changing in climate also affect emerging infectious diseases (EIDs), including increasing viral cross-species transmission and fungal infections. A recent modeling study on 3000 mammalian species predicts that if global temperatures rise by 2 °C, migration patterns could lead to over 4000 instances of viral spillover in the next 50 years, Which factors render climate change a predominant catalyst for zoonotic disease transmission (Carlson et al., 2022; Grobusch & Grobusch, 2022).

5.6 Threatened Ecosystems and Vulnerable Species

Climate change is progressively transforming ecosystems worldwide, endangering several habitats and species. Increasing temperatures, altered precipitation patterns, ocean acidification, and habitat destruction are causing ecosystem deterioration and species extinctions. This chapter examines the ecosystems most impacted by climate change and the animals that are especially susceptible to these environmental alterations. Species are currently affected by anthropogenic climate change, and its swift progression is constraining the capacity of numerous species to acclimatize to their surroundings (Foden & Young, 2016). Table 1 presents the impact of climate change of several critically endangered, endangered, and vulnerable species.

5.7 Forests (Tropical, Temperate, and Boreal)

Forests are essential for preserving world biodiversity and controlling climate, although they are significantly susceptible to climate change. Tropical forests, characterized by the highest levels of biodiversity, are seeing accelerated deforestation, extended droughts, and heightened occurrences of wildfires. These pressures diminish carbon sequestration and jeopardize species like orangutans, jaguars, and several amphibians (Domke et al., 2023). Temperate forests encounter changing seasonal patterns, resulting in discrepancies in species interactions and heightened insect infestations. Boreal forests, located in northern areas, are especially imperilled by escalating temperatures, resulting in permafrost thaw, heightened wildfire occurrences, and habitat degradation for animals like the boreal caribou and Siberian tiger. The destruction of forest ecosystems disturbs carbon storage equilibrium and hastens climate change EPA, 2021).

Table 1 Impact of climate change on Flora and Fauna (IUCN, 2024)

Species affected	Observed effects	Cause of change	Conservation efforts
FLORA			
Amazon Rainforest Trees (Multiple species)	Increased tree mortality, reduced carbon sequestration	Rising temperatures, deforestation	Reforestation, REDD+ (Reducing Emissions from Deforestation and Forest Degradation) programs
Baobab Tree (*Adansonia* spp.)	Higher mortality rates, failing to reproduce	Extreme droughts, changing rainfall patterns	Conservation through seed banking, habitat protection
Coral Reefs (*Zooxanthellae* algae & Coral species)	Coral bleaching, loss of marine biodiversity	Ocean warming, acidification	Marine protected areas, reef restoration projects
Sundari Tree (*Heritiera* fomes)—Sundarbans	Declining population, increased salinity tolerance	Rising sea levels, coastal erosion	Mangrove restoration, afforestation
Shisham (*Dalbergia sissoo*)	Increased vulnerability to pests and diseases	Erratic rainfall, temperature fluctuations	Research on resistant varieties, agroforestry initiatives
Sandalwood (*Santalum album*)	Decreasing natural populations due to habitat loss	Rising temperatures, overexploitation	Conservation plantations, sustainable harvesting policies
Tea (*Camellia sinensis*)—Assam & Darjeeling	Reduced yield, changes in aroma and quality	Temperature shifts, erratic rainfall	Research on climate-adaptive varieties, sustainable farming practices
Buransh (*Rhododendron* arboreum)	Flowering occurring earlier than usual, reduced flowering in some areas	Rising temperatures, altered precipitation patterns	Afforestation, conservation programs, seed banking
Oak (*Quercus leucotrichophora*)	Declining population, replacement by pine species	Increasing droughts, temperature rise	Sustainable forest management, assisted regeneration
Brahma Kamal (*Saussurea obvallata*)	Shifting to higher altitudes, reduced habitat availability	Glacial retreat, habitat loss	Protection under the Wildlife Protection Act, cultivation in controlled conditions
Deodar (*Cedrus deodara*)	Reduced regeneration, increased vulnerability to pests	Rising temperatures, altered rainfall patterns	Sustainable forestry practices, reforestation programs
Himalayan Yew (*Taxus wallichiana*)	Population decline due to overexploitation and climate change	Unsustainable harvesting, habitat destruction	Conservation in protected areas, regulation on harvesting
Chir Pine (*Pinus roxburghii*)	Expanding at the cost of oak forests, increased forest fires	Higher temperatures, reduced humidity, changing rainfall	Fire control measures, oak regeneration programs

(continued)

Table 1 (continued)

Species affected	Observed effects	Cause of change	Conservation efforts
Himalayan Birch (*Betula utilis*)	Migration to higher altitudes, reduced growth rates	Temperature rise, habitat fragmentation	Afforestation programs, habitat monitoring
Juniper (*Juniperus indica*)	Reduced growth rates, habitat loss	Climate-induced stress, soil degradation	Protection of alpine zones, afforestation efforts
FAUNA			
Royal Bengal Tiger (*Panthera tigris tigris*)	Habitat shrinking, prey depletion	Deforestation, rising temperatures	Project Tiger, national parks & reserves
Indian Elephant (*Elephas maximus indicus*)	Human-elephant conflicts, loss of migration corridors	Deforestation, erratic rainfall	Project Elephant, wildlife corridors
One-horned Rhinoceros (*Rhinoceros unicornis*) - Assam	Flood-related deaths, habitat loss	Increased flooding, habitat fragmentation	Kaziranga National Park protection efforts
Ganges River Dolphin (*Platanista gangetica*)	Population decline due to river pollution	Water pollution, temperature increase	Dolphin conservation projects, clean Ganga initiatives
Great Indian Bustard (*Ardeotis nigriceps*)	Habitat loss, reduced breeding success	Desertification, erratic weather patterns	Conservation breeding programs, habitat restoration
Indian Pangolin (*Manis crassicaudata*)	Declining population due to habitat destruction	Deforestation, climate stress	Anti-poaching laws, conservation projects
Snow Leopard (*Panthera uncia*)	Shrinking habitat, food scarcity, increased human-wildlife conflict	Warming temperatures, human encroachment	Snow Leopard Conservation Project, habitat protection measures
Himalayan Monal (*Lophophorus impejanus*)	Altered migration patterns, population decline	Habitat degradation, changing climate conditions	Biodiversity conservation programs, eco-tourism initiatives
Musk Deer (*Moschus chrysogaster*)	Population decline due to habitat loss	Rising temperatures, deforestation	Protection in wildlife sanctuaries, anti-poaching efforts
Himalayan Brown Bear (*Ursus arctos isabellinus*)	Shifting habitat to higher altitudes, food scarcity	Glacier melting, changing vegetation	Community-based conservation, habitat monitoring
Asiatic Black Bear (*Ursus thibetanus*)	Increased human-wildlife conflict, altered hibernation patterns	Temperature rise, habitat disturbance	Wildlife conflict mitigation programs, habitat conservation
Himalayan Griffon Vulture (Gyps himalayensis)	Declining population due to reduced food availability	Habitat destruction, climate variability	Vulture breeding programs, reducing pesticide use
Himalayan Tahr (*Hemitragus jemlahicus*)	Habitat fragmentation, increasing human-animal conflicts	Temperature changes, grazing land loss	Community-driven conservation, habitat restoration

5.8 Coral Reefs and Marine Life

As a result of climate change, coral reefs, sometimes referred to as the "rainforests of the sea," are among the most endangered ecosystems. Rising water temperatures have caused widespread coral bleaching, a process whereby corals drive out the symbiotic algae that provide them with vital nutrients, leading to high mortality rates (Heron et al., 2017). Coral skeletons are compromised and marine food webs are disrupted by ocean acidification, which is brought on by increased absorption of carbon dioxide. Because they depend on coral reefs for habitat and food, marine animals like clownfish, sea turtles, and reef sharks are especially vulnerable to reef damage. Furthermore, pollution and overfishing put additional stress on marine life, endangering the balance of entire oceanic ecosystems (Obura, 2017).

5.9 Freshwater Ecosystems and Wetlands

Freshwater habitats, encompassing rivers, lakes, and wetlands, are essential for biodiversity and human sustenance. Climate change has modified water temperatures, flow patterns, and seasonal cycles, jeopardizing aquatic life. Elevated evaporation rates and extended droughts diminish water availability, adversely affecting fish populations, amphibians, and migratory birds (Vose et al., 2018). Wetlands, vital carbon sinks, are diminishing due to elevated temperatures and anthropogenic activity, including land conversion and water extraction. Freshwater turtles, amphibians, and wetland birds are facing habitat degradation and population declines, necessitating conservation initiatives for these vulnerable ecosystems (Domke et al., 2020).

5.10 Arctic and Alpine Biodiversity

The Arctic and alpine regions are among the most swiftly transforming habitats as a result of global warming. Increasing temperatures in the Arctic have resulted in the depletion of sea ice, endangering animals such as polar bears, walruses, and Arctic foxes that rely on ice-covered environments for their survival. The changing temperature is modifying the distribution of tundra flora, affecting herbivores such as caribou and their dependent carnivores (Salimi et al., 2021). Alpine ecosystems, located in elevated places, are undergoing alterations in snowlines and temperature variations, compelling animals such as snow leopards, pikas, and mountain goats to migrate to greater altitudes. The scarcity of appropriate habitats renders these species exceedingly susceptible to extinction (Kramer et al., 1997).

5.11 Disruption of Carbon Sequestration Processes

One essential ecosystem service that controls atmospheric carbon dioxide levels and slows down climate change is carbon sequestration. Oceans, wetlands, and forests are important carbon sinks that store large amounts of CO_2. However, the capacity of these natural systems to store carbon is being weakened by climate change-related disturbances like wildfires, deforestation, and ocean acidification. Warming temperatures and prolonged droughts degrade forests, leading to more frequent and intense wildfires that release stored carbon into the atmosphere. Similarly, methane, a potent greenhouse gas, is being released by the thawing of permafrost in Arctic regions, exacerbating global warming. A feedback loop that further destabilizes global ecosystems is created when carbon sequestration capacity declines, speeding up climate change. Over the past 250 years, human combustion of fossil fuels has significantly increased carbon dioxide (CO_2) concentrations above 380 parts per million, which has led to global warming. Sea level rise, extreme weather events (wildfires, floods, droughts, and storms), altered precipitation patterns, and ecological disruptions are some of the negative consequences that result from this. Additionally, increased absorption of CO_2 by saltwater is causing ocean acidification, endangering marine life, including coral reefs and plankton. Implementing technically and financially viable methods that reduce CO_2 emissions and enhance CO_2 extraction from the atmosphere is crucial to addressing these challenges (Sundquist et al., 2008).

5.12 Loss of Genetic Resources for Agriculture and Medicine

Biodiversity functions as an essential reservoir of genetic resources that underpin agriculture, medicine, and ecological resilience. Climate change is causing habitat destruction and animal extinction, resulting in diminished genetic diversity. Numerous wild relatives of crops, which exhibit characteristics for drought resilience, pest resistance, and climatic adaptability, are vanishing due to habitat fragmentation and severe weather phenomena (Tester & Langridge, 2010). This loss diminishes the capacity to cultivate robust crop varieties essential for global food security. Furthermore, numerous medicinal plants and organisms utilized in pharmaceuticals are endangered due to habitat damage linked to climate change. The reduction in genetic resources constrains future prospects for the development of novel pharmaceuticals and sustainable agricultural methods, jeopardizing both human health and food security.

5.13 *Overexploitation of Natural Resources*

The unsustainable extraction of natural resources for sustenance, energy, and raw materials significantly contributes to biodiversity decline. Overfishing, deforestation, and hunting have resulted in the reduction of numerous species, affecting food webs and ecological services. The demand for agricultural land has propelled the proliferation of intensive farming practices that deteriorate soil health and water quality. Excessive groundwater extraction and dam construction disrupt hydrological cycles, affecting freshwater species and wetland habitats. Mining operations devastate ecosystems, emit hazardous contaminants, and lead to soil degradation. To alleviate these effects, it is essential to implement sustainable resource management techniques, enforce conservation legislation, and encourage responsible consumption.

6 Adaptation and Mitigation Strategies

Alterations in temperature and precipitation patterns force species to migrate to suitable habitats; those unable to adapt or relocate face increased extinction risks. For instance, climate change is projected to threaten approximately 5% of Earth's species with extinction under a 2 °C warming scenario, escalating to 16% under a 4.3 °C rise. Biodiversity enhances ecosystem resilience by enabling ecosystems to maintain functionality amid disturbances. However, climate change can overwhelm this capacity, leading to reduced ecosystem stability. For example, increase in frequency and intensity of extreme weather events can disrupt ecological balance, resulting in habitat loss and species decline (EPA, 2021). Ecosystems with greater plant diversity exhibit lower sensitivity to temperature variability, highlighting the stabilizing role of biodiversity. Conversely, loss of biodiversity due to climate change can alter energy flow and material circulation, further destabilizing ecosystems. The loss or migration of species impacts the availability of essential resources such as food, water, fuelwood, fiber, biochemicals, and genetic materials. While ecosystems may continue to function, their ability to provide these services can be significantly reduced. Climate change directly influences an ecosystem's ability to regulate climate, but it also has indirect effects by altering land cover and plant composition. One critical function of ecosystems is carbon dioxide absorption. Studies suggest that terrestrial ecosystems' net carbon uptake is expected to peak before the middle of the century and then weaken or even reverse, intensifying climate change (IPCC, 2007a). Forests play a vital role in climate regulation due to their high evapotranspiration rates, which contribute to atmospheric moisture and influence precipitation patterns. For instance, the decline of the Amazon rainforest, driven by changing rainfall patterns, could severely disrupt this regulatory function (Millennium Ecosystem Assessment, 2005). Other regulatory services, such as disease control, are also affected by shifts in land cover and vegetation. Warmer temperatures and increased rainfall can expand the range of disease carriers like mosquitoes, leading to the spread of vector-borne illnesses.

Cultural ecosystem services are closely linked to biodiversity. If species are lost or ecosystems are altered, the ability to provide cultural benefits, such as recreation and aesthetic value, may decline. A prime example is coral reef ecosystems, which are under threat from rising sea surface temperatures. An increase of 1–3 °C is projected to cause more frequent coral bleaching events and widespread coral mortality unless corals can adapt or acclimatize (Hoegh-Guldberg et al., 2022). Higher atmospheric CO_2 levels may boost primary productivity in some species, but biodiversity declines could reduce both the effectiveness and stability of these fundamental ecosystem services (IPBES, 2019; IPCC, 2022. Ecosystem changes and biodiversity loss resulting from climate change can further impact regional and global climates. This occurs through shifts in greenhouse gas absorption and release, leading to either positive or negative climate feedback loops (Arora et al., 2020; McDowell et al., 2020). Following are some key points for adaptation and mitigation Strategies for Climate Change.

6.1 Preservation and Rehabilitation of Natural Ecosystems

Efforts for preservation and rehabilitation of Natural Ecosystems include reforestation, afforestation, wetland restoration, and marine ecosystem conservation (Brooks et al., 2005). Protecting ecosystems like mangroves, coral reefs, and grasslands helps sequester carbon, enhance water retention, and provide habitat for diverse species. Restoration projects often incorporate native plant species to maintain ecological balance and improve ecosystem services.

6.2 Climate-Resilient Agriculture and Forestry

Climate-resilient agriculture and forestry focus on sustainable practices that minimize climate risks and improve productivity. Some techniques like conservation tillage, agroforestry, cover cropping, and Integrated Pest Management (IPM) enhance soil health and reduce greenhouse gas emissions. Drought-resistant crop varieties, precision irrigation, and organic farming contribute to sustainability (Smit & Wandel, 2006). In forestry, sustainable logging, afforestation with resilient species, and fire management help maintain forest ecosystems against climate change threats.

6.3 *Assisted Migration and Species Conservation Programs*

Assisted migration involves relocating species to suitable habitats to protect them from climate-induced changes. This approach is crucial for species facing habitat loss due to rising temperatures, shifting precipitation patterns, or increased natural disasters. Conservation programs include seed banks, captive breeding, and habitat corridors that facilitate genetic diversity and species survival (Dow & Downing, 2006). These strategies require careful planning to prevent ecological imbalances in new environments.

6.4 *Role of Indigenous Knowledge and Community Participation*

Indigenous knowledge and community participation are crucial for climate adaptation strategies. Traditional knowledge in ecology which has been passed down through generations, offers insights into sustainable resource management, weather prediction, and ecosystem conservation. Community-based adaptation programs integrate local practices with modern science to enhance resilience. Participatory approaches empower communities to implement nature-based solutions, disaster risk management, and sustainable livelihoods, ensuring long-term climate mitigation benefits.

7 Conclusion

Climate change presents a substantial threat to global biodiversity, affecting ecosystems, modifying species ranges, and heightening extinction risks. Increasing temperatures, habitat destruction, ocean acidification, and severe weather phenomena are hastening the decrease of biodiversity. Immediate measures are required to diminish greenhouse gas emissions, enhance protected area networks, and execute nature-based remedies. The future of biodiversity relies on prompt mitigation initiatives and enduring sustainable development policies. If not addressed, climate change would further accelerate biodiversity loss, resulting in permanent ecological repercussions.

8 Future Perspectives

Future research should focus on developing climate-resilient conservation strategies, such as assisted migration, genetic adaptation, and ecosystem-based approaches. Strengthening global policies, promoting reforestation, and reducing carbon emissions are essential steps to mitigate biodiversity loss. Advanced technologies like remote sensing and AI can aid in monitoring species responses to climate change. Collaboration between governments, scientists, and local communities is vital to ensuring a sustainable and biodiverse future.

References

AbdElgawad, H., Farfan-Vignolo, E. R., de Vos, D., & Asard, H. (2015). Elevated CO_2 mitigates drought and temperature-induced oxidative stress differently in grasses and legumes. *Plant Science, 231*, 1–10. https://doi.org/10.1016/j.plantsci.2014.11.001

Abdisa, A., Dirba, K., & Muktar, M. (2017). *Impacts of various ENSO phases on cereal crop productivity in the Upper Awash Basin, Central Highland of Ethiopia*. Haramaya University.

Adger, W. N. (2000). Social and ecological resilience: Are they related? *Progress in Human Geography, 24*(3), 347–364.

Ahmed, H., Tessema, Z., Adugna, T., & Diriba, K. (2017). *Interconnection between El-Niño-southern oscillation induced rainfall variability, livestock population dynamics and pastoralists adaptation strategies in Eastern Ethiopia*. Haramaya University.

Allemand, D., & Osborn, D. (2019). Ocean acidification impacts on coral reefs: From sciences to solutions. *Regional Studies in Marine Science, 28*, 100558. https://doi.org/10.1016/j.rsma.2019.100558

Arora, V. K., Katavouta, A., Williams, R. G., et al. (2020). Carbon–concentration and carbon–climate feedbacks in CMIP6 models, and their comparison to CMIP5 models. *Biogeosciences, 17*(16), 4173–4222. https://doi.org/10.5194/bg-17-4173-2020

Atkinson, A., Siegel, V., Pakhomov, E. A., & Rothery, P. (2004). Long-term decline in krill stock and increase in salps within the Southern Ocean. *Nature, 432*(7013), 100–103.

Barnett, K. L., & Facey, S. L. (2016). Grasslands, invertebrates, and precipitation: A review of the effects of climate change. *Frontiers in Plant Science, 7*, 1196. https://doi.org/10.3389/fpls.2016.01196

Bellard, C., Bertelsmeier, C., Leadley, P., Thuiller, W., & Courchamp, F. (2012). Impacts of climate change on the future of biodiversity. *Ecology Letters, 15*(4), 365–377.

Brooks, T. M., Mittermeier, R. A., Mittermeier, C. G., et al. (2002). Habitat loss and extinction in the hotspots of biodiversity. *Conservation Biology, 16*, 909–923.

Brooks, N., Adger, W. N., & Kelly, P. M. (2005). The determinants of vulnerability and adaptive capacity at the national level and the implications for adaptation. *Global Environmental Change, 15*(2), 151–163. https://doi.org/10.1016/j.gloenvcha.2004.12.006

Burkle, L. A., & Alarcon, R. (2011). The future of plant-pollinator diversity: Understanding interaction networks across time, space, and global change. *Botany, 90*(3), 528–538.

Burney, J., & Ramanathan, V. (2014). Recent climate and air pollution impacts on Indian agriculture. *Proceedings of the National Academy of Sciences of the United States of America, 111*(46), 16319–16324. https://doi.org/10.1073/pnas.1317275111

Burns, F., et al. (2016). Agricultural management and climatic change are the major drivers of biodiversity change in the UK. *PLoS One, 11*(3), e0151595. https://doi.org/10.1371/journal.pone.0151595

Carlson, C. J., Albery, G. F., Merow, C., et al. (2022). Climate change increases cross-species viral transmission risk. *Nature, 607*(7919), 555–562.

CBD. (2010). *Global Biodiversity Outlook 3* (94p). Secretariat of the Convention on Biological Diversity.

Chaparro-Pedraza, P. C., & de Roos, A. M. (2019). Environmental change effects on life-history traits and population dynamics of anadromous fishes. *The Journal of Animal Ecology, 88*(8), 1178–1190.

Chater, C., Peng, K., Movahedi, M., Dunn, J. A., Walker, H. J., Liang, Y. K., et al. (2015). Elevated CO_2-induced responses in stomata require ABA and ABA signaling. *Current Biology, 25*, 2709–2716. https://doi.org/10.1016/j.cub.2015.09.013

Chen, I. C., Hill, J. K., Ohlemüller, R., Roy, D. B., & Thomas, C. D. (2011). Rapid range shifts of species associated with high levels of climate warming. *Science, 333*(6045), 1024–1026.

Colinet, H., et al. (2015). Insects in fluctuating thermal environments. *Annual Review of Entomology, 60*, 123–140.

Comeau, S., Gorsky, G., Jeffree, R., Teyssie, J.-L., & Gattuso, J.-P. (2009). Impact of oceanacidification on a key Arctic pelagic mollusc (Limacina helicina). *Biogeosciences, 6*, 1877–1882.

Convention on Biological Diversity (CBD). (1992). *Convention on biological diversity*. Retrieved from https://www.cbd.int

Convention on Biological Diversity (CBD). (2003). *Interlinkages between biological diversity and climate change*. CBD technical series 10.

Corbeels, M., Graaff, J., Ndah, T. H., et al. (2014). Understanding the impact and adoption of conservation agriculture in Africa: A multi-scale analysis. *Agriculture, Ecosystems & Environment, 187*, 155–170. https://doi.org/10.1016/j.agee.2013.10.011

Dicks, L. V., et al. (2015). How much flower-rich habitat is enough for wild pollinators? Answering a key policy question with incomplete knowledge. *Ecological Entomology, 40*, 22–35.

Domke, G. M., Oswalt, S. N., Walters, B. F., & Morin, R. S. (2020). Tree planting has the potential to increase carbon sequestration capacity of forests in the United States. *Proceedings of the National Academy of Sciences, 117*(40), 24649–24651. https://doi.org/10.1073/pnas.2010840117

Domke, G. M., et al. (2023). Chapter 7: Forests. In *Fifth national climate assessment* (pp. 7–16). U.S. Global Change Research Program.

Doney, S. C., Fabry, V. J., Feely, R. A., & Kleypas, J. A. (2009). Ocean acidification: The other CO_2 problem. *Annual Review of Marine Science, 1*, 169–192.

Dow, K., & Downing, T. E. (2006). *The atlas of climate change*. Earthscan.

Duarte, C. M., Losada, I. J., Hendriks, I. E., Mazarrasa, I., & Marbà, N. (2013). The role of coastal plant communities for climate change mitigation and adaptation. *Nature Climate Change, 3*(11), 961–968.

Dupont, A. (2008). The strategic implications of climate change. *Survival, 50*(3), 29–54. https://doi.org/10.1080/00396330802173107

EEA. (2006). *Transport and the environment: Facing a dilemma*. European Environment Agency. EEA Report No. 3/2006.

EPA. (2021). *Climate change indicators: Streamflow*. Retrieved 21 Mar 2022

Ewers, R. M., & Didham, R. K. (2006). Confounding factors in the detection of species responses to habitat fragmentation. *Biological Reviews, 81*(1), 117–142.

Fabry, V. J. (1990). Shell growth rates of pteropod and heteropod mollusks and aragonite production in the open ocean: Implications for the marine carbonate system. *Journal of Marine Research, 48*, 209–222.

Fahrig, L. (2003). Effects of habitat fragmentation on biodiversity. *Annual Review of Ecology, Evolution, and Systematics, 34*, 487–515.

Foden, W. B., & Young, B. E. (Eds.). (2016). *IUCN SSC guidelines for assessing species' vulnerability to climate change*. IUCN Species Survival Commission.

Fox, R. (2013). The decline of moths in Great Britain: A review of possible causes. *Insect Conservation and Diversity, 6*, 5–19.

Franks, S. J., & Weis, A. E. (2009). Climate change alters reproductive isolation and potential gene flow in an annual plant. *Evolutionary Applications, 2*(4), 481–488. https://doi.org/10.1111/j.1752-4571.2009.00073.x

Franzen, M., & Ockinger, E. (2012). Climate-driven changes in pollinator assemblages during the last 60 years in an Arctic mountain region in northern Scandinavia. *Journal of Insect Conservation, 16*, 227–238. https://doi.org/10.1007/s10841-011-9410-y

Frei, C., Scholl, R., Fukutome, S., Schmidli, J., & Vidale, P. L. (2006). Future change of precipitation extremes in Europe: Intercomparison of scenarios from regional climate models. *Journal of Geophysical Research, 111*, D06105. https://doi.org/10.1029/2005JD005965

Frungillo, L., Skelly, M. J., Loake, G. J., Spoel, S. H., & Salgado, I. (2014). S-nitrosothiols regulate nitric oxide production and storage in plants through the nitrogen assimilation pathway. *Nature Communications, 5*, 5401. https://doi.org/10.1038/ncomms6401

Fuchs, R., et al. (2015). Gross changes in reconstructions of historic land cover/use for Europe between 1900 and 2010. *Global Change Biology, 21*, 299–313.

Gadgil, M., Berkes, F., & Folke, C. (1993). Indigenous knowledge for biodiversity conservation. *Ambio, 22*(2–3), 151–156.

Gazeau, F., Quiblier, C., Jansen, J., Gattuso, J. P., Middelburg, J., & Heip, C. H. R. (2007). Impact of elevated CO_2 on shellfish calcification. *Advancing Earth and Space Sciences, 34*(7). https://doi.org/10.1029/2006GL028554

Gerard, F., et al. (2010). Land cover change in Europe between 1950 and 2000 determined employing aerial photography. *Progress in Physical Geography, 34*(2), 183–205.

Gladstone-Gallagher, R., et al. (2019). Linking traits across ecological scales determines functional resilience. *Trends in Ecology & Evolution, 34*(12), 1080–1091.

Goldewijk, K. K. (2001). Estimating global land use change over the past 300 years: The HYDE database. *Global Biogeochemical Cycles, 15*(2), 417–433.

González-Varo, J. P., Biesmeijer, J. C., Bommarco, R., et al. (2013). Combined effects of global change pressures on animal-mediated pollination. *Trends in Ecology & Evolution, 28*(9), 524–530.

Goulson, D., Nicholls, E., Botías, C., & Rotheray, E. L. (2015). Bee declines driven by combined stress from parasites, pesticides, and lack of flowers. *Science, 347*(6229), 1435–1446.

Grobusch, L. C., & Grobusch, M. P. (2022). A hot topic at the environment–health nexus: Investigating the impact of climate change on infectious diseases. *International Journal of Infectious Diseases, 116*, 7–9. https://doi.org/10.1016/j.ijid.2021.12.004

Gupta, K. J., & Igamberdiev, A. U. (2016). Reactive nitrogen species in mitochondria and their implications in plant energy status and hypoxic stress tolerance. *Frontiers in Plant Science, 7*, 369. https://doi.org/10.3389/fpls.2016.00369

Hallmann, C. A., Sorg, M., Jongejans, E., et al. (2017). More than 75 percent decline over 27 years in total flying insect biomass in protected areas. *PLoS One, 12*(10), e0185809.

Hall-Spencer, J., Rodolfo-Metalpa, R., Martin, S., et al. (2008). Volcanic carbon dioxide vents show ecosystem effects of ocean acidification. *Nature, 454*, 96–99. https://doi.org/10.1038/nature07051

Hansen, A. J., & Cramer, W. (2015). Global biodiversity indicators: Measuring and monitoring biodiversity change. In T. E. Lovejoy & L. Hannah (Eds.), *Biodiversity and climate change: Transforming the biosphere* (pp. 229–241). Yale University Press.

Hawkins, B. A., Field, R., Cornell, H. V., et al. (2003). Energy, water, and broad-scale geographic patterns of species richness. *Ecology, 84*, 3105–3117.

Heron, S. F., Maynard, J. A., van Hooidonk, R., & Eakin, C. M. (2017). *Impacts of climate change on world heritage coral reefs: A first global scientific assessment*. UNESCO World Heritage Centre.

Hoegh-Guldberg, O., Poloczanska, E. S., Skirving, W., & Dove, S. (2022). Coral reefs. In H.-O. Pörtner et al. (Eds.), *Climate change 2022: Impacts, adaptation and vulnerability. Contribution of working group II to the sixth assessment report of the intergovernmental panel on climate change* (pp. 1239–1312). Cambridge University Press.

Hughes, T. P., Barnes, M. L., Bellwood, D. R., et al. (2017). Coral reefs in the Anthropocene. *Nature, 546*(7656), 82–90.

Huntley, B., Green, R. E., Collingham, Y. C., et al. (2004). The performance of models relating species geographical distributions to climate is independent of trophic level. *Ecology Letters, 7*, 417–426.

IPBES. (2019). *Global assessment report on biodiversity and ecosystem services*. Intergovernmental Science-Policy Platform on Biodiversity and Ecosystem Services. https://doi.org/10.5281/zenodo.3831673

IPCC. (2007a). *Climate change 2007: The physical science basis. Summary for policymakers*. Cambridge University Press.

IPCC. (2007b). *Climate change 2007: Mitigation of climate change. Summary for policymakers*. Cambridge University Press.

IPCC. (2014). *Climate change 2014: Mitigation of climate change. Contribution of working group III to the fifth assessment report of the intergovernmental panel on climate change*. Cambridge University Press.

IPCC. (2021). *Climate change 2021: The physical science basis*. Intergovernmental Panel on Climate Change.

IPCC. (2022). Summary for policymakers. In H.-O. Pörtner et al. (Eds.), *Climate change 2022: Impacts, adaptation and vulnerability. Contribution of working group II to the sixth assessment report of the intergovernmental panel on climate change* (pp. 3–33). Cambridge University Press.

Ishimatsu, A., Hayashi, M., & Kikkawa, T. (2008). Fishes in high-CO_2, acidified oceans. *Marine Ecology Progress Series, 373*, 295–302. https://doi.org/10.3354/MEPS07823

IUCN. (2024). The IUCN Red List of Threatened Species. Version 2024-1.

Jaeger, J. (2005). *Was verträgt unsere Erde noch? Wege in die Nachhaltigkeit*. Fischer Verlag.

Jhun, I., Coull, B. A., Zanobetti, A., & Koutrakis, P. (2015). The impact of nitrogen oxides concentration decreases on ozone trends in the USA. *Air Quality, Atmosphere and Health, 8*, 283–292. https://doi.org/10.1088/1748-9326/10/8/084009

Kramer, P. J., Boyer, J. S., & Carlson, W. C. (1997). Water relations of plants and soils. *Forest Science, 43*, 151–152.

Kromp-Kolb, H., & Formayer, H. (2005). *Schwarzbuch Klimawandel. Wie viel Zeit bleibt uns noch?* Ecowin Verlag.

Kudo, G., & Ida, T. Y. (2013). Early onset of spring increases the phenological mismatch between plants and pollinators. *Ecology, 94*(10), 2311–2320.

Latif, M. (2007). *Bringen wir das Klima aus dem Takt*. Forum für Verantwortung, Fischer.

Malhi, Y., Roberts, J. T., Betts, R. A., et al. (2009). Climate change, deforestation, and the fate of the Amazon. *Science, 326*(5957), 1350–1351.

McDowell, N. G., Allen, C. D., Anderson-Teixeira, K., et al. (2020). Pervasive shifts in forest dynamics in a changing world. *Science, 368*(6494), eaaz9463. https://doi.org/10.1126/science.aaz9463

Miles, L., Grainger, A., & Phillips, O. (2004). The impact of global climate change on tropical forest biodiversity in Amazonia. *Global Ecology and Biogeography, 13*, 553–565.

Millennium Ecosystem Assessment. (2005). *Ecosystems and human well-being: Biodiversity synthesis*. World Resources Institute.

Mittler, R., & Blumwald, E. (2010). Genetic engineering for modern agriculture: Challenges and perspectives. *Annual Review of Plant Biology, 61*, 443–462. https://doi.org/10.1146/annurev-arplant-042809-112116

NASA (2022). National Aeronautics and Space Administration. *Global climate change: Vital signs of the planet*. https://climate.nasa.gov/

Noss, R. F. (1990). Indicators for monitoring biodiversity: A hierarchical approach. *Conservation Biology, 4*(4), 355–364.

Obura, D. (2017). Refilling the coral reef glass. *Science, 357*(6357), 1215.

Parmesan, C. (2006). Ecological and evolutionary responses to recent climate change. *Annual Review of Ecology, Evolution, and Systematics, 37*, 637–669.

Parmesan, C., & Yohe, G. (2003). A globally coherent fingerprint of climate change impacts across natural systems. *Nature, 421*(6918), 37–42.

Patz, J. A., Campbell-Lendrum, D., Holloway, T., et al. (2005). Impact of regional climate change on human health. *Nature, 438*(7066), 310–317.

Potts, S. G., Biesmeijer, J. C., Kremen, C., et al. (2010). Global pollinator declines: Trends, impacts and drivers. *Trends in Ecology & Evolution, 25*(6), 345–353.
Powney, G. D., Carvell, C., Edwards, M., et al. (2019). Widespread losses of pollinating insects in Britain. *Nature Communications, 10*(1), 1018.
Rana, N., Manish, K., & Pandit, M. K. (2025). Effect of climate change on the flowering phenology of *Rhododendron arboreum* Sm. In the Western Himalaya. *Journal of Asia-Pacific Biodiversity, 18*(1), 197–203. https://doi.org/10.1016/j.japb.2024.10.004
Ranasinghe, R., Ruane, A. C., Vautard, R., Arnell, N. W., Coppola, E., Cruz, F. A., et al. (2022). *Climate change 2022: Impacts, adaptation and vulnerability. Contribution of working group II to the sixth assessment report of the intergovernmental panel on climate change* (pp. 45–54). Cambridge University Press.
Reynolds, J. F., Stafford Smith, D. M., Lambin, E. F., Turner, B. L., Mortimore, M., Batterbury, S. P., et al. (2007). Global desertification: Building a science for dryland development. *Science, 316*(5826), 847–851. https://doi.org/10.1126/science.1131634
Rustad, L. E., Campbell, J. L., Marion, G. M., Norby, R. J., Mitchell, M. J., Hartley, A. E., Cornelissen, J. H. C., & Gurevitch, J. (2001). A meta-analysis of the response of soil respiration, net nitrogen mineralization, and aboveground plant growth to experimental ecosystem warming. *Oecologia, 126*(4), 543–562.
Salimi, S., Almuktar, S. A. A. A., & Scholz, M. (2021). Impact of climate change on wetland ecosystems: A critical review of experimental wetlands. *Science of the Total Environment, 775*, 145788. https://doi.org/10.1016/j.scitotenv.2021.145788
Scheffers, B. R., De Meester, L., Bridge, T. C., Hoffmann, A. A., Pandolfi, J. M., Corlett, R. T., et al. (2016). The broad footprint of climate change from genes to biomes to people. *Science, 354*, aaf7671. https://doi.org/10.1126/science.aaf7671
Seibold, S., Gossner, M. M., Simons, N. K., Blüthgen, N., Müller, J., Ambarlı, D., et al. (2019). Arthropod decline in grasslands and forests is associated with landscape-level drivers. *Nature, 574*(7780), 671–674. https://doi.org/10.1038/s41586-019-1684-3
Semenza, J. C., & Paz, S. (2021). Climate change and infectious disease in Europe: Impact, projection, and adaptation. *The Lancet Regional Health - Europe, 9*, 100230. https://doi.org/10.1016/j.lanepe.2021.100230
Sgro, C. M., Lowe, A. J., & Hoffmann, A. A. (2011). Building evolutionary resilience for conserving biodiversity under climate change. *Evolutionary Applications, 4*(2), 326–337. https://doi.org/10.1111/j.1752-4571.2010.00157.x
Smit, B., & Wandel, J. (2006). Adaptation, adaptive capacity, and vulnerability. *Global Environmental Change, 16*(3), 282–292. https://doi.org/10.1016/j.gloenvcha.2006.03.008
Steffen, W., Sanderson, A., Tyson, P. D., Jäger, J., Matson, P. A., Moore, B., et al. (2004). *Global change and the earth system: A planet under pressure*. Springer.
Steffen, W., Richardson, K., Rockström, J., Cornell, S. E., Fetzer, I., Bennett, E. M., et al. (2018). Trajectories of the earth system in the Anthropocene. *Proceedings of the National Academy of Sciences of the United States of America, 115*(33), 8252–8259. https://doi.org/10.1073/pnas.1810141115
Stirling, I., & Derocher, A. E. (2012). Effects of climate warming on polar bears: A review of the evidence. *Global Change Biology, 18*(9), 2694–2706. https://doi.org/10.1111/j.1365-2486.2012.02753.x
Sundquist, E., Burruss, R., Faulkner, S., Gleason, R., Harden, J., Kharaka, Y., Tieszen, L., & Waldrop, M. (2008). *U.S. geological survey, Fact sheet 2008-3097*.
Tester, M., & Langridge, P. (2010). Breeding technologies to increase crop production in a changing world. *Science, 327*(5967), 818–822. https://doi.org/10.1126/science.1183700
Thuiller, W., Lavorel, S., Araújo, M. B., Sykes, M. T., & Prentice, I. C. (2005). Climate change threats to plant diversity in Europe. *PNAS, 102*(23), 8245–8250.
Tol, R. S. J., Bohn, M., Downing, T. E., Guillerminet, M. L., Hizsnyik, E., Kasperson, R., et al. (2006). Adaptation to five metres of sea level rise. *Journal of Risk Research, 9*, 467–482. https://doi.org/10.1080/13669870600717632

Underwood, E., et al. (2017). Pollinator initiatives in EU member states: Success factors and gaps. *Science of the Total Environment, 775*, 145788. https://doi.org/10.1016/j.scitotenv.2021.145788

UNEP (2021). From pollution to solution: *A global assessment of marine litter and plastic pollution*. Nairobi: UNEP. Available at: https://www.unep.org/

United Nations Framework Convention on Climate Change (UNFCCC). (2015). *The Paris agreement*. https://unfccc.int

Van der Sluijs, J. P., & Turkenburg, W. (2006). Climate change and the precautionary principle. In E. Fisher, J. Jones, & R. von Schomberg (Eds.), *Implementing the precautionary principle: Perspectives and prospects*. Edward Elgar.

Vasiliev, D., & Greenwood, S. (2020). Pollinator biodiversity and crop pollination in temperate ecosystems, implications for national pollinator conservation strategies: Mini review. *Science of the Total Environment, 744*, 140880. https://doi.org/10.1016/j.scitotenv.2020.140880

Vasiliev, D., & Greenwood, S. (2021). The role of climate change in pollinator decline across the northern hemisphere is underestimated. *Science of the Total Environment, 775*, 145788. https://doi.org/10.1016/j.scitotenv.2021.145788

Vellinga, M., & Wood, R. A. (2002). Global climatic impacts of a collapse of the Atlantic thermohaline circulation. *Climatic Change, 54*, 251–267. https://doi.org/10.1023/A:1016168827653

Vose, J. M., Peterson, D. L., Domke, G. M., Fettig, C. J., Joyce, L. A., Prestemon, J. P., et al. (2018). Forests. In *Impacts, risks, and adaptation in the United States: Fourth national climate assessment* (Vol. II, p. 246). U.S. Global Change Research Program. https://doi.org/10.7930/NCA4.2018.CH6

Waldbusser, G. G., Voigt, E. P., Bergschneider H. G., Mark A., & Newell, R. I. E (2011). Seawater carbonate chemistry and calcification rate of eastern oyster Crassostrea virginica. *Pangaea*. https://doi.org/10.1594/PANGAEA.758181.

Weiner, C. N., Werner, M., Linsenmair, K. E., & Blüthgen, N. (2014). Land-use impacts on plant–pollinator networks: Interaction strength and specialization predict pollinator declines. *Ecology, 95*(2), 466–474. https://doi.org/10.1890/13-0436.1

Williams, S. E., Bolitho, E. E., & Fox, S. (2003). Climate change in Australian tropical rainforests: An impending environmental catastrophe. *Proceedings of the Royal Society of London - Series B: Biological Sciences, 270*, 1887–1892. https://doi.org/10.1098/rspb.2003.2464

Williams, J. M., Scott, S. E., Galbraith, C. A., Martay, B., Macphie, K. H., Hereward, H. F. R., et al. (2023). *Climate change and migratory species: A review of impacts, conservation actions, indicators and ecosystem services. Summary for policy makers*. JNCC. ISBN 978-0-86139-000-7. jncc.gov.uk/climate-migratory-species-report

Woods, J. K., et al. (2008). Influence of climate on butterfly community and population dynamics in Western Ohio. *Environmental Entomology, 37*(3), 696–706. https://doi.org/10.1603/0046-225X(2008)37[696:IOCOTC]2.0.CO;2

Impacts of Climate Change on the Biological Communities in the Sikkim Himalaya: An Opinionated Review

Laavanya Nilesh Rathi, Kumar Manish, and Mohd Tariq

Abstract The Himalaya, one of the world's most biodiverse regions, are habitat to an extensive range of flora and fauna, with the majority of them being native to the area. Millions of people rely on the Himalaya's biodiversity for their livelihoods; therefore, safeguarding ecological balance and sustaining biodiversity becomes extremely important. Climate change has become the biggest threat to the sustenance of the biodiversity of the Himalaya. The region is warming at three times the rate of the global average. Declining cryosphere, changing land use and altering vegetation patterns, extreme weather events have led to high biodiversity losses including species extinction and extermination. The changed biological equilibrium of the area, due to climate change, causes a domino effect that leads to numerous other negative consequences. The migration of animals to higher altitudes results in the loss of unique ecosystems and the services they offer. At the same time, there are some generalist species that are benefiting from climate change. The purpose of the current review is to collect all related information on the impacts of climate change, including changes in ecology, behaviour, biology, phenology, and distribution on different taxa in the Himalaya and synthesize the overall impacts. The methodology of the study involves a comprehensive literature review that uses the Scopus and

L. N. Rathi
Jindal School of Environment and Sustainability, O.P. Jindal Global University, Sonipat, India

K. Manish (✉)
Jindal School of Environment and Sustainability, O.P. Jindal Global University, Sonipat, India

Centre for Biodiversity and Conservation, O.P. Jindal Global University, Sonipat, India
e-mail: kumar.manish@jgu.edu.in

M. Tariq
Faculty of Allied, Health and Life Sciences, Sumandeep Vidyapeeth Deemed to be University, Vadodara, Gujarat, India

Department of Biotechnology, Graphic Era (Deemed to be University), Dehradun, Uttarakhand, India

S. Siddiqui et al. (eds.), *Frontiers in Sustainability*,
https://doi.org/10.1007/978-3-032-19740-5_6

Google Scholar databases to ensure a transparent and meticulous synthesis of previous research on the impact of climate change on biological communities in Sikkim Himalaya. The study attempts to collect and analyse data on biodiversity patterns, ecosystem services, and changes in species distribution in response to changes in the climate in a logical and structured manner.

Keywords Biodiversity · Climate change · Ecosystem · Extinction · Himalaya

1 Introduction

The Himalaya are one of the youngest and the highest mountain chains of the world (Pandit, 2017). The region is home to 1.4 billion people with extremely high biodiversity, geo-hydrological and cultural values (Pandit et al., 2014). The mountain system is geo-ecologically divided into two parts: Eastern and Western Himalaya. Sikkim, one of the smallest states of India forms a part of the Eastern Himalayan range and is recognized due to its exceptional ecological diversity (Manish et al., 2017). Sikkim encompasses nine protected areas covering 47% of the state's total geographical area and 11 Important Bird Areas (IBA). Sikkim is known for its unique terrain and climatic fluctuations mainly because it is home to Kanchenjunga, the third-highest peak in the world. This region also supports a range of ecosystems, including temperate forests and high-altitude wetlands. In addition to a wide variety of flowers and medicinal plants, the state is home to rare and endemic species, making it a hotspot for biodiversity worldwide. Due to the unique blend of its ecological and cultural significance, Sikkim harbours India's first "Mixed Heritage" site and a "World Heritage Site." The rich and unique biodiversity of Sikkim, however, is being threatened by a range of natural and anthropogenic drivers, including deforestation, hydropower development, urbanization, tourism, climate change, etc. in recent years (Pandit, 2017). Of all the threats, climate change is emerging as the biggest threat in Sikkim because of its varied impacts on biodiversity, hydrology, glacial run-off, agriculture, livestock, ecosystem services, etc. (Bhattacharya et al., 2012).

The Himalaya are warming at a much higher rate than the global average (Dimri & Dash, 2012; Shrestha et al., 1999). Shrestha et al. (2012) pointed out that the Himalaya are warming at a rate three times more than the global average. Overall, the temperature has increased by 1.5 °C, and annual precipitation has increased by 16.3 cm between 1982 and 2006 in the Himalaya (Shrestha et al., 2012). Climate change has caused accelerated melting of glaciers, shifts in intensity, duration and timing of monsoonal rainfall, altered hydrological regimes, etc. in the Himalaya (Xu et al., 2009). In Sikkim, various studies have reported similar results. Rahman et al. (2012) reported an increase in mean minimum temperature by 1.95 °C and an increase in rainfall by 12.4 cm between 1981 and 2010 in Tadong (Sikkim). Luitel et al. (2012) reported that the glaciers in Sikkim had receded by 460 m between 1976 and 2009. The rate of climate change has been reported to be greater in the higher elevations in Sikkim as compared to the lower elevations (Chaudhary & Bawa, 2011). Because of these reasons, the delicate equilibrium between ecology and biodiversity is reportedly being disturbed in the region (Banerjee et al., 2019).

There have been reports of changes in forest composition and structure, ecosystem services, biodiversity losses, shifts in suitable habitats and geographical ranges of species, behavioural changes, biological invasions, decreased crop productivity, changes in sowing and harvesting seasons, etc. due to climate change in Sikkim (Arrawatia & Tambe, 2012). The purpose of this study is to synthesize the available literature and analyse the impacts of climate change with respect to different taxa (plants, birds, amphibians and insects) in Sikkim. Furthermore, the study aims to propose a policy framework for conservation actions in order to mitigate the impacts of climate change on the biodiversity of Sikkim. The overarching goal was to document the biological response to climate change in the Himalaya and suggest ways to conserve biodiversity in the era of ensuing climate change.

2 Literature Review

In order to evaluate the effects of climate change on the biological communities in the Sikkim Himalaya, this study adopted the systematic review methodology and snowball approach. Systematic review involved following the broad PRISMA (Preferred Reporting Items for Systematic Reviews and Meta-Analyses) guidelines with some modifications (Moher et al., 2010). This offered a strong foundation for carrying out and reporting meta-analyses and systematic reviews. PRISMA not only ensures transparency but also accuracy, and thoroughness. It helps in identifying, screening, and choosing suitable studies (Moher et al., 2010).

The present study primarily utilized two databases, Scopus and Google Scholar. Scopus database was used to search for peer-reviewed research articles while Google Scholar was used to find grey literature, such as reports, theses, and articles that were not listed in Scopus (Fig. 1). Grey literature can be useful for these kinds of studies since it helps identify and fill certain gaps in the system. It often includes up-to-date findings or useful, unpublished data that may not yet be accessible and available from peer-reviewed journals. Reports from governmental organizations, non-profits, or academic institutions provide thorough, regionally specific information that may be useful when doing such studies. Peer-reviewed research can occasionally take longer to publish, which might keep some significant information from being missed. Some grey literature, such as government websites and studies (Sikkim government websites and reports), were included in this study. In both Scopus and Google Scholar, the following keywords were used for searching: "Sikkim," "Himalaya," "Species," and "Climate change." To get thorough results, keywords were combined using Boolean operators (AND, OR). For instance, one of the search queries consisted of the following terms "Climate change "AND "Species" AND "Sikkim" AND "Himalaya" (Fig. 1).

In order to ensure the quality and relevancy of searches, inclusion and exclusion criteria were created. Studies that looked at the biological impacts of climate change in the Sikkim Himalaya or nearby regions met the inclusion criteria. The study also concentrated on peer-reviewed research, English-language articles, and relevant

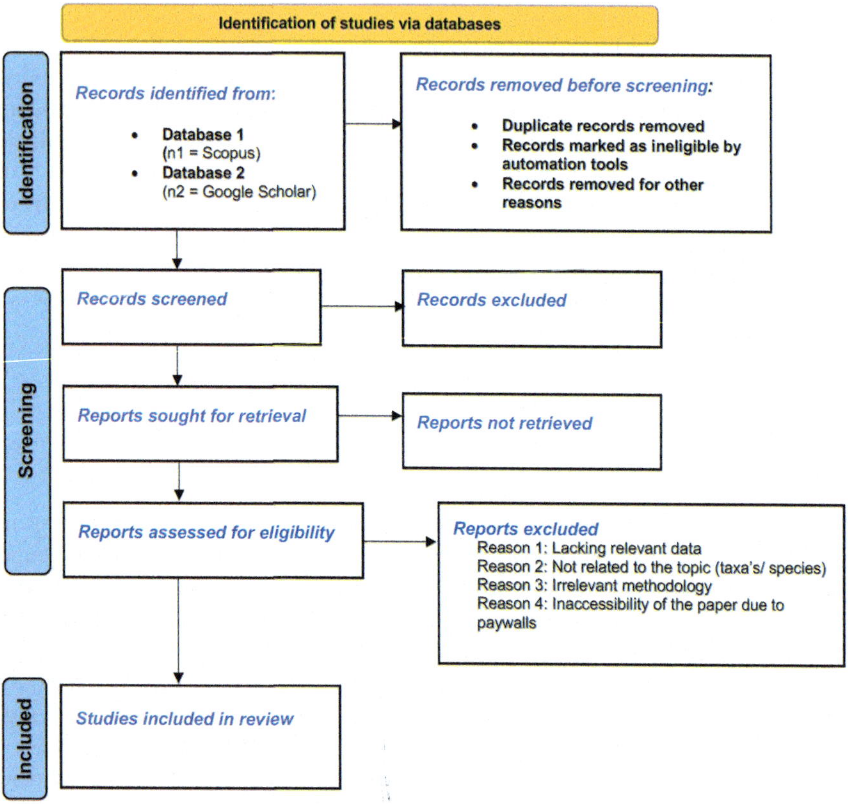

Fig. 1 Broad outline of the methodology used for literature review in the present study

grey literature. The exclusion criteria were centred on studies that had little to do with the biological effects of climate change or the Sikkim Himalaya. Moreover, articles with no quantitative or empirical support were rejected. Specifically, according to PRISMA criteria, the research articles were selected in two stages: (i) Identification: From Google Scholar and Scopus, a total of 37 records were obtained and their abstracts were screened for selection/exclusion (ii) Screening and elimination: 16 records were eliminated after abstracts and titles were screened because they were irrelevant to the study's goals. A total of 21 studies were finally selected for inferences regarding the impacts of climate change on different taxa in the current study. We further used the snowball technique of literature review where we looked for additional studies pertaining to the scope of the present study in the references section of the 21 studies that we selected. Overall, we limited our search to four taxa (plants, birds, amphibians and insects) with respect to the impacts of climate change on biodiversity in Sikkim. After synthesizing the impacts, we classified them into three different categories: (i) Impacts on habitat distribution, (ii) Impacts on reproductive biology, and (iii) Impacts on the performance of Protected Areas (PAs) in conservation.

3 Impacts on Habitat Distribution

The existing literature revealed that the majority of the species across the plant, bird, amphibian, and insect taxa shows a prominent upward and northward shift towards higher elevations in the Sikkim Himalayan region. Telwala et al. (2013) reported a shift of 23–998 m in the upper elevation limit of the sub-alpine and alpine endemic plant species of Sikkim Himalaya. The authors reported a mean upward displacement rate of 27.53 m/decade and a median upward shift of 240 m towards the northward direction for the endemic species (Telwala et al., 2013). Amongst 124 endemic species, six species showed 50% contraction in their elevational ranges (*Rheum nobile*, *Saussurea stella*, *Rhodiola bupleuroides*, *Saussurea uniflora*, *Saussurea leontodontoides*, *Ponerorchis chusua*) while four species (*Microgynaecium tibeticum*, *Meconopsis simplicifolia*, *Pedicularis trichoglossa*, *Lancea tibetica*) showed 100% contraction. Native plant species in particular are expected to experience maximum decline in their habitat distribution in future climate conditions (Manish et al., 2016; Fig. 2). Kumar (2012) reported that as a result of climate change,

Fig. 2 Photographs of some native Himalayan plant species of Sikkim: (**a**) *Rhododendron cinnabarinum* subsp. *cinnabarinum*; (**b**) *Rhododendron campylocarpum*; (**c**) *Meconopsis horridula*; (**d**) *Primula atrodentata*. Photo credits: Kumar Manish

Rhododendrons in Sikkim will experience a reduction in suitable habitats and a decline in populations. Manish et al. (2016) reported that climate change in Sikkim could lead to habitat loss for 16–18% of the endemic plant species by 2050 and 2070, respectively. The study also showed that shrublands will likely expand and will gain 1–3% of the geographical area that is currently occupied by herbaceous meadows by 2070. Thus, climate change will likely favour the expansion of woody plants and herbaceous species will experience a decline in their populations. This is likely to impact the medicinal plant base of the highlands as herbaceous meadows harbour a large proportion of medicinal plants (Manish et al., 2016). In a later study, Manish (2022) showed that climate change is likely to lead to habitat loss in 13–16% of medicinal plants in Sikkim (Fig. 3). The regions harbouring the maximum number of medicinal plant species are expected to shift by 200–400 m by 2050–2070 (Manish, 2022). Anand and Garg (2024) reported that *Ephedra gerardiana*, a valuable Himalayan medicinal plant will lose significant habitat and shift to higher elevations as a result of climate change. Verma et al. (2023) reported that the invasive

Fig. 3 Photographs of some medicinal plant species of Sikkim: (**a**) *Hypericum patulum*; (**b**) *Angelica archangelica*; (**c**) *Selinum wallichianum*; (**d**) *Bauhinia variegata*. Photo credits: Kumar Manish

species have started to spread to the higher elevations in Sikkim. Within 3 years, *Ageratina adenophora* showed significant expansion in the distribution in Skkim particularly at 2400–2700 m where they were previously absent or present in low numbers (Verma et al., 2023). Similar to the plants, many bird species in Sikkim are also shifting their ranges upwards along the elevation gradient in response to climate change (Fig. 4; Acharya & Chettri, 2012). Prominent species showing such a shift include Blood Pheasant (*Ithaginis cruentus*), Ashy Wood Pigeon (*Columba pulchricollis*), Hodgson's Hawk Cuckoo (*Hierococcyx fugax*), Ibisbill (*Ibidorhyncha struthersii*), Peregrine Falcon (*Falco peregrinus*), and Fire-tailed Sunbird (*Aethopyga ignicauda*) (Acharya & Chettri, 2012). The authors reported a mean elevational range expansion/shift of 456 m across 25 species in Sikkim due to climate change (Acharya & Chettri, 2012). A study on high-elevation frogs in the Eastern Himalaya, including Sikkim, highlights the importance of considering realistic migration constraints when projecting future distributions (Subba et al., 2018). Without migration, amphibian species are projected to experience a significant decrease in

Fig. 4 Photographs of some unique bird species of Sikkim: (**a**) Blue-throated Barbet (*Psilopogon asiaticus*); (**b**) Green-Tailed Sunbird (*Aethopyga nipalensis*); (**c**) White-capped water redstart (*Phoenicurus leucocephalus*); (**d**) Yellow-billed Blue Magpie (*Urocissa flavirostris*). Photo credits: Kumar Manish

Fig. 5 Photographs of some unique butterfly and moth species in Sikkim: (**a**) *Milionia pulchrinervis*; (**b**) Krishna Peacock (*Papilio krishna*); (**c**) Metallic Cerulean (*Jamides alecto*); (**d**) Common Punchinello (*Zemeros flegyas*). Photo credits: Kumar Manish

distribution by 2080 (Subba et al., 2018). Some of the butterfly species have shown shifts in their elevation range due to climate change in Sikkim (Fig. 5). These include Common Birdwing (*Troides helena*), Yellow Orangetip (*Ixias pyrene*), Golden Sapphire (*Heliophorus brahma*), Common Earl (*Tanaecia julii*), etc. (Acharya & Chettri, 2012). Species adapted to cooler climates (e.g., high-altitude butterflies like *Parnassius* spp.) may face habitat shrinkage. Species like the Kaiser-i-Hind (*Teinopalpus imperialis*) may struggle with habitat loss (Acharya & Chettri, 2012). The upward shift in species distributions is likely to lead to increased interspecific competition in the upper elevations. It is likely that due to climate change, there will be a preponderance of generalist species with the ability to survive in wide environmental conditions while the specialist species adapted to specific elevations and climates will likely suffer a decline in populations. In a nutshell, climate change is causing a progressive change in species assemblages, composition and structure in various ecosystems (particularly in the sub-alpine and alpine ecosystems) and altered ecosystem functioning.

4 Impacts on Reproductive Biology

In plants, climate change has been reported to cause phenological changes including early blooms, disruption in cycles of seed germination and maturity and decreased crop productivity in Sikkim (Arrawatia & Tambe, 2012). Sharma et al. (2016) reported that the production of large cardamom (*Amomum subulatum*) had decreased in Sikkim due to climate change. Traditional farming practices are also affected, with shifts in sowing and harvesting periods and increased pest and disease incidence (Ravindranath et al., 2012; Sharma & Rai, 2012). Rising temperatures and unpredictable rainfall patterns have led to increased incidences of insect pests and diseases, particularly affecting crops like large cardamom, a significant agricultural product in Sikkim (Shukla et al., 2016). Climate change, particularly warming temperatures, facilitates the geographic expansion of insect pests, increases their overwintering survival, and allows for more generations per year. This can lead to more frequent and severe pest outbreaks (Pureswaran et al., 2018). In forested areas, climate change can lead to increased outbreaks of pests like bark beetles, which are influenced by warmer temperatures. These outbreaks can have significant impacts on forest health and biodiversity (Jactel et al., 2019). Acharya and Vijayan (2009) reported that the Common Tailorbird (*Orthotomus sutorius*) showed breeding activity for only 2–3 months in Sikkim, a marked departure from the year-round breeding that the species generally shows. Acharya and Chettri (2012) reported delayed nest-building activities, laying of eggs and hatchlings emergence in various bird species in Sikkim such as Ashy Drongo (*Dicrurus leucophaeus*), Black Bulbul (*Hypsipetes leucocephalus*), White-capped Water Redstart (*Chaimarrornis leucocephalus*), and so on. Acharya and Chettri (2012) also reported cases of breeding failure in species that prefer water habitats such as Ruddy Shelduck (*Tadorna ferruginea*), Ibisbill (*Ibidorhyncha struthersiii*), Common Redshank (*Tringa totanus*), and Black-necked Crane (*Grus nigricolis*) in Sikkim. There are also reports of early breeding and changed breeding activities, upward migration of species towards higher elevations and biased sex ratios towards females in the amphibians due to climate change in Sikkim (Acharya & Chettri, 2012).

5 Impacts on the Performance of Protected Areas (PAs) in Conservation

The major issue with the shift in the distribution of species and their habitats as a result of climate change is the fact that the current Protected Area (PA) network would be rendered ineffective in the conservation of plant species in future. Manish and Pandit (2019) showed that there is a need to add 896 square kilometres to the existing PA network in Sikkim Himalaya in order to effectively conserve plant species in future climate change scenarios. Specifically, there will be a need to create three new PAs and expand the boundaries of the existing PAs in the region. The authors suggested that for effective conservation of plant species in Sikkim in future climate change scenarios, it will be more prudent to design large PAs that encompass a myriad of habitats and elevations instead of designing small PAs that are focussed on the flagship species (Manish & Pandit, 2019).

6 Conclusions and Policy Implications

The results of the present study suggest that climate change is having a profound impact on biodiversity in the Sikkim Himalayan region. Though the responses are taxa-specific, some generalized patterns do emerge. For example, all the taxa including plants, birds, amphibians and insects (butterflies) are showing or are expected to show a prominent shift in their distributions towards upper elevations due to climate change in the region (Fig. 6). Many birds and amphibians in Sikkim depend on insects for food; reduced insect populations can disrupt entire food chains. Some pests, like crop-damaging aphids, may have longer active periods, increasing agricultural risks. Warmer temperatures cause some insects (e.g., moths, bees) to emerge earlier, leading to mismatches with food sources (flowers, host plants). Changes in climate can disrupt the synchrony between insects and their host plants, potentially leading to mismatches that affect both pest and plant populations. Additionally, the effectiveness of natural predators and biological control methods may be reduced, complicating pest management strategies (Harrington et al., 2001). Effective management and adaptation strategies are crucial to mitigate these impacts and ensure the sustainability of local biodiversity and agricultural productivity. The drying of streams, which serve as crucial habitats for amphibians, has been noted as a significant factor in their population decline (Acharya & Chettri, 2012). Long dry spells as a result of climate change particularly in the low-elevation regions have exacerbated this issue, leading to habitat loss. However, models that incorporate realistic migration rates show different outcomes, emphasizing the need for accurate migration modelling in conservation planning (Subba et al., 2018). There is likely to be a preponderance of generalist species amongst all taxa. Species that are adapted to wide climate distributions and can tolerate wide climate fluctuations are likely to be the most dominant in future climates while endemic species with narrow tolerances are likely to experience decline in their populations. Slowly, the generalist species will push the specialist species to extinction across all taxa as a result of climate change. The shifts in habitat distributions are likely to render the existing PAs ineffective in conserving species in future climate change scenarios, and there is an urgent need to create new PAs in the region. Incorporating climate change into PA planning is essential. This includes using species distribution models and conservation planning tools to identify priority areas for conservation and climate refugia.

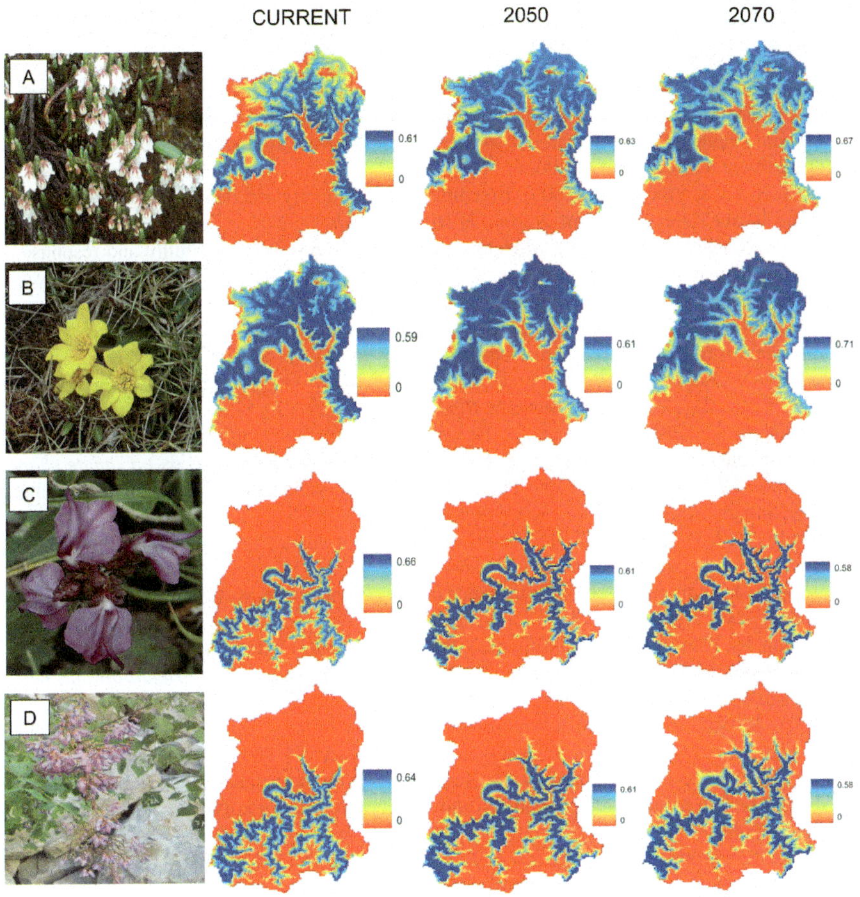

Fig. 6 Outputs of Species Distribution Modelling (SDM) of some selected plant species of Sikkim: (**a**) *Cassiope fastigiata*; (**b**) *Caltha palustris*; (**c**) *Pedicularis megalantha*; (**d**) *Desmodium elegans*. The results of SDM have been sourced from Manish et al. (2016). Photo credits: Kumar Manish

Expanding and strategically planning PAs in Sikkim, along with adaptive management and international cooperation, are crucial steps to mitigate the impacts of climate change on biodiversity in the region.

Acknowledgements The authors would like to thank the Dean, Jindal School of Environment and Sustainability and O.P. Jindal Global University for providing necessary support during the course of the study.

References

Acharya, B. K., & Chettri, B. (2012). Effect of climate change on birds, herpetofauna and butterflies in Sikkim Himalaya: A preliminary investigation. In M. L. Arrawatia & S. Tambe (Eds.), *Climate change in Sikkim: Patterns, impacts and initiatives*. Information and Public Relations Department, Government of Sikkim.

Acharya, B. K., & Vijayan, L. (2009). Breeding bird community and their nesting characteristics in the Teesta Valley of Sikkim. *Journal of Hill Research, 22*, 1–8.

Anand, A., & Garg, V. K. (2024). Modeling the species occurrence probability and response of climate change on Himalayan Somalata plant under different shared socioeconomic pathways. *Environmental Monitoring and Assessment, 196*, 647.

Arrawatia, M. L., & Tambe, S. (2012). *Climate change in Sikkim: Patterns, impacts and initiatives*. Information and Public Relations Department, Government of Sikkim.

Banerjee, A., Chakraborty, P., & Bandopadhyay, R. (2019). Urgent conservation needs in the Sikkim Himalaya biodiversity hotspot. *Biodiversity, 20*, 88–97.

Bhattacharya, S., Krishnaswamy, S., & Rao, C. K. (2012). Vulnerability of Sikkim to climate change and strategies for adaptation. In M. L. Arrawatia & S. Tambe (Eds.), *Climate change in Sikkim: Patterns, impacts and initiatives*. Information and Public Relations Department, Government of Sikkim.

Chaudhary, P., & Bawa, K. S. (2011). Local perceptions of climate change validated by scientific evidence in the Himalayas. *Biology Letters, 7*, 767–770.

Dimri, A. P., & Dash, S. K. (2012). Wintertime climatic trends in the western Himalayas. *Climatic Change, 111*, 775–800.

Harrington, R., Fleming, R. A., & Woiwod, I. P. (2001). Climate change impacts on insect management and conservation in temperate regions: Can they be predicted? *Agricultural and Forest Entomology, 3*, 233.

Jactel, H., Koricheva, J., & Castagneyrol, B. (2019). Responses of forest insect pests to climate change: Not so simple. *Current Opinion in Insect Science, 35*, 103–108.

Kumar, P. (2012). Assessment of impact of climate change on rhododendrons in Sikkim Himalayas using Maxent modelling: Limitations and challenges. *Biodiversity and Conservation, 21*, 1251–1266.

Luitel, K. K., Shrestha, D. G., Sharma, N. P., & Sharma, R. K. (2012). Impact of climate change on east-Rathong glacier in Rangit basin, West Sikkim. In M. L. Arrawatia & S. Tambe (Eds.), *Climate change in Sikkim: Patterns, impacts and initiatives*. Information and Public Relations Department, Government of Sikkim.

Manish, K. (2022). Medicinal plants in peril due to climate change in the Himalaya. *Ecological Informatics, 68*, 101546.

Manish, K., & Pandit, M. K. (2019). Identifying conservation priorities for plant species in the Himalaya in current and future climates: A case study from Sikkim Himalaya, India. *Biological Conservation, 233*, 176–184.

Manish, K., Telwala, Y., Nautiyal, D. C., & Pandit, M. K. (2016). Modelling the impacts of future climate change on plant communities in the Himalaya: A case study from eastern Himalaya, India. *Modeling Earth Systems and Environment, 2*, 1–12.

Manish, K., Pandit, M. K., Telwala, Y., Nautiyal, D. C., Koh, L. P., & Tiwari, S. (2017). Elevational plant species richness patterns and their drivers across non-endemics, endemics and growth forms in the eastern Himalaya. *Journal of Plant Research, 130*, 829–844.

Moher, D., Liberati, A., Tetzlaff, J., Altman, D. G., & Prisma Group. (2010). Preferred reporting items for systematic reviews and meta-analyses: The PRISMA statement. *International Journal of Surgery, 8*, 336–341.

Pandit, M. K. (2017). *Life in the Himalaya: An ecosystem at risk*. Harvard University Press.

Pandit, M. K., Manish, K., & Koh, L. P. (2014). Dancing on the roof of the world: Ecological transformation of the Himalayan landscape. *Bioscience, 64*, 980–992.

Pureswaran, D. S., Roques, A., & Battisti, A. (2018). Forest insects and climate change. *Current Forestry Reports, 4*, 35–50.

Rahman, H., Karuppaiyan, R., Senapati, P. C., Ngachan, S. V., & Kumar, A. (2012). An analysis of past three decade weather phenomenon at the mid-hills of Sikkim and strategies for mitigating possible impact of climate change on agriculture. In M. L. Arrawatia & S. Tambe (Eds.), *Climate change in Sikkim: Patterns, impacts and initiatives*. Information and Public Relations Department, Government of Sikkim.

Ravindranath, N. H., Sagadevan, A., Jayaraman, M., & Munsi, M. (2012). Climate change and its impacts on forests of Sikkim. In M. L. Arrawatia & S. Tambe (Eds.), *Climate change in Sikkim: Patterns, impacts and initiatives*. Information and Public Relations Department, Government of Sikkim.

Sharma, G., & Rai, L. K. (2012). Climate change and sustainability of agrodiversity in traditional farming of the Sikkim Himalaya. In M. L. Arrawatia & S. Tambe (Eds.), *Climate change in Sikkim: Patterns, impacts and initiatives*. Information and Public Relations Department, Government of Sikkim.

Sharma, G., Partap, U., Dahal, D. R., Sharma, D. P., & Sharma, E. (2016). Declining large-cardamom production systems in the Sikkim Himalayas: Climate change impacts, agroeconomic potential, and revival strategies. *Mountain Research and Development, 36*, 286–298.

Shrestha, A. B., Wake, C. P., Mayewski, P. A., & Dibb, J. E. (1999). Maximum temperature trends in the Himalaya and its vicinity: An analysis based on temperature records from Nepal for the period 1971–94. *Journal of Climate, 12*, 2775–2786.

Shrestha, U. B., Gautam, S., & Bawa, K. S. (2012). Widespread climate change in the Himalayas and associated changes in local ecosystems. *PLoS One, 7*, e36741.

Shukla, G., Kumar, A., Pala, N. A., & Chakravarty, S. (2016). Farmers perception and awareness of climate change: A case study from Kanchandzonga biosphere reserve, India. *Environment, Development and Sustainability, 18*, 1167–1176.

Subba, B., Sen, S., Ravikanth, G., & Nobis, M. P. (2018). Direct modelling of limited migration improves projected distributions of Himalayan amphibians under climate change. *Biological Conservation, 227*, 352–360.

Telwala, Y., Brook, B. W., Manish, K., & Pandit, M. K. (2013). Climate-induced elevational range shifts and increase in plant species richness in a Himalayan biodiversity epicentre. *PLoS One, 8*, e57103.

Verma, A. K., Nayak, R., Manika, N., Bargali, K., Pandey, V. N., Chaudhary, L. B., & Behera, S. K. (2023). Monitoring the distribution pattern and invasion status of *Ageratina adenophora* across elevational gradients in Sikkim Himalaya, India. *Environmental Monitoring and Assessment, 195*, 152.

Xu, J., Grumbine, R. E., Shrestha, A., Eriksson, M., Yang, X., Wang, Y. U., & Wilkes, A. (2009). The melting Himalayas: Cascading effects of climate change on water, biodiversity, and livelihoods. *Conservation Biology, 23*, 520–530.

Biodiversity in Response to Global Change

Navneeti Chamoli, Bharat Hamal, Deepti Prabha, Jai Singh Chauhan, Jitender Kumar, Priyanka Sharma, Nitesh Kumar, Rajeev Kumar, and Arti Sharma

Abstract Biodiversity serves as our strongest natural defence against climate change, playing a crucial role in maintaining ecosystem balance, sequestering carbon, and enhancing resilience to environmental changes. However, climate change is drastically affecting biodiversity by altering species distributions, degrading habitats, and increasing the frequency of extreme weather events. These disruptions threaten the stability of ecosystems, accelerate species extinctions, and weaken the essential services that biodiversity provides, including climate regulation and disaster mitigation. Biodiversity is vital for limiting climate change, as diverse ecosystems act as carbon sinks, stabilize climatic patterns, and support species adaptation to environmental shifts. The loss of biodiversity diminishes nature's ability to regulate greenhouse gases, exacerbating global warming. This chapter examines the intricate relationship between climate change and biodiversity, detailing the adverse impacts and the urgent need for conservation efforts. Strategies such as habitat pro-

---

N. Chamoli · D. Prabha · J. S. Chauhan
Department of Seed Science & Technology, Hemvati Nandan Bahuguna Garhwal University, Srinagar, Uttarakhand, India

B. Hamal
Department of Forestry, Hemvati Nandan Bahuguna Garhwal University, Srinagar, Uttarakhand, India

J. Kumar
Department of Plant Sciences, Central University of Himachal Pradesh, Kangra, Himachal Pradesh, India

P. Sharma
Department of Bio Sciences, Career Point University, Hamirpur, Himachal Pradesh, India

A. Sharma (✉)
Faculty of Agriculture, Abhilashi University, Mandi, Himachal Pradesh, India

N. Kumar
Department of Biosciences, Himachal Pradesh University, Shimla, Himachal Pradesh, India

R. Kumar
School of Bioengineering and Biosciences, Lovely Professional University, Phagwara, Punjab, India

S. Siddiqui et al. (eds.), *Frontiers in Sustainability*,
https://doi.org/10.1007/978-3-032-19740-5_7

tection, sustainable land use, ecological restoration, and climate-resilient policies are essential to mitigate biodiversity loss and strengthen nature's capacity to counteract climate change. Integrating scientific research, policy frameworks, and community participation will be key to ensuring the sustainability of biodiversity amid ongoing global challenges.

Keywords Biodiversity · Climate change · Ecosystem · Global warming · Climate-resilient

1 Introduction

Deforestation, habitat fragmentation, pollution, intensive agriculture and livestock farming, urbanization, greenhouse gas emissions, ocean acidification, industrial overfishing, and the large-scale transportation of species—human activities are exerting unprecedented pressures on the planet's biodiversity (UNEP,2021). These rapid environmental changes are reshaping ecosystems, altering species distributions, and disrupting ecological interactions, leading to biodiversity loss at an alarming rate. How do species and ecological communities respond to these pressures? Can we predict their dynamics in an era of accelerating global change? Understanding biodiversity in response to environmental shifts requires an exploration of the key drivers of species adaptation and survival (CBD, 2010). Some species exhibit resilience through rapid reproduction and genetic adaptation, while others struggle to cope with habitat degradation and shifting climatic conditions. The balance between generalist and specialist species is shifting, with profound consequences for ecological stability (Kromp-Kolb & Formayer, 2005; Latif, 2007). In this chapter, we will examine the mechanisms by which biodiversity responds to global change. The role of genetic diversity, reproductive strategies, dispersal mechanisms, and phenotypic plasticity in determining species survival will be explored and we will also discuss how ecosystem shifts, species invasions, and altered food webs influence biodiversity at local and global scales. By understanding these patterns, we can better anticipate the future of biodiversity and develop strategies for conservation and sustainable management. This chapter will provide a comprehensive overview of the challenges and opportunities for biodiversity in the face of ongoing global change.

2 The Keys of Adaptation to Change

Since the time of Darwin, it has been well understood that species undergo change over time due to the process of natural selection (Darwin, 1859). This mechanism filters out individuals within a population based on their ability to survive and reproduce, favouring those who are better suited to their environment (Mayr, 1982). More specifically, individuals who successfully pass on their genetic traits to future generations shape the course of evolution. This continuous process of adaptation allows species to respond to both gradual and sudden environmental changes, operating over different time scales (Futuyma & Kirkpatrick, 2017). Over the past 60 years, research in ecology and evolutionary biology has provided extensive evidence that species are not only adapted to the usual variations in their living conditions but also to the relative frequency of extreme or catastrophic events that occur in their habitats (Hughes et al., 2017). Organisms evolve strategies to cope with periodic environmental fluctuations such as seasonal changes, resource availability, and predator-prey dynamics. However, they must also develop resilience to rare but significant disturbances, such as droughts, wildfires, hurricanes, or disease outbreaks, which can drastically impact populations (Gould & Eldredge, 1977). For example, some plant species in fire-prone ecosystems, like certain pine trees, have evolved serotinous cones that only release seeds in response to the heat of a wildfire, ensuring regeneration after destruction. Similarly, desert amphibians like the spadefoot toad have developed the ability to remain dormant underground for years, emerging only after heavy rains to breed rapidly before the environment becomes hostile again.

The ability of species to endure both predictable and extreme environmental challenges depends on a combination of genetic diversity, phenotypic plasticity, and adaptive behaviours (Teyssèdre, 2004).While, gradual environmental changes may allow populations to adjust over generations, more frequent or intense catastrophic events exacerbated by human-induced climate change can push species beyond their adaptive limits, leading to population declines or even extinctions. Understanding these mechanisms is essential for biodiversity conservation, as it helps predict which species are most vulnerable to future environmental disruptions and guides efforts to protect and sustain ecosystems (Latour, 2017).

2.1 Niche Width and Phenotypic Plasticity

The concept of an ecological niche, first introduced by Hutchinson (1957) describes the unique role a species plays in its ecosystem. This niche is defined by a combination of physical, chemical, and biological factors that determine where and how a species can survive and reproduce (Odling-Smee et al., 2003). Since these factors are shaped by natural selection over generations, a species' niche continuously evolves along with its physiological and behavioural traits. One key characteristic of a species' niche is its width, or the range of environmental conditions it can tolerate. Species with a narrow niche width (Stenotypic species) are specialized for stable conditions, while those with a broad niche width (Eurytopic species) can survive in a wide range of environments (Smith, 2007). For example:

1. Pelagic (open ocean) fish and invertebrates are usually stenohaline, meaning they can only tolerate a narrow range of salinity because the salinity of the open ocean remains relatively constant (~32 g/L).

2. Estuarine species, such as fish that live where rivers meet the sea, must be euryhaline, capable of surviving large fluctuations in salinity due to the mixing of freshwater and seawater.

To cope with environmental variability, organisms develop phenotypic plasticity, which refers to their ability to modify their morphology, physiology, or behaviour in response to environmental changes (MEA, 2005). This adaptability can occur in two ways:

Developmental plasticity: Changes that occur during an organism's growth and are often irreversible (e.g., a plant growing longer roots in dry soil).

Reversible plasticity: Adjustments that can be made repeatedly throughout an organism's life in response to environmental shifts (e.g., seasonal fur colour changes in Arctic foxes).

Niche construction, as described by Odling-Smee, Laland, and Feldman (2003), highlights the dynamic relationship between organisms and their environment, emphasizing that species not only modify their surroundings but also influence their own evolutionary pathways. Through various activities and interactions, organisms actively shape their ecological niche, creating environmental changes that, in turn, exert selection pressures on their future adaptation. This concept challenges the traditional view of evolution as a one-way process where species merely adapt to pre-existing conditions. Instead, niche construction underscores how species themselves play a role in shaping the selective forces acting upon them, leading to co-evolutionary changes over time (IPBES, 2019).A clear example of this process can be observed in the interaction between butterflies and flowers. When a butterfly forages on a flower, it extracts nectar, depleting a resource that will no longer be available to itself or other insects for the rest of the day (Steffen et al., 2015). Simultaneously, the butterfly collects pollen, which may be deposited on another flower of the same species, facilitating fertilization. These actions represent immediate, short-term impacts on both the butterfly's environment and the broader ecological network in which it participates (Ceballos et al., 2017). The depletion of nectar influences the behaviour of future pollinators, while pollen transfer contributes to plant reproduction and genetic diversity (Dirzo et al., 2014). Over the long term, these interactions drive evolutionary changes in both species involved, as they exert mutual selective pressures on one another. Plants that produce more nectar and develop more effective signalling mechanisms—such as distinct odours, colours, or petal shapes—are more likely to attract pollinators, increasing their chances of reproduction (Fenster et al., 2004). At the same time, butterflies with better sensory abilities, such as enhanced visual or olfactory recognition of flowers, have an advantage in locating high-quality nectar sources (Raguso, 2008). Morphological adaptations occur, where the shape of a butterfly's proboscis and the structure of a flower's corolla evolve in response to each other, leading to a specialized co-evolutionary relationship.

This mutual adaptation illustrates how niche construction influences not only individual survival and reproduction but also broader evolutionary trends. Rather than passively responding to environmental pressures, species actively shape their own adaptive landscapes through their behaviours and interactions (Waser et al., 1996). In this way, niche construction serves as a powerful mechanism driving biodiversity, ecosystem complexity, and the intricate interdependencies observed in nature.

2.1.1 Phenotypic Adaptations and Environmental Change

Natural selection favours individuals that can adjust to regular environmental fluctuations, giving them a survival advantage in dynamic ecosystems (Ghalambor et al., 2007). The ability to modify traits in response to environmental changes—a process known as phenotypic plasticity—is crucial for survival (West-Eberhard, 2003). However, when environmental changes occur too rapidly or are too extreme, such as those caused by climate change or habitat destruction, they may exceed the adaptive capacity of many species, increasing the risk of population decline or extinction (Urban, 2015).

2.1.1.1 Irreversible Phenotypic Adaptations

Some adaptations are permanent and occur only during an organism's development. These changes, once established, cannot be reversed, even if environmental conditions shift (Tollrian, 1990). Examples include:

1. Daphnia (water fleas) and predator-induced defence mechanisms: When daphnia larvae detect the presence of predatory fish in their habitat, they develop protective helmets or elongated spines. This adaptation enhances their survival by making them harder to consume. However, once developed, this defensive structure remains throughout their lifetime.
2. Caste differentiation in social insects: In species like ants, termites, and bees, larvae develop into different castes—workers, soldiers (in termites), or reproductive individuals—depending on the food they receive from their colony. For example, in honeybees, a larva fed royal jelly becomes a queen, while those given regular food develop into workers. This caste specialization is irreversible and determines the role each individual will play in the colony.

2.1.1.2 Reversible Phenotypic Plasticity

When environmental changes are cyclical or seasonal, organisms may develop reversible adaptations that allow them to adjust to periodic changes in climate, food availability, or other ecological factors (Piersma & Drent, 2003).

1. Deciduous Trees and Seasonal Cycles: In temperate regions, trees like oak, maple, and beech respond to seasonal changes by shedding their leaves in

autumn. This process, triggered by day length (photoperiod), helps them conserve energy during winter when water and sunlight are scarce. They enter a state of dormancy and only begin new growth in spring when conditions improve.
2. Hibernation and Torpor in Animals: Many terrestrial animals, including mammals, reptiles, and amphibians, enter a state of dormancy to survive harsh winter conditions (Malhi et al., 2009). Bears, hedgehogs, and ground squirrels hibernate, lowering their metabolic rate to conserve energy. Frogs and some reptile species enter diapause (a suspended developmental state) to withstand freezing temperatures. Unlike these groups, birds do not hibernate due to their high metabolic demands, but they have evolved alternative survival strategies.
3. Migration and Food Storage in Birds: Birds have adapted to seasonal food shortages in different ways: Seed-eating birds like jays, nuthatches, and nutcrackers store food during late summer and autumn to consume in winter (Atkinson et al., 2004).Insectivorous birds like swallows and swifts cannot survive without insects, which become scarce in cold weather. Instead, they accumulate fat reserves and migrate to warmer regions, returning to their breeding grounds in summer.
4. Circadian Rhythms and Hormonal Regulation: The 24-h activity cycle (circadian rhythm) in vertebrates is another example of phenotypic plasticity. This rhythm is regulated by melatonin, a hormone produced by the brain in response to the length of day and night. It influences sleep patterns, feeding behaviour, and body temperature, ensuring that animals remain in sync with Earth's natural light-dark cycles.

2.1.2 Behavioural Flexibility in Response to Unexpected Changes

Organisms are not rigidly programmed to react only to predictable environmental cues. Many species also display behavioural flexibility, allowing them to respond to unforeseen environmental challenges. For example, Birds altering migration routes due to unexpected climate changes, Mammals shifting foraging patterns in response to food scarcity while Insects adjusting reproductive timing based on seasonal variability. This behavioural elasticity enables species to maximize survival and reproduction even in unpredictable environments (Teyssèdre, 2006). However, extreme or rapid environmental changes—such as global warming, habitat destruction, and pollution—can outpace organisms' ability to adapt, threatening biodiversity.

Phenotypic plasticity plays a vital role in species survival by allowing organisms to adjust to environmental fluctuations. While some adaptations are irreversible and shaped during development, others are reversible and help individuals cope with cyclical changes. On the other hand, many species exhibit behavioural flexibility, allowing them to respond to unexpected challenges. However, with the increasing speed of climate change and human-induced environmental pressures, many species may struggle to adapt quickly enough, highlighting the need for conservation efforts to preserve biodiversity.

2.1.3 Adaptive Responses in Plants and Animals

Just like animals, plants also exhibit adaptive responses to both regular (cyclical) and unexpected environmental changes. Despite lacking a nervous system or sensory organs, plants can still perceive environmental signals and adjust their behaviour accordingly. Plants respond to daily and seasonal changes, such as light intensity and temperature, but they also react to sudden threats like predation, infection, or competition. Some common plant responses include:

1. Flower Orientation and Opening: Many plants adjust the position or timing of flower opening based on light exposure to maximize pollination success. For example, sunflowers track the sun throughout the day, a behaviour known as heliotropism.
2. Production of Defensive Compounds: When attacked by herbivores or pathogens, plants can produce toxins or deterrent chemicals to protect themselves. For instance, mustard plants release glucosinolates, which are toxic to many insects.
3. Chemical Signalling for defence: Plants communicate with their neighbours by releasing volatile organic compounds (VOCs) when under attack. These chemicals serve two main functions. First is to attract beneficial species (some plants release pheromones that attract predatory insects (like ladybugs) to eat herbivorous pests and second is warning nearby plants (neighbouring plants of the same species can detect these signals and activate their own defence mechanisms before being attacked).

Despite the lack of a central nervous system, plants have evolved sophisticated biochemical and physiological responses to cope with environmental stress, demonstrating a form of behavioural plasticity at the biochemical level.

2.1.4 Behavioural Plasticity in Animals

While plants show adaptive behaviours through chemical signalling and physiological changes, behavioural plasticity is far more pronounced in animals, especially those with higher cognitive abilities, complex social structures, and greater mobility (West-Eberhard, 2003).Some invertebrates display remarkable intelligence and problem-solving skills. For example, Honeybees can learn to associate flower colours with food rewards and even communicate through complex dance movements to share information about food locations. Similarly, Octopuses, despite being non-social, demonstrate problem-solving abilities, such as unscrewing jars to access food or using tools in the wild while birds and mammals exhibit the highest levels of learning capacity, social interactions, and behavioural flexibility (Gould & Gould, 1988; Dyer, 2002). Crows and parrots have demonstrated tool use, problem-solving skills, and even the ability to recognize themselves in mirrors (Mather, 2008; Zylinski & How, 2012; Emery & Clayton, 2004; Pepperberg, 2008) Primates, such as chimpanzees, can modify their hunting techniques, use tools for extracting termites, and develop unique behaviours based on their local environment (Whiten et al., 1999; Boesch & Boesch-Achermann, 2000).To survive in ever-changing environments, organisms rely on three major adaptive strategies:

1. Physiological Tolerance: This refers to an organism's inherent ability to withstand extreme environmental conditions, such as temperature fluctuations, salinity changes, or oxygen variations. For example, Desert plants like cacti store water and have thick, waxy coatings to prevent dehydration.
2. Morphological or Physiological Adjustments: Organisms can modify their physical traits to better adapt to environmental conditions. For example, Arctic foxes grow a thick white coat in winter for warmth and camouflage then shed it in summer for a thinner brown coat.
3. Behavioural Flexibility: Animals can alter their behaviour based on changing conditions, which may involve learning, problem-solving, or social cooperation. For example, Elephants change migration routes based on seasonal water availability and pass this knowledge down through generations.

In the past, species adapted to natural environmental fluctuations caused by climate cycles, predators, and resource availability. However, today's environmental changes are largely driven by human activities, such as:

1. Deforestation and habitat destruction, leading to habitat loss.
2. Climate change, causing rising temperatures, shifting rainfall patterns, and extreme weather events.
3. Pollution and chemical contamination, affecting species' survival and reproduction.

Because many of these changes occur faster than natural adaptation can keep up, many species struggle to survive. Conservation efforts, including habitat protection, reforestation, and reducing carbon emissions, are essential to maintaining biodiversity and allowing species to adapt at a sustainable pace. Both plants and animals have evolved impressive mechanisms to cope with environmental changes, from chemical signalling in plants to complex learning in animals. The ability to tolerate, adjust, and change behaviour is critical for survival. However, rapid human-driven changes present a major challenge to natural adaptation. Understanding and supporting these adaptive mechanisms is key to biodiversity conservation and ecological sustainability.

3 Habitat Stability and Demogenetic Strategies

Over multiple generations, species must adjust not just at the individual level but also demographically and genetically to cope with long-term environmental fluctuations. This process, known as demogenetic adaptation, involves both population dynamics (demography) and genetic evolution, ensuring that species can survive and reproduce despite environmental uncertainties. Pioneering research by MacArthur and Wilson (1967) laid the foundation for understanding how species evolve in response to habitat stability and disturbance. They proposed that species adopt one of two major demogenetic strategies, depending on the predictability and intensity of environmental changes over generations:

3.1 Strategy ('Quantity-Based Strategy')

Prioritizing rapid reproduction in unpredictable environments, adapting to Unstable or Temporary Habitats. Species that follow the r-strategy thrive in highly variable, unpredictable environments, where the likelihood of catastrophic disturbances is high. These species rely on rapid reproduction and widespread dispersal to compensate for high mortality rates.

3.1.1 Key Traits of r-Species

According to Pianka (1970), the r-strategy is characterized by:

1. Early reproduction: These species mature quickly and reproduce at a young age.
2. Short lifespan and small body size: They maximize reproductive output before environmental pressures take a toll.
3. High fecundity: They produce a large number of offspring to increase survival chances.
4. Early and widespread dispersal: Seeds, eggs, or larvae are scattered widely to colonize new habitats, often through passive (wind, water) or active (migration) mechanisms.
5. Minimal parental care: Offspring are typically independent soon after birth.

3.1.2 The 'r' Strategy: Resilience Through Rapid Multiplication and Dispersal

The 'r' strategy, characterized by rapid reproduction and high dispersal, is a key survival mechanism for species living in unstable, temporary, or unpredictable habitats. This strategy is commonly observed in microorganisms, fungi, algae, annual plants, insects, and some small vertebrates (Reznick et al., 2002; Beaumont et al., 2009; Botero et al., 2015). Instead of relying on long-term adaptation to specific environments, r-strategists focus on sheer numbers and rapid colonization, ensuring that at least some individuals will survive environmental changes.

3.1.3 r-Strategy in Microorganisms and Viruses

Microorganisms, including bacteria, protists, fungi, and viruses, thrive by adopting the r-strategy. They inhabit temporary or fluctuating environments are present in Table 1.

Since these environments can be short-lived, r-strategists multiply rapidly before conditions become unfavourable. When local populations experience episodic decimation (e.g., antibiotic treatment, immune responses, environmental shifts), a fraction of individuals with genetic mutations may survive and repopulate quickly (Starrfelt & Kokko, 2012).

When exposed to antibiotics, most bacterial cells die, but a few may carry mutations conferring resistance. These survivors reproduce rapidly, creating a new population resistant to the antibiotic. Over time, resistant bacterial strains become

Table 1 Species following r-strategist strategies (Zheng et al., 2024)

Species	Category	Characteristics	Example Environments
Bacteria	Microorganisms	Rapid reproduction, short generation time, high mutation rate	Soil, water, host organisms
Fruit flies (*Drosophila* spp.)	Insects	Short lifespan, high fecundity, rapid development	Decaying fruit, temperate regions
Mosquitoes (*Anopheles*, *Aedes*)	Insects	High reproductive output, small size, minimal parental care	Stagnant water, tropical regions
Houseflies (*Musca domestica*)	Insects	Fast life cycle, large number of offspring	Urban environments, organic waste
Locusts (*Schistocerca gregaria*)	Insects	Boom-and-bust population cycles, high dispersal ability	Grasslands, arid regions
Rabbits (*Oryctolagus cuniculus*)	Mammals	High reproductive rate, early sexual maturity	Meadows, farmlands
Rodents (*Mus musculus*, *Rattus rattus*)	Mammals	Short gestation period, frequent breeding	Urban, agricultural environments
Frogs (*Rana* spp.)	Amphibians	Many eggs, external fertilization, high juvenile mortality	Wetlands, ponds
Fish (*Cod*, *Tilapia*, *Carp*)	Aquatic vertebrates	High egg production, little parental care	Rivers, oceans, freshwater bodies
Weeds (*Amaranthus*, *Chenopodium*)	Plants	Fast-growing, high seed output, early maturity	Disturbed habitats, agricultural fields
Algae (*Chlamydomonas*, *Spirogyra*)	Aquatic plants	Rapid reproduction, quick colonization	Ponds, lakes, marine environments
Coral spawning species	Marine invertebrates	Release of large numbers of gametes, external fertilization	Coral reefs, coastal waters

dominant, leading to public health challenges like multidrug-resistant tuberculosis (MDR-TB) and MRSA (Methicillin-resistant *Staphylococcus aureus*). This ability to rapidly evolve and adapt to environmental stress makes bacteria formidable in the face of climate change, pollution, and medical interventions.

3.1.4 r-Strategy in Plants: Fast Growth and Wide Dispersal

Many annual plants, grasses, and pioneer species follow the r-strategy. These plants produce vast quantities of seeds and pollen, often dispersed by wind or animals and grow quickly in disturbed or open habitats. Also, these plants have short life cycles, ensuring rapid turnover. For example, Dandelions (*Taraxacum* spp.) produce light, wind-dispersed seeds, quickly colonizing open ground. Some weeds like *Amaranthus* spp. (pigweed) have evolved herbicide resistance, allowing them to thrive in agricultural fields despite control efforts while grasses in fire-prone areas regenerate quickly after burning, taking advantage of temporarily cleared spaces. These species proliferate after disturbances (e.g., fire, grazing, drought), making them highly resilient in dynamic ecosystems.

3.1.5 r-Strategy in Animals: Rapid Reproduction and High Dispersal

Many invertebrates, fish, and small mammals exhibit r-selected traits, such as early maturity (short time to reproductive age), high reproductive output (many offspring per breeding cycle) rapid dispersal of eggs, larvae, or juveniles and short lifespans and high mortality rates. For example, Insects (*Drosophila*, mosquitoes, and locusts reproduce in large numbers and quickly spread across regions), Fish (Species like cod and anchovies produce millions of eggs, ensuring that at least some will survive predation), Rodents (Mice and rats) have short gestation periods, large litters, and high reproductive rates, allowing populations to rebound quickly after population crashes. Unlike birds and large mammals, which invest heavily in offspring survival, these species rely on quantity over quality, ensuring some individuals survive even under extreme environmental pressures.

For example, overuse of insecticides has led to the rise of resistant insect populations, such as mosquitoes resistant to DDT and pyrethroids, challenging malaria control efforts, Colorado potato beetles resistant to multiple pesticides, making them a persistent agricultural pest, Resistant individuals survive chemical treatments, reproduce rapidly, and spread their resistance genes, making pest management more difficult.

3.1.6 The r-Strategy as Strength in a Changing World

The intrinsic growth rate and genetic diversity of r-strategists allow them to recover quickly from population declines (e.g., after pesticide application, droughts, or habitat destruction), take advantage of newly available habitats (e.g., agricultural fields, deforested land, warming oceans) and adapt rapidly to environmental stress through mutation and natural selection. Following are some emerging threats due to r-Strategists-

1. Resistant pathogens: Bacteria, fungi, and viruses are evolving resistance to antibiotics, antifungals, and antivirals.
2. Invasive species: Highly reproductive plants and animals (e.g., kudzu vine, zebra mussels) outcompetes native species in new ecosystems.
3. Pest outbreaks: Climate change is expanding the range of r-strategist pests, such as desert locusts, leading to agricultural losses.

The r-strategy of rapid reproduction and dispersal is a powerful evolutionary response to unstable or temporary habitats. This strategy allows species to recover from population declines, colonize new areas, and adapt rapidly to environmental changes. However, it also accelerates the evolution of resistant pathogens, invasive species, and agricultural pests, posing significant global challenges (Bongers, 1999). Thus, while r-strategists have an evolutionary advantage in changing environments, they also present challenges for biodiversity conservation, agriculture, and public health.

3.2 K-Strategy ('Quality-Based Strategy')

Favouring long-term survival and competition in stable habitats K-Strategists. In contrast, K-strategists dominate in stable, resource-limited habitats, where competition is intense, and population sizes are near the carrying capacity (K) of the environment (Teyssèdre, 2018). These species emphasize quality over quantity, investing in long-term survival and ensuring the success of fewer offspring (Table 2).

3.2.1 Key Traits of K-Strategists

1. K-strategists are typically long-lived species with slow reproductive rates, investing heavily in offspring survival rather than producing large numbers of offspring. Their key traits include:
2. Late sexual maturity (e.g., elephants reproduce after 10–15 years).
3. Low fecundity (e.g., whales and apes produce only one offspring per breeding cycle).
4. Extended parental care (e.g., primates nurse and protect their young for years).
5. Stable populations in saturated, competitive environments (e.g., large mammals in forests and savannahs).
6. High dependency on intact ecosystems for survival.

Table 2 List of organisms following K-strategies (Zheng et al., 2024)

Category	Examples	Characteristics
Mammals	Elephants, whales, humans, lions	Long gestation, few offspring, high parental care, slow growth
Birds	Eagles, albatrosses, owls	Few eggs, extended parental care, long lifespan
Reptiles	Tortoises, crocodiles	Long lifespan, slow growth, parental investment in some species
Amphibians	Giant salamander	Large body size, few offspring, longer development
Fish	Sharks, sturgeons	Slow reproduction, long lifespan, fewer offspring than r-strategists
Insects	Termites, some beetles (e.g., dung beetles)	Longer life cycle, social structures, parental investment
Plants	Oak trees, coconut palms	Long lifespan, slow growth, produce fewer seeds but with high survival rates

7. While this strategy maximizes offspring survival, it limits population recovery after disturbances, making K-strategists highly susceptible to rapid environmental changes.

K-strategists thrive in stable ecosystems but struggle with sudden environmental disturbances, making them more vulnerable to habitat destruction and climate change.

3.2.2 The Disadvantage of K-Strategy in Habitat Disruption

Since K-strategists have low genetic diversity and slow reproduction, they struggle to adapt quickly to sudden changes in their habitat. This can lead to local or regional population declines (e.g., declining orangutan populations due to deforestation), fragmentation-driven extinction (e.g., habitat loss isolates populations, reducing genetic exchange), and increased risk of inbreeding (e.g., cheetahs have low genetic diversity due to past population bottlenecks). Following are some case studies of K species.

3.2.2.1 Case Study: The Decline of Large Mammals

1. Elephants: Poaching and habitat fragmentation have reduced their numbers. With long reproductive cycles (one calf every 3–4 years), they struggle to recover.
2. Tigers: Due to habitat loss and poaching, tiger populations have declined. With low reproductive rates, their ability to rebound is limited.

3. Whales: Overfishing, ocean noise pollution, and climate change have reduced whale populations. With slow reproductive rates, they face long-term population decline.

These examples highlight the vulnerability of K-strategists when faced with rapid environmental changes.

3.2.2.2 K-Strategists with High Fecundity

Some K-strategists, like large pelagic fish (e.g., tuna and cod), deviate from the typical pattern by having high fecundity but limited parental care:

1. Tuna and Cod: Despite being K-strategists due to their size and long lifespan, they lay millions of eggs to compensate for the high mortality rate of larvae.
2. Sea Turtles: Although long-lived, sea turtles lay thousands of eggs, relying on sheer numbers rather than parental care.

This demonstrates evolutionary flexibility, where certain K-strategists adopt elements of r-strategy to survive.

3.2.2.3 Eusocial Species: A Special K-Strategy Case

Eusocial species, such as honeybees, termites, and naked mole rats, combine low reproductive rates (few breeding individuals) with cooperative brood care. These species are more K-like because they rely on social structure, cooperation, and predictable environments rather than sheer reproductive output.

1. Honeybees: A single queen lays all the eggs, while worker bees ensure colony stability.
2. Naked Mole Rats: A dominant queen breeds while others assist, ensuring colony persistence despite environmental fluctuations.

Such species increase their environmental predictability through social cooperation, which helps buffer against habitat disruptions.

4 r/K Strategies in Response to Global Change

As global climate change alters habitat stability, species with different demogenetic strategies respond in different ways-

1. r-strategists may benefit from disturbances like deforestation, agricultural expansion, and urbanization, as they can quickly colonize disturbed areas. This is why many pest species and invasive plants spread rapidly.

2. K-strategists are at higher risk, as their slow reproductive rates and habitat specialization make it difficult to adapt quickly to rapid environmental changes. Many endangered species, such as elephants and tigers, follow the K-strategy.

5 Intermediate Strategies and Adaptation

Not all species fit neatly into r or K categories. Some species exhibit a mix of both traits, adapting their reproductive strategies based on environmental conditions. For example, Sea turtles: Lay many eggs (r-strategy), but individuals that survive live long and mature slowly (K-strategy while some species of Rodents reproduce rapidly when resources are abundant but shift to K-like behaviours in stable environments. Species adapt to long-term environmental fluctuations through demographic (r/K strategies) and genetic evolution. While r-strategists flourish in unpredictable environments, K-strategists dominate in stable, competitive habitats. However, as climate change and human activity disrupt ecosystems; K-selected species face greater threats, emphasizing the need for conservation efforts to protect biodiversity.

6 The 'K' Strategy: Adaptation and Vulnerability in a Changing World

Unlike the r-strategy, which thrives in unstable and unpredictable environments through rapid reproduction and dispersal, the K-strategy is characterized by low fecundity, slow development, and strong competition for resources. While this strategy ensures stability in predictable and resource-saturated ecosystems, it also makes K-strategists highly vulnerable to rapid environmental disruptions, such as climate change, habitat fragmentation, and human activity.

6.1 Adaptation Potential of Species to Global Change

Three key factors that influence species adaptation:

1. Genetic Diversity: Higher diversity increases adaptability to environmental changes.
2. Intrinsic Growth Rate: High reproduction rates improve population resilience.
3. Phenotypic Plasticity: Behavioural and physiological flexibility allows adaptation to new conditions.

Since K-strategists generally have low genetic diversity and slow growth, they rely more on behavioural plasticity to cope with global change.

6.2 *Expansion of Generalist Species*

Certain K-strategists with high adaptability (generalists) have expanded in human-modified ecosystems:

1. Wild boars and deer: Populations are increasing due to deforestation and agriculture.
2. Olive baboons: Thriving in human settlements in central Africa.
3. Crows and pigeons: Expanded globally due to urban environments.

These species benefit from human activity, unlike specialist K-strategists, which decline due to habitat loss.

6.3 *Displacement, Decline, and Extinction in K-Species*

While some species adapt, others struggle to survive.

6.3.1 Species Moving to Favourable Habitats

Both r- and K-strategists are shifting their ranges due to climate change. For example, Bark beetles and forest pests are migrating towards higher altitudes, Birds and insects are shifting their breeding patterns to match temperature changes. However, migration does not guarantee survival, especially for habitat specialists.

6.3.2 Decline and extinction of Specialist K-Species

K-strategists that require specific conditions are particularly vulnerable. For example, large mammals (e.g., tigers, rhinos, and elephants) struggle due to habitat fragmentation, Coral reef fish and invertebrates are declining due to ocean acidification, and many tropical forest species are disappearing as deforestation accelerates. Without rapid adaptation or conservation intervention, these species face irreversible declines. While r-strategists thrive in changing environments through rapid reproduction, K-strategists are more vulnerable due to slow adaptation rates. Their survival depends on behavioural flexibility, conservation efforts, and the rate of environmental change. The future of biodiversity will depend on how species adapt, migrate, or decline in response to human activity and climate change.

7 Dynamics of Communities and Ecological Networks

Ecosystems are dynamic and complex systems composed of networks of co-adapted species that interact with each other and their physical environment over many generations. Human activities, such as habitat conversion, overexploitation, pollution, and climate change, are significantly altering these ecosystems, leading to homogenization of biodiversity, shifts in community composition, and potential ecosystem collapses.

7.1 The Homogenization of Biodiversity

Human activities modify natural habitats, often favouring generalist species over specialists. This results in the homogenization of regional faunas and floras and a decline in global biodiversity, unless sufficient diversification occurs through speciation in newly emerging habitats (McKinney & Lockwood, 1999; Clavel et al., 2010; Thomas, 2015). For instance, in Europe, agricultural intensification has caused an expansion of generalist bird species, such as chickadees, starlings, carrion crows, and wood pigeons, which can thrive in degraded fields, towns, villages, and forests. Conversely, specialists of agricultural environments, such as linnets, hoopoes, grey partridges, and crested lapwings, have experienced significant population declines. Researchers have developed a community specialization index to track these changes, revealing an increasing homogenization of France's bird populations over the past two decades (McKinney & Lockwood, 1999; Clavel, 2011).

7.2 The Fragility of Ecological Networks

Ecosystems are structured as interconnected networks where species depend on one another for survival. However, rapid environmental changes, driven by climate change and human activity, are disrupting these networks (Julliard et al., 2006). Species have different rates of adaptation, leading to desynchronization in interactions and increasing the risk of ecosystems shifting to alternative, often less favourable, stable states.

7.3 Case Study: Mediterranean Region Ecosystem Shifts

In the Mediterranean region, centuries of deforestation and overgrazing have led to widespread ecosystem shifts (Watson & Pauly, 2001; Folke et al., 2004; Cardinale et al., 2012):

1. Forest ecosystems have transitioned to bushy scrublands (e.g., maquis vegetation) due to soil degradation and loss of tree cover.
2. Scrublands, with low evapotranspiration, have become poor climate regulators, further intensifying arid conditions.
3. Continued pressure through overgrazing and climate change increases the risk of desertification.

Such shifts reduce ecosystem productivity, biodiversity, and resilience, negatively impacting both wildlife and human populations reliant on these ecosystems.

7.4 The Impact of Human Activities on Ecological Communities

Human societies continually convert biodiverse habitats into more impoverished landscapes, such as cultivated fields, overfished seas, and polluted rivers. This has resulted in a 25% reduction in total available resources for wildlife, leading to declines in abundance and biomass of animal and plant communities, decreased species richness, as fewer species can survive in degraded environments and shortened food webs, increasing the risk of ecosystem collapse and shifts to new, less productive stable states (Folke et al., 2004; Cardinale et al., 2012). Here are some points to understand Species, Environmental Change, and Ecosystem Dynamics-

- Species Shape Their Environment: Every species influences its ecosystem, constructing its ecological niche over generations.
- Role of Environmental Engineers: Some species, such as corals, beavers, and humans, have a disproportionate impact on habitat structure and ecosystem function.
- Responses to Environmental Change: A species' ability to adapt depends on external factors (e.g., magnitude and frequency of changes) and internal traits (e.g., behavioural plasticity and reproductive fecundity).
- Specialists vs. Generalists: Specialists have narrow ecological niches and struggle with habitat modifications, while generalists can adapt more easily.
- Demogenetic Strategies and Global Change: Species are adapted to the natural frequency of extreme environmental events. However, human-driven climate change and habitat destruction are exceeding these natural thresholds.
- Ecosystem Resilience is limited: Although ecosystems are adaptive, beyond a certain level of disturbance, they transition to a new stable state, often with reduced biodiversity and ecosystem services.
- Current Global Trends: Many ecosystems worldwide are undergoing transformation due to human activity, resulting in biodiversity loss and negative consequences for human societies.

The dynamics of ecological communities and networks are being reshaped by human-induced global change. The widespread shift towards simplified, low-diversity ecosystems poses significant risks for both biodiversity and human well-being. Understanding these dynamics is crucial for developing conservation strategies and mitigating the negative impacts of ecosystem degradation (Haberl, 2007; Gaston et al., 2003; Smil, 2011; Teyssèdre & Couvet, 2007; Pauly et al., 1998). Without urgent action, the homogenization of biodiversity, ecosystem collapses, and the loss of essential ecological functions will continue to accelerate, endangering the stability of life on Earth.

8 Conclusion

Biodiversity plays a fundamental role in maintaining ecological balance, ecosystem resilience, and overall environmental stability. The response of organisms to global change is largely governed by their reproductive and survival strategies, with r-strategists adapting quickly to disturbances and K-strategists maintaining population stability in competitive environments. However, the increasing impact of climate change, habitat destruction, pollution, and invasive species has led to significant biodiversity loss, particularly among K-strategists with slower reproductive rates. As ecosystems undergo rapid transformations, understanding these survival strategies is crucial for designing effective conservation measures and promoting sustainable biodiversity management. By integrating ecological, genetic, and conservation-based approaches, it is possible to mitigate biodiversity loss and ensure the continued survival of species in a changing world.

9 Future Perspective

The future of biodiversity conservation depends on the integration of scientific research, technological advancements, and policy-driven initiatives to enhance species resilience and ecosystem sustainability. The application of genomics and molecular tools can help identify and preserve genetic diversity, especially among endangered K-strategists. Developing climate-resilient conservation models will be essential to protect species from the adverse effects of global change. Sustainable agricultural practices, such as the conservation of traditional landraces, organic farming, and improved seed systems, will contribute to both biodiversity preservation and food security. Additionally, long-term ecological monitoring will provide critical insights into species adaptability and ecosystem shifts, enabling more effective conservation planning. Strengthening community involvement, environmental education, and policy enforcement will further support biodiversity protection at local and global levels. A holistic and multidisciplinary approach is necessary to mitigate biodiversity loss and promote sustainable coexistence between human activities and natural ecosystems.

References

Atkinson, A., Siegel, V., Pakhomov, E. A., & Rothery, P. (2004). Long-term decline in krill stock and increase in salps within the Southern Ocean. *Nature, 432*(7013), 100–103.

Beaumont, H. J. E., Gallie, J., et al. (2009). Experimental evolution of bet hedging. *Nature, 462*, 90–92.

Boesch, C., & Boesch-Achermann, H. (2000). *The chimpanzees of the Taï forest: Behavioural ecology and evolution*. Oxford University Press.

Bongers, T. (1999). The Maturity Index, the evolution of nematode life history traits, adaptive radiation and cp-scaling. *Plant and Soil, 212*, 13–22. https://doi.org/10.1023/A:1004571900425

Botero, C. A., Weissing, F. J., Wright, J., & Rubenstein, D. R. (2015). Evolutionary tipping points in the capacity to adapt to environmental change. *Proceedings of the National Academy of Sciences of the United States of America, 112*, 184–189.

Cardinale, B., et al. (2012). Biodiversity loss and its impact on humanity. *Nature, 486*, 59–67.

CBD (2010) Secretariat of the Convention on Biological Diversity, Global Biodiversity Outlook 3. Montréal, 94p.

Ceballos, G., Ehrlich, P. R., & Dirzo, R. (2017). Biological annihilation via the ongoing sixth mass extinction signaled by vertebrate population losses and declines. *Proceedings of the National Academy of Sciences of the United States of America, 114*(30), E6089–E6096. https://doi.org/10.1073/pnas.1704949114

Clavel, J. (2011) *L'Homogénéisationbiotique*. Regards et débats sur la biodiversité, SFE, regard n°16, April 2011.

Clavel, J., Julliard, R., & Devictor, V. (2010). Worldwide decline of specialist species: Toward a global functional homogenization? *Frontiers in Ecology and the Environment, 2*, 222–228.

Darwin, C. (1859). *On the origin of species by means of natural selection*. John Murray.

Dirzo, R., Young, H. S., et al. (2014). Defaunation in the Anthropocene. *Science, 345*, 401–406.

Dyer, A. G. (2002). The biology of floral scents and colour perception in bees. *Naturwissenschaften, 89*(10), 525–531.

Emery, N. J., & Clayton, N. S. (2004). The mentality of crows: Convergent evolution of intelligence in corvids and apes. *Science, 306*(5703), 1903–1907.

Fenster, C. B., Armbruster, W. S., Wilson, P., Dudash, M. R., & Thomson, J. D. (2004). Pollination syndromes and floral specialization. *Annual Review of Ecology, Evolution, and Systematics, 35*, 375–403.

Folke, C., Carpenter, S., et al. (2004). Regime shifts, resilience and biodiversity in ecosystem management. *Annual Review of Ecology, Evolution, and Systematics, 35*, 557–581.

Futuyma, D. J., & Kirkpatrick, M. (2017). *Evolution* (4th ed.). Sinauer Associates.

Gaston, K. J., Blackburn, T. M., & Klein Goldewijk, K. (2003). Habitat conversion and global avian biodiversity loss. *Proceedings of the Royal Society of London B, 270*, 1293–1300.

Ghalambor, C. K., McKay, J. K., Carroll, S. P., & Reznick, D. N. (2007). Adaptive versus non-adaptive phenotypic plasticity and the potential for contemporary adaptation in new environments. *Functional Ecology, 21*(3), 394–407.

Gould, S. J., & Eldredge, N. (1977). Punctuated equilibria: The tempo and mode of evolution reconsidered. *Paleobiology, 3*(2), 115–151.

Gould, J. L., & Gould, C. G. (1988). *The honey bee*. Scientific American Library.

Haberl, H. (2007). Quantifying and mapping the human appropriation of net primary production in earth's terrestrial ecosystems. *Proceedings of the National Academy of Sciences of the United States of America, 104*, 12944–12947.

Hughes, T. P., Barnes, M. L., Bellwood, D. R., et al. (2017). Coral reefs in the Anthropocene. *Nature, 546*(7656), 82–90.

Hutchinson, G. E. (1957). Concluding remarks. *Cold Spring Harbor Symposia on Quantitative Biology, 22*, 415–427.

IPBES. (2019). *IPBES report 2019*.

Julliard, R., Clavel, J., et al. (2006). Spatial segregation of bird specialists and generalists in bird communities. *Ecology Letters, 9*, 1237–1244.
Kromp-Kolb, H., & Formayer, H. (2005). *SchwarzbuchKlimawandel. Wie viel Zeit bleibtunsnoch?* Ecowin Verlag.
Latif, M. (2007). *Bringenwir das Klima ausdem Takt*. Forum für Verantwortung, Fischer.
Latour, B. (2017). *Oùatterrir?* La Découverte.
MacArthur, R. H., & Wilson, E. O. (1967). *The theory of Island biogeography*. Princeton University Press.
Malhi, Y., Roberts, J. T., Betts, R. A., et al. (2009). Climate change, deforestation, and the fate of the Amazon. *Science, 326*(5957), 1350–1351.
Mather, J. A. (2008). Cephalopod consciousness: Behavioural evidence. *Consciousness and Cognition, 17*(1), 37–48.
Mayr, E. (1982). *The growth of biological thought: Diversity, evolution, and inheritance*. Harvard University Press.
McKinney, M. L., & Lockwood, J. L. (1999). Biotic homogenization: A few winners replacing many losers in the next mass extinction. *Trends in Ecology & Evolution, 14*, 450–453.
MEA. (2005). *Millennium Ecosystem Assessment.*
Odling-Smee, F., Laland, K., & Feldman, W. (2003). *Niche construction*. Princeton University Press.
Pauly, D., Christensen, V., Dalsgaard, J., Froese, R., & Torres, F. S. B. (1998). Fishing down marine food webs. *Science, 279*, 860–863.
Pepperberg, I. M. (2008). *The Alex studies: Cognitive and communicative abilities of grey parrots*. Harvard University Press.
Pianka, E. R. (1970). On r- and K-selection. *The American Naturalist, 104*, 592–597.
Piersma, T., & Drent, J. (2003). Phenotypic flexibility and the evolution of organismal design. *Trends in Ecology & Evolution, 18*(5), 228–233.
Raguso, R. A. (2008). Wake up and smell the roses: The ecology and evolution of floral scent. *Annual Review of Ecology, Evolution, and Systematics, 39*, 549–569.
Reznick, D., Bryant, M. J., & Bashey, F. (2002). R- and K-selection revisited: The role of population regulation in life history evolution. *Ecology, 83*(6), 1509–1520.
Smil, V. (2011). Harvesting the biosphere: The human impact. *Population and Development Review, 37*(4), 613–636.
Smith, B. D. (2007). The ultimate ecosystem engineers. *Science, 315*, 1797–1798.
Starrfelt, J., & Kokko, H. (2012). Bet hedging: A triple trade-off between means, variances, and correlations. *Biological Reviews, 87*, 742–755.
Steffen, W., Richardson, K., Rockström, J., et al. (2015). Planetary boundaries: Guiding human development on a changing planet. *Science, 347*, 736–746.
Teyssèdre, A. (2004). Chapter II: Versunesixièmegrande crise d'extinctions? In R. Barbault & B. Chevassus-au-Louis (Eds.), *Biodiversité et changementsglobaux* (pp. 24–49). ADPF.
Teyssèdre, A. (2006). *Les clés de la Communication animale*. Delachaux et Niestlé.
Teyssèdre, A. (2018). *Les mammifères face aux changementsglobaux.* Regards et débats sur la biodiversité, SFE, regard R80b, June 2018.
Teyssèdre, A., & Couvet, D. (2007). Expected impact of agriculture expansion on the world avifauna. *Comptes Rendus Biologies, 330*, 247–254.
Thomas, C. D. (2015). Rapid acceleration of plant speciation during the Anthropocene. *Trends in Ecology & Evolution, 30*, 448–455.
Tollrian, R. (1990). Predator-induced helmet formation in *Daphnia cucullata* (SARS). *Archiv für Hydrobiologie, 119*(2), 191–196.
Urban, M. C. (2015). Accelerating extinction risk from climate change. *Science, 348*(6234), 571–573.
Waser, N. M., Chittka, L., Price, M. V., Williams, N. M., & Ollerton, J. (1996). Generalization in pollination systems, and why it matters. *Ecology, 77*(4), 1043–1060.
Watson, R., & Pauly, D. (2001). Systematic distortions in world fisheries catch trends. *Nature, 414*(6863), 534–536.

West-Eberhard, M. J. (2003). *Developmental plasticity and evolution*. Oxford University Press.
Whiten, A., Goodall, J., McGrew, W. C., Nishida, T., Reynolds, V., Sugiyama, Y., et al. (1999). Cultures in chimpanzees. *Nature, 399*(6737), 682–685.
Zheng, W., Fan, X., Chen, H., Ye, M., Yin, C., Wu, C., & Liang, Y. (2024). The response patterns of r- and K-strategist bacteria to long-term organic and inorganic fertilization regimes within the microbial food web are closely linked to rice production. *Science of the Total Environment, 942*, 173681. https://doi.org/10.1016/j.scitotenv.2024.173681. Epub 2024 Jun 4.
Zylinski, S., & How, M. J. (2012). Dynamic camouflage in cephalopods. *BioinspirBiomim, 7*(2), 025006.

Nature-Based Solutions for Climate Change Mitigation

Priya Dubey and Aditi Roy

Abstract Climate change is profoundly impacting both natural and urban environments, disrupting ecosystems and threatening their stability. Implementing nature-based solutions (NBS) in cities offers a promising approach to mitigating these effects while enhancing adaptation strategies. In addition to climate resilience, NBS contributes to ecosystem service provision, improving urban sustainability and liability. To effectively assess the impact of NBS on climate change mitigation and adaptation, Key Performance Indicators (KPIs) can serve as valuable tools for quantifying benefits to both humans and the environment. This chapter presents a comprehensive literature review aimed at identifying relevant KPIs for evaluating climate adaptation and mitigation strategies and exploring the role of NBS in this context. The review follows a structured methodology, beginning with the identification of major climate change drivers and their associated effects. It then establishes a set of KPIs to measure performance against specific climate impacts. Furthermore, the study links these KPIs to relevant Sustainable Development Goals (SDGs) and Ecosystem Services (ES), ensuring a holistic evaluation framework. Finally, potential NBS interventions are identified at multiple scales, from individual buildings to district and city-wide applications, highlighting their effectiveness in addressing climate challenges.

Findings indicate that NBS play a crucial role in climate adaptation, particularly at urban and district levels. Moreover, KPIs are widely integrated into environmental assessment frameworks, reinforcing their applicability in measuring NBS effectiveness. This study underscores the potential of NBS as a sustainable, multifunctional approach to tackling climate change in urban environments.

Keyword Mitigation · Climate change · Natural Climate Solutions (NCS) · Climate Resilience

---

P. Dubey (✉)
Department of Biosciences, Integral University, Lucknow, India

A. Roy
Department of Botany, University of Lucknow, Lucknow, India

S. Siddiqui et al. (eds.), *Frontiers in Sustainability*,
https://doi.org/10.1007/978-3-032-19740-5_8

1 Introduction

As urban populations expand and climate change accelerates, cities face growing risks from climate-related hazards. These hazards, intensified by urban environments, threaten human well-being and demand urgent adaptation and mitigation efforts (Osaka et al., 2021). City residents must either reduce their exposure and mitigate harmful effects or, in extreme cases, relocate. However, cities also offer opportunities for climate adaptation, as they concentrate essential services, institutions, and resources. To enhance urban resilience, policymakers and planners are increasingly turning to nature-based solutions (NbS), which restore or mimic natural ecosystems to mitigate climate risks (Chausson et al., 2020). NbS provide adaptable and sustainable alternatives to conventional engineering approaches, incorporating strategies such as expanding green spaces, planting trees, and developing infrastructure that supports natural water management, including stormwater ponds, bioswales, green roofs, and riparian buffers. Additionally, restoring coastal habitats strengthens urban resilience, reinforcing NbS as vital tools for sustainable urban development (Chausson et al., 2020).

In recent years, various stakeholders have introduced innovative climate change solutions, particularly those addressing human-induced impacts. Among these, NbS and natural climate solutions (NCS) have gained prominence for leveraging ecosystems in conservation, restoration, and sustainable resource management (Osaka et al., 2021). Research suggests that NbS could contribute up to 30% of cost-effective carbon reductions necessary to limit global warming below 2 °C (Griscom et al., 2017). Furthermore, an analysis of national climate commitments under the Paris Agreement reveals that nearly two-thirds of participating nations have incorporated NbS into their policies (Seddon et al., 2020). Although NbS and NCS have gained attention relatively recently (Seddon et al., 2020), the idea of using ecosystems for climate mitigation and adaptation has historical roots. Indigenous and traditional knowledge systems have long acknowledged these benefits. In modern discourse, the concept of "ecosystem services" emerged in the 1980s (Osaka et al., 2021) and gained wider recognition in the 1990s and early 2000s (Lele et al., 2013). The United Nations' Millennium Ecosystem Assessment highlighted the role of ecosystems in providing essential services like clean water, food security, and disaster risk reduction (Millennium Ecosystem Assessment, M. E. A., 2005).

By the late 2000s, the term "nature-based solutions" began appearing in environmental policy and management discussions (Nesshöver et al., 2017). This shift moved beyond viewing ecosystems as passive service providers to recognizing their active role in addressing environmental and societal challenges (Cohen-Shacham et al., 2016). Major environmental organizations, including the International Union for Conservation of Nature (IUCN), the World Wildlife Fund (WWF), and the World Bank, have promoted NbS as sustainable alternatives to traditional engineering-based solutions (Cohen-Shacham et al., 2016). While NbS broadly tackle climate change, food security, and disaster resilience, the term "natural climate solutions" (NCS) emerged in the late 2010s, emphasizing emissions reduction in alignment with the Paris Agreement goals (Fargione et al., 2018; Griscom et al., 2017). Despite their differences, both NbS and NCS benefit from being framed as "natural." Studies indicate that policies perceived as "natural" receive stronger public support, whereas artificial interventions often face skepticism (Osaka et al., 2021).

Some of the key indicators for evaluating the effectiveness of NbS are categorized into four main aspects: Integrated Environmental Performance, Health and Well-being, Citizen Involvement, and Transferability and Monitoring (Fig. 1). Integrated environmental performance considers factors such as environmental regulation, biodiversity, and the revitalization of degraded areas, while also accounting for potential negative ecosystem impacts. Health and well-being indicators assess the influence of NbS on physical and mental health, as well as their impact on quality of life, happiness, and employment opportunities. Citizen involvement is crucial, encompassing engagement in green projects, a sense of ownership, and the adoption of NbS within communities. Lastly, transferability and monitoring focus on governance, long-term viability, budget considerations, and the ability to replicate successful initiatives elsewhere. These indicators collectively provide a comprehensive framework for assessing the effectiveness and sustainability of nature-based solutions. The KPIs in Fig. 1 were identified through the structured literature review (Sect. 2.5) by cross-referencing commonly cited indicators in NbS assessment frameworks with SDG targets and ecosystem service categories. Their relevance was determined based on three criteria: (a) measurability using existing data or monitoring tools (e.g., air quality sensors, remote sensing, citizen science), (b) applicability across multiple NbS types and hazard contexts, and (c) policy alignment with urban resilience and sustainability goals. Operational usefulness was

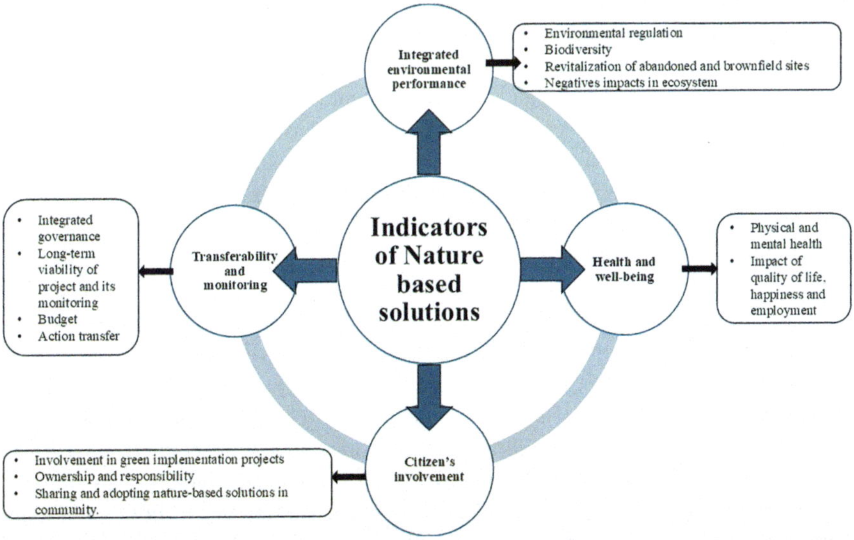

Fig. 1 Possible indicators for evaluating the effectiveness of nature-based solutions (NbS) in climate change adaptation and mitigation, along with their associated co-benefits

evaluated by examining case studies where KPIs such as surface temperature reduction, water retention capacity, or biodiversity indices demonstrated quantifiable benefits against specific hazards.

2 Exposure and Intensification of Climate Change Hazards in Urban Areas

Urban areas are particularly susceptible to climate change due to their high population densities and extensive infrastructure, often located in vulnerable regions. Cities exhibit characteristics that can intensify climate-related hazards, such as widespread impervious surfaces that worsen heat accumulation and water management issues. The urban populations will face significant impacts from climate change, including rising sea levels, increasing temperatures, reduced precipitation leading to water scarcity, and a growing frequency of extreme weather events such as heavy rainfall, heat waves, coastal storms, flooding, and droughts (Hobbie & Grimm, 2020). Among these, extreme weather events are anticipated to have the most immediate and widespread consequences.

2.1 Rising Sea Levels and Coastal Storms

The rate of sea-level rise is accelerating, with projections estimating an increase of 0.3–1.3 m by the end of the century (Holden et al., 2022). Many major cities are situated in low-lying coastal areas, making them increasingly vulnerable to flooding, storm surges, coastal erosion, and saltwater intrusion (Bosserelle et al., 2022). For example, in nine Asian megadeltas, significant populations live at elevations of less than 10 m above sea level, with rapid urban expansion further amplifying these risks. Some regions, such as South Florida in the United States, actively regulate groundwater levels to prevent saltwater intrusion into freshwater reserves (Czajkowski et al., 2018). However, rising groundwater reduces soil storage capacity, heightening the risk of surface flooding and potentially necessitating the relocation of communities from highly exposed coastal areas. Coastal cities also face growing threats from intensifying storms. Rising sea levels contribute to higher storm surges, increasing the likelihood of catastrophic flooding (Rosenzweig & Solecki, 2014). A notable example is Superstorm Sandy in 2012, which devastated New York City when storm surges coincided with high tide, leading to widespread inundation and severe erosion along the coastline. In regions such as East Asia's Pearl River Delta, Tokyo, and Manila, tropical cyclones pose an even greater risk, with strong winds and heavy rainfall causing extensive damage to both natural and built environments (Sundermann et al., 2013). Even inland areas can suffer from extreme storms, which can disrupt infrastructure and worsen air pollution by dispersing harmful pollutants, exacerbating health concerns.

2.2 *Extreme Heat and the Urban Heat Island Effect*

Since 1880, global temperatures have risen by approximately 0.8 °C and are expected to continue increasing (Arnfield, 2003). Urban areas experience an amplified warming effect due to the urban heat island (UHI) phenomenon, where cities are significantly warmer than their surrounding rural areas (Arnfield, 2003). In the United States, for instance, cities are heating up 1.5 times faster than rural regions due to factors such as reduced vegetation, heat retention by buildings, restricted airflow, and excess heat from human activities. Rising temperatures pose several challenges to urban environments. Extreme heat can hinder human activity and infrastructure functionality, particularly in tropical and subtropical cities. The UHI effect, along with higher nighttime temperatures and prolonged warm seasons, increases the risk of heat stress for urban populations (Zhou et al., 2014). Additionally, energy demand spikes due to the heightened reliance on cooling systems, while drought conditions intensify water shortages. Air quality also deteriorates as high temperatures exacerbate pollution and prolong thermal inversion events. Heat waves are projected to become more frequent and severe, particularly in hot and humid climates (Zhou et al., 2014; Oleson et al., 2015). The risk of heat stress is influenced by factors such as temperature, humidity, and individual physiological responses. Studies suggest that even moderate temperatures in humid conditions can be lethal (Epstein & Moran, 2006). For example, by 2100, Jakarta, Indonesia, could face between 117 and 365 days per year of potentially fatal heat and humidity levels, compared to 9 to 50 such days annually in New York City, depending on emission scenarios (Mora et al., 2017).

2.3 *Water Security, Inland Storms, and Urban Flooding*

Unlike temperature trends, the effects of climate change on water availability will vary significantly by region. Many urban centers depend on distant watersheds or slowly replenishing groundwater for their water supply, making them particularly vulnerable to shifts in precipitation patterns (Barron et al., 2013). Climate change is expected to alter not only the overall volume of rainfall but also its seasonal distribution and intensity, leading to significant fluctuations in water accessibility. Drought conditions in cities are aggravated by increasing water demand for purposes such as landscape irrigation and cooling. Among all natural disasters, riverine flooding impacts the largest number of people globally, with approximately 379 million individuals exposed to flood risks (Sundermann et al., 2013). Given that many cities are built near rivers, the threat of flooding and erosion remains a major concern. Historically, cities have relied on engineered flood defenses, such as levees and channelization, which can create a false sense of security as climate change shifts flood probabilities. Urbanization exacerbates flood risks through the proliferation of impervious surfaces, such as roads and buildings, which prevent water from infiltrating the ground. This leads to increased surface runoff and contributes to pluvial flooding—flooding that occurs when rainfall overwhelms drainage systems (Markolf et al., 2018). Unlike coastal or riverine flooding, pluvial floods can develop in locations far from water bodies and are notoriously difficult to predict. Existing flood risk maps often fail to account for this type of flooding, leaving many low-lying urban communities—especially those with limited adaptive capacity—highly exposed (Grimm et al., 2017).

2.4 Additional Climate-Related Hazards

Beyond extreme heat and water-related challenges, cities are also increasingly vulnerable to other climate-driven hazards, including wildfires, tornadoes, hurricanes, landslides, and even seismic activity, some of which may be worsened by climate change (Grimm et al., 2017). The way human settlements expand plays a critical role in determining exposure to these risks. While urban fires were once a major threat in the early twentieth century, modern building regulations have significantly reduced their occurrence. However, as more people settle in areas near forests and grasslands, wildfire risk is escalating (Radeloff et al., 2018). Climate change is expected to intensify wildfire activity, particularly in regions experiencing prolonged droughts. Apart from direct destruction of property and loss of life, wildfires severely impact urban air quality, increasing health risks for city residents. A striking example is the 2010 Russian heat wave, during which wildfires near Moscow led to hazardous air pollution levels, significantly contributing to heat-related deaths (Shaposhnikov et al., 2014).

2.5 Scientiometric Analysis

To provide a systematic and evidence-based understanding of NbS research in the context of climate change mitigation and adaptation, this chapter employed a scientometric analysis approach. Scientometrics is widely used to evaluate the structure, development, and trends of scientific knowledge by applying quantitative methods to bibliographic data. Bibliographic records were retrieved from Scopus for the period 2015–2025 to capture the emergence and evolution of NbS as a research field. This yielded 2438 publications, including research articles, reviews, and proceedings in the past decade. Keywords included: "Nature-based Solutions" OR "NbS" OR "Natural Climate Solutions" AND "climate change" OR "adaptation" OR "mitigation" AND "ecosystem services" OR "SDGs" AND "Key Performance

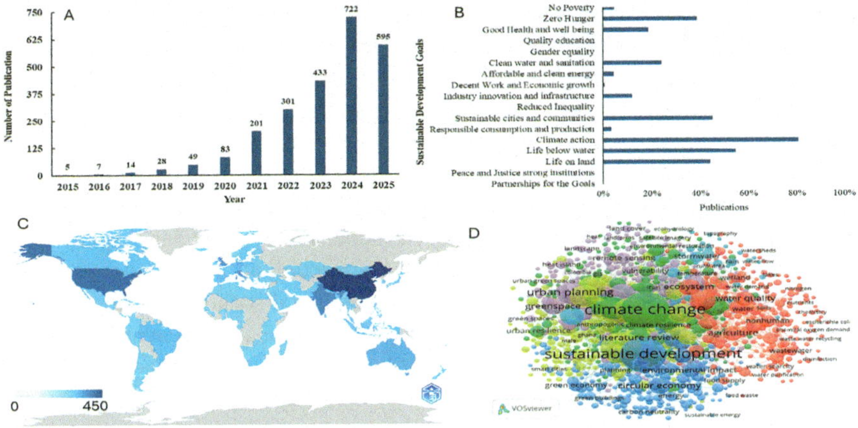

Fig. 2 Global research trends for NbS in the context of climate change mitigation and SDGs: (**a**) annual publication growth (2015–2025), (**b**) % of publications for SDGs, (**c**) geographic distribution of publications, and (**d**) keyword co-occurrence network highlighting thematic clusters in sustainability-related research

Indicators (KPI)". Records were included if they (a) were peer-reviewed journal articles, book chapters, or authoritative policy reports; (b) explicitly addressed NbS in relation to climate hazards, mitigation/adaptation strategies, ecosystem services, or KPIs; and (c) were published in English. Grey literature, conference abstracts, and non-English records were excluded to ensure quality and comparability. The bibliographic data were exported in CSV format and analyzed using the bibliometrix package in R and its web-based application Biblioshiny. These tools enabled the conversion of bibliographic data into structured datasets and facilitated quantitative analyses (Fig. 2).

3 Climate Change Impacts on Social, Ecological, and Technical Components in Cities

3.1 Social Impacts

Urban populations are increasingly vulnerable to extreme temperatures, exacerbated by the urban heat island (UHI) effect (Giridharan & Emmanuel, 2018). These rising temperatures contribute to various heat-related health issues, including heatstroke, dehydration, heat exhaustion, and increased mortality rates. The most at-risk groups include the elderly, children, socially isolated individuals, outdoor workers, those experiencing homelessness, and people without access to air conditioning. Additionally, lower-income communities are disproportionately affected due to reduced access to cooling infrastructure and limited green spaces that provide natural temperature regulation (Hondula et al., 2015). Heavy rainfall and subsequent urban flooding pose further health risks. Runoff from rainfall can carry heavy metals, nutrients, salts, and pathogenic bacteria into stormwater systems, leading to high pollution levels downstream. This contamination threatens both public health and aquatic ecosystems. In some cities, combined sewer overflows (CSOs) occur when excessive rainwater overwhelms sewer systems, resulting in untreated sewage being discharged into surface waters (Kjellstrom et al., 2016). This issue is expected to worsen with climate change, heightening health and sanitation concerns. Beyond immediate physical risks, climate change also contributes to human displacement. It can directly cause injury, mortality, and the destruction of homes, but it also indirectly drives migration through economic, political, and social disruptions (Pandit et al., 2016). Low- and middle-income countries will bear the heaviest burden of these changes, as many urban residents in these regions already face heightened risks from sea-level rise, coastal storms, and riverine flooding. Informal settlements, which often lack resilient infrastructure, clean water, and reliable healthcare, are particularly susceptible to these hazards. In the absence of significant adaptation efforts, extreme heat will render some regions uninhabitable. Even if global warming is limited to 1.5 °C above preindustrial levels, nearly 40% of megacities, primarily in Africa and Asia, will experience dangerously high heat indices annually (Hall et al., 2017). At 4 °C of warming, nearly 80% of the world's largest cities will face lethal heat conditions. Mental health is another significant concern linked to climate change-related hazards. Weather-related disasters can lead to post-traumatic stress disorder (PTSD), depression, and anxiety. Prolonged exposure to extreme heat has been associated with increased aggression, crime rates, suicides, mood disorders, and cognitive decline, including dementia. Disruptions to infrastructure, livelihoods, and forced migration further strain social structures, reducing the adaptive capacity of affected communities (McMichael et al., 2010).

3.2 *Ecological Impacts*

Urban ecosystems are particularly vulnerable to climate change due to their interaction with other stressors such as pollution, habitat fragmentation, and the spread of pests and pathogens. Rapid urban expansion is expected to reshape regional biodiversity, with plant diversity being especially sensitive to shifting climatic conditions (Seto et al., 2012). Studies suggest that cities in warmer regions tend to have lower plant species richness, whereas milder winter climates may promote the growth of spontaneous species, including trees in North America. The combined effects of UHI, restricted rooting space, compacted soils, and contamination from heavy metals and chemicals will intensify heat- and drought-related stress on urban plant life (Zhou et al., 2014). Areas with high levels of impervious surfaces and elevated temperatures will see the greatest impact (Ordóñez & Duinker, 2015). Additionally, climate change is expected to facilitate the expansion of pests and pathogens, which can further threaten urban plant communities. Many tree species currently planted in cities may not be well-suited to future climate conditions, leading to reductions in urban tree cover. The nursery trade also increases the risk of introducing novel pests and diseases, while the low diversity of urban forests can accelerate their spread. Severe storms pose another threat to urban trees, as their root systems are often inadequate to withstand strong winds. Urban soils, which tend to be compacted, waterlogged, or drought-stressed, further compromise tree stability. Damage from construction, mowing, and trenching exacerbates these vulnerabilities, increasing the risk of tree loss during extreme weather events (Brandt et al., 2016). Climate change will also impact urban water bodies. Rising water temperatures, combined with high nutrient loads, will encourage harmful algal blooms and cyanobacteria growth in lakes and ponds (Paerl et al., 2016). These conditions will extend water stratification periods, reducing oxygen levels and leading to hypoxic events that can devastate aquatic life. Additionally, reduced water circulation can trigger anaerobic processes, releasing more nutrients from sediments and further accelerating eutrophication (Ladwig et al., 2018). More stable and warmer surface waters will promote harmful cyanobacteria, increasing the risk of toxin exposure for both humans and animals. Urban streams, many of which have been heavily modified through channelization and burial, will also be adversely affected by climate change. These waterways are already struggling with increased temperatures, higher peak flows during extreme weather events, and greater pollution loads from stormwater and CSOs. Combined with urbanization, climate change will continue to degrade stream ecosystems, threatening fish populations and biodiversity (Nelson et al., 2013).

3.3 Technical Impacts

Infrastructure systems in cities are designed to provide essential services such as electricity, water supply, transportation, and waste management. Additionally, protective infrastructure such as stormwater systems, seawalls, and climate-controlled buildings is crucial for safeguarding urban populations (Chester et al., 2019). However, climate change is increasingly challenging the resilience of these systems due to factors such as aging infrastructure, inadequate design standards, and the increasing frequency and intensity of extreme weather events (Gilroy & McCuen, 2012). Urban infrastructure is particularly susceptible to climate-related hazards due to its high density and interconnected nature. Storms that impact cities tend to cause more damage compared to rural areas because of the concentration of infrastructure. Moreover, critical systems such as power lines, sewer networks, and water supply systems are often located along public roads and near water bodies, making them more vulnerable to flooding. When multiple infrastructure systems fail simultaneously, the consequences can be severe. For instance, in some cities, stormwater and wastewater treatment systems are integrated. During heavy rainfall, these systems can become overwhelmed, leading to CSOs and the discharge of untreated pollutants into water bodies. Infrastructure designed for historical climate conditions is increasingly failing in the face of more frequent and intense weather events. Many systems were built to withstand specific risk thresholds based on past climate patterns, such as the widely used 1% probability standard for extreme events (Kim et al., 2020). However, this approach is no longer adequate in an era of climate uncertainty, where past probabilities do not necessarily predict future risks. To address these challenges, recent efforts to incorporate nonstationary flood-frequency analysis into infrastructure planning are showing promise for developing more resilient urban systems.

4 Nature-Based Solutions for Climate Change Mitigation

Nature-Based Solutions (NbS) are increasingly recognized as essential strategies for climate change mitigation, biodiversity conservation, and sustainable resource management (Fig. 3). While significant progress has been made, future advancements in research, policy, and governance are necessary to maximize their impact. Scientific research should focus on long-term carbon sequestration potential, improved monitoring and verification technologies, and synergies with technological innovations. Scaling up NbS requires integrating them into national climate policies, expanding financial mechanisms such as green bonds and carbon markets, and fostering multi-stakeholder collaborations. However, challenges such as land-use conflicts, policy fragmentation, and inadequate funding must be addressed through clear land tenure rights, evidence-based decision-making, and capacity-building initiatives. The future of NbS in climate action depends on continuous

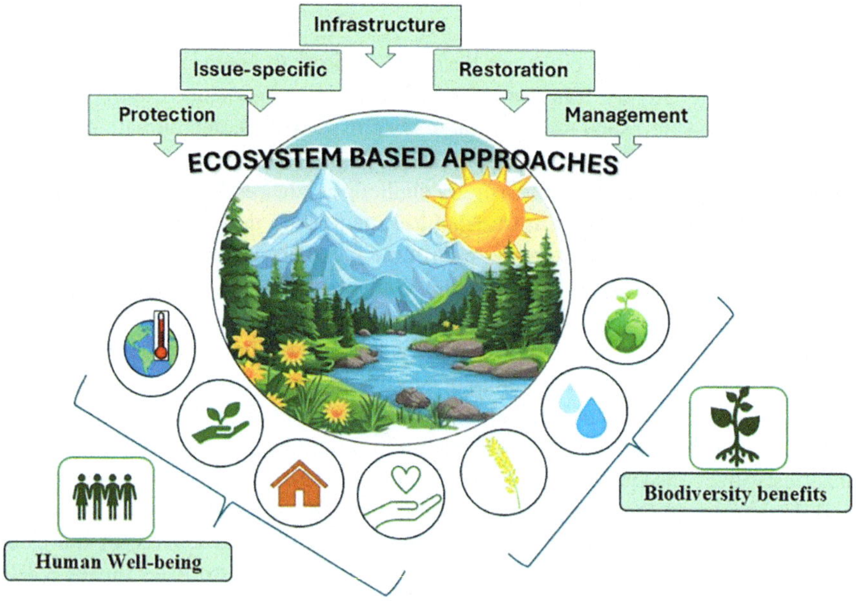

Fig. 3 Nature-based solutions chart outlined by IUCN

scientific innovation, stronger policy commitments, and adaptive governance frameworks. As research trends indicate growing academic interest, ensuring the long-term success of NbS will require a concerted effort from governments, institutions, and communities to integrate them into mainstream climate strategies for a sustainable and resilient future.

4.1 Forest-Based Solutions

Forest-based solutions are vital for climate change mitigation, particularly through carbon sequestration, biodiversity conservation, and ecosystem restoration. Avoided deforestation, which prevents the loss of existing forests, is one of the most cost-effective strategies to reduce greenhouse gas emissions (Busch & Engelmann, 2017). Deforestation contributes to nearly 10% of global CO_2 emissions, making forest conservation a critical component of climate action (IPCC, 2019). The UN's Reducing Emissions from Deforestation and Forest Degradation (REDD+) initiative supports developing countries in conserving and sustainably managing forests while providing financial incentives for carbon storage and emission reductions (Angelsen et al., 2018). In addition to reducing emissions, REDD+ programs contribute to biodiversity conservation, livelihood improvement, and climate resilience (Seymour & Busch, 2016). However, the success of REDD+ and avoided deforestation efforts depends on effective governance, land tenure security, and equitable benefit-sharing mechanisms (Nepstad et al., 2014). Integrating these strategies with afforestation, reforestation, and sustainable forest management can enhance their long-term impact on climate change mitigation.

4.2 Wetlands and Peatlands Conservation

Wetlands and peatlands play a crucial role in climate change mitigation by serving as long-term carbon sinks, storing vast amounts of organic carbon in their waterlogged soils. Peatlands, which cover only about 3% of the Earth's land surface, store nearly twice as much carbon as all the world's forests combined, making their conservation vital for preventing carbon release into the atmosphere (Leifeld & Menichetti, 2018). Wetlands also support biodiversity, regulate hydrological cycles, and act as natural buffers against extreme weather events such as floods and droughts (Davidson, 2014). However, drainage, land-use change, and agricultural expansion threaten these ecosystems, leading to significant carbon emissions. Restoration efforts, such as rewetting drained peatlands, preventing land conversion, and implementing sustainable management practices, can enhance their carbon sequestration potential while improving ecosystem resilience (Griscom et al., 2017). International initiatives, including the Ramsar Convention and the Global Peatlands Initiative, aim to promote wetland and peatland protection through policy frameworks, financial incentives, and community engagement (Joosten et al., 2012). Strengthening these conservation efforts is essential for maintaining the carbon storage capacity of wetlands and peatlands and mitigating global climate change.

4.3 Coastal and Marine Ecosystems

Coastal and marine ecosystems play a significant role in climate change mitigation by acting as highly efficient carbon sinks. Mangroves, seagrasses, and coral reefs are particularly important for carbon sequestration, collectively storing vast amounts of atmospheric CO_2 in their biomass and sediments. Mangroves, found along tropical and subtropical coastlines, can sequester carbon at rates up to four times higher than terrestrial forests due to their dense root systems that trap organic material in anaerobic soils (Donato et al., 2011). Seagrass meadows, covering less than 0.2% of the ocean floor, store approximately 10–18% of the ocean's total carbon in their sediments (Fourqurean et al., 2012). Coral reefs, while not direct carbon sinks, contribute to climate regulation by supporting biodiversity, protecting coastlines from erosion, and promoting carbon cycling within marine ecosystems (Woodhead et al., 2019). However, these ecosystems are increasingly threatened by human activities, such as deforestation, coastal development, and ocean acidification, making their conservation critical for long-term carbon storage and climate resilience.

Blue carbon initiatives focus on the protection, restoration, and sustainable management of coastal and marine ecosystems to enhance their carbon sequestration capacity and mitigate climate change. The Blue Carbon Initiative, led by Conservation International, IUCN, and UNESCO, promotes scientific research, policy development, and community-driven conservation efforts to safeguard blue carbon ecosystems (Herr & Landis, 2016). International agreements, such as the Paris Agreement, recognize the role of blue carbon, encouraging nations to include coastal ecosystem conservation in their Nationally Determined Contributions (NDCs) (Macreadie et al., 2019). Efforts such as mangrove afforestation, seagrass meadow restoration, and the designation of marine protected areas (MPAs) help to enhance blue carbon stocks while providing co-benefits such as biodiversity protection and improved livelihoods for coastal communities. Strengthening governance frameworks, securing financial incentives for conservation, and integrating traditional knowledge are crucial for ensuring the long-term success of blue carbon initiatives.

4.4 Regenerative Agriculture and Soil Health

Regenerative agriculture focuses on restoring soil health, enhancing biodiversity, and improving ecosystem resilience while sequestering atmospheric carbon in soils. Agroforestry, a key component of regenerative agriculture, integrates trees and shrubs into farming systems, improving soil structure, reducing erosion, and increasing organic carbon storage in the soil (Griscom et al., 2017). Sustainable farming practices, such as cover cropping, crop rotation, reduced tillage, and compost application, enhance soil microbial activity, boost nutrient cycling, and increase soil organic matter, which enhances carbon sequestration and long-term agricultural productivity (Lal, 2020). Healthy soils act as significant carbon sinks, with the potential to store up to 2–3 billion tons of CO_2 annually through improved land management practices (Paustian et al., 2016). By reducing reliance on chemical fertilizers and pesticides, regenerative agriculture also minimizes greenhouse gas emissions and promotes ecosystem services such as water retention and habitat restoration. Scaling up these practices requires supportive policies, financial incentives for farmers, and increased research on region-specific approaches to soil carbon sequestration.

4.5 Urban Green Infrastructure

Urban green infrastructure plays a crucial role in enhancing climate resilience, improving air quality, and mitigating the urban heat island effect. Green roofs, which involve vegetation-covered rooftops, provide insulation, reduce stormwater runoff, and lower urban temperatures by absorbing heat and carbon dioxide (Oberndorfer et al., 2007). Urban forests, consisting of trees and green spaces integrated into city landscapes, sequester carbon, filter air pollutants, and provide shade, thereby reducing energy demand for cooling (Nowak et al., 2013). Public parks and green corridors contribute to biodiversity conservation, improve mental and physical well-being, and create natural cooling zones in densely populated areas. These nature-based solutions not only enhance urban aesthetics but also strengthen the resilience of cities against climate extremes such as heatwaves, heavy rainfall, and flooding.

Sustainable urban planning integrates climate adaptation and mitigation strategies to create low-carbon, resilient, and livable cities. Compact and mixed-use urban development reduces dependence on private vehicles, thereby lowering greenhouse gas emissions and promoting energy-efficient public transport systems (Newman & Kenworthy, 2015). The incorporation of pedestrian-friendly infrastructure, bike lanes, and electric vehicle charging stations supports sustainable mobility and reduces air pollution. Additionally, permeable surfaces, rain gardens, and urban wetlands improve stormwater management and mitigate the risk of urban flooding (Gill et al., 2007). By integrating green infrastructure and renewable energy solutions, sustainable urban planning enhances climate resilience while fostering economic growth and social well-being. Strong policy frameworks, community engagement, and investment in green technologies are essential for achieving climate-smart urban development.

In addition to these broad categories, specific forms of urban green infrastructure provide hazard-targeted benefits that can be monitored through measurable indicators while also advancing key SDGs. Green roofs mitigate the urban heat island effect by lowering rooftop and ambient temperatures (SDG 11: Sustainable Cities and Communities), while their capacity to retain stormwater reduces the burden on drainage systems during heavy rainfall (SDG 6: Clean Water and Sanitation). They also enhance biodiversity by creating habitats in dense urban cores (SDG 15: Life on Land). KPIs include reductions in surface temperature, percentage of stormwater retained, annual CO_2 absorbed per square meter, and species richness indices. Urban forests act as natural cooling systems, significantly reducing heat stress during extreme heat waves (SDG 3: Good Health and Well-being), while also filtering pollutants (SDG 11) and sequestering substantial amounts of carbon (SDG 13: Climate Action). Their role in storm protection and soil stabilization further enhances resilience (SDG 15). Relevant KPIs include canopy cover percentage, pollutant removal efficiency, annual carbon sequestration, and reductions in heat-related hospitalizations. Bioswales, shallow vegetated depressions designed to capture and filter runoff, address hazards linked to flooding and water pollution by improving infiltration and nutrient absorption (SDG 6) while contributing to drought resilience through groundwater recharge and localized cooling (SDG 13). KPIs for bioswales include infiltration rates (mm/hour), nutrient load reduction percentages, annual groundwater recharge volumes, and maintenance cost-effectiveness. Together, these targeted interventions illustrate how distinct NbS types reduce hazard exposure, deliver ecosystem services, and simultaneously advance multiple SDGs when evaluated through differentiated and quantifiable KPIs.

5 Co-Benefits of Nature-Based Solutions

NBS offer a holistic approach to addressing climate change, environmental degradation, and socioeconomic challenges while simultaneously providing a wide array of co-benefits (Fig. 4). These solutions, which include conservation, restoration, and sustainable management of natural ecosystems, not only contribute to climate resilience but also enhance biodiversity, improve water and air quality, and strengthen local economies and livelihoods. By integrating ecological processes into urban, rural, and coastal landscapes, NBS foster a sustainable relationship between humans and nature, ultimately supporting long-term environmental and economic stability (Cohen-Shacham et al., 2016).

One of the most significant co-benefits of nature-based solutions is their role in enhancing biodiversity. Ecosystems such as wetlands, forests, grasslands, and coral reefs are home to diverse species, many of which are threatened by habitat destruction, pollution, and climate change. Implementing NBS, such as reforestation, wetland restoration, and sustainable agriculture, helps protect and restore these ecosystems, providing vital habitats for wildlife (Seddon et al., 2020). For example, mangrove restoration projects along coastlines not only sequester carbon but also create breeding grounds for fish, crabs, and mollusks, supporting marine biodiversity (Alongi, 2012). Similarly, urban green spaces and green roofs contribute to

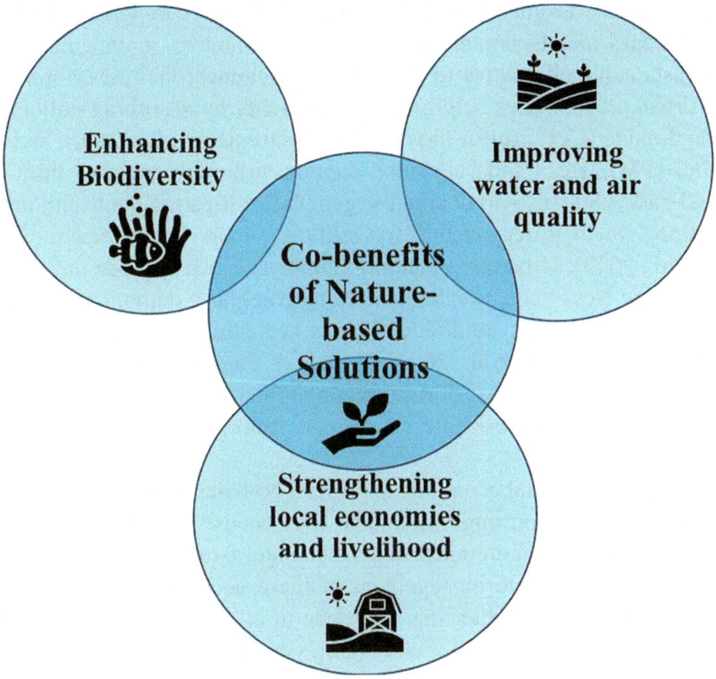

Fig. 4 Co-benefits of nature-based solutions

biodiversity by offering refuge for pollinators, birds, and small mammals in otherwise built environments. Moreover, biodiversity plays a crucial role in ecosystem services that benefit human well-being. Diverse ecosystems are more resilient to environmental changes and disruptions, as they provide genetic variation and ecological redundancy that help maintain ecosystem stability. Agroforestry, a nature-based approach that integrates trees with crops and livestock, enhances soil fertility, reduces erosion, and fosters habitat connectivity, which supports pollinators and natural pest controllers (Jose, 2009). By maintaining and restoring biodiversity, NBS contribute to ecosystem health, ensuring that nature continues to provide essential services such as food production, climate regulation, and disease control.

Nature-based solutions also play a crucial role in improving water and air quality, benefiting both ecosystems and human communities. Wetlands, forests, and green infrastructure act as natural filters that help remove pollutants from water bodies and air, reducing the adverse effects of human activities on the environment (Mitsch & Gosselink, 2015). Wetlands, including marshes, swamps, and peatlands, function as natural water purifiers by trapping sediments, absorbing excess nutrients, and breaking down contaminants through microbial processes. Restoring and protecting wetlands can significantly improve water quality in rivers and lakes by reducing nitrogen and phosphorus runoff from agriculture, which can otherwise lead to harmful algal blooms and oxygen depletion (Verhoeven et al., 2006). For instance, constructed wetlands have been successfully used in wastewater treatment to filter out heavy metals, organic pollutants, and pathogens, making them an effective and sustainable alternative to conventional treatment facilities. Similarly, forests and urban green spaces help improve air quality by absorbing pollutants such as carbon dioxide (CO_2), sulfur dioxide (SO_2), nitrogen oxides (NO_x), and particulate matter (PM). Trees and vegetation capture airborne pollutants through their leaves and bark while also producing oxygen, thus mitigating the health impacts of air pollution, which is responsible for millions of premature deaths worldwide (Nowak et al., 2013). Urban tree-planting initiatives, such as those implemented in major cities like New York, London, and Singapore, have demonstrated significant improvements in air quality and reductions in heat stress, contributing to healthier urban environments (Gill et al., 2007). Moreover, coastal mangroves and seagrass meadows act as buffers against storm surges and filter sediments, preventing coastal erosion and protecting freshwater resources from saltwater intrusion (Duarte et al., 2013).

NbS provide substantial economic and social co-benefits by generating employment opportunities, supporting sustainable livelihoods, and fostering ecotourism. Many rural and coastal communities depend on natural resources for their economic well-being, and sustainable management of these ecosystems ensures long-term economic stability while enhancing resilience to climate change (United Nations Environment Programme, 2019). Ecotourism, which relies on well-preserved natural environments, is a major economic driver in many regions, generating revenue and employment opportunities for local communities. National parks, marine protected areas, and restored wetlands attract tourists interested in wildlife observation, hiking, and water-based activities, providing an economic incentive for conservation efforts (Honey, 2008). For example, Costa Rica's investment in protected areas and reforestation has boosted ecotourism, contributing significantly to the country's GDP while conserving biodiversity and mitigating climate change. In addition to tourism, NbS promote sustainable agriculture and fisheries, benefiting farmers, fishers, and Indigenous communities. Agroforestry and regenerative agriculture improve soil fertility and increase crop yields, reducing dependence on chemical fertilizers and pesticides while enhancing food security (Pretty et al., 2006). Sustainable fisheries management, supported by marine protected areas and mangrove conservation, helps replenish fish stocks, ensuring a stable income for coastal communities. A successful example is the restoration of oyster reefs in the United States, which has led to improved water quality, increased fish populations, and revitalized local fishing economies (Beck et al., 2011).

Furthermore, NBS create job opportunities in conservation, restoration, and sustainable land management. Tree planting, wetland restoration, and green infrastructure development require skilled labor, contributing to employment generation, especially in regions where economic opportunities are limited. The European Union's Green Deal, for example, has invested in nature-based solutions to create green jobs, improve urban resilience, and promote sustainable development across member states (Fetting, 2020). Investing in NBS also reduces the economic burden associated with climate change-related disasters, such as floods, storms, and heatwaves. Natural infrastructure, such as mangroves and floodplains, provides cost-effective protection against extreme weather events, reducing the need for expensive artificial defenses like seawalls and drainage systems. Studies have shown that every dollar invested in ecosystem-based adaptation can yield multiple benefits, including reduced disaster recovery costs, improved public health, and enhanced economic stability (Seddon et al., 2020).

The co-benefits of nature-based solutions extend far beyond their primary role in climate mitigation and adaptation. By enhancing biodiversity, improving water and air quality, and strengthening local economies and livelihoods, NBS provide a multifaceted approach to sustainable development that balances environmental conservation with economic and social well-being. As governments, businesses, and communities increasingly recognize the value of working with nature rather than against it, investments in NBS are gaining momentum worldwide. However, scaling up these solutions requires integrated policies, financial incentives, and collaboration among stakeholders to ensure their long-term success. By prioritizing nature-based solutions, societies can build a more resilient, equitable, and sustainable future for both people and the planet.

5.1 Nexus among NbS Types, SDGs, Ecosystem Services, and KPIs

The integration of NbS with SDGs, ES, and KPIs provides a multidimensional framework for evaluating their effectiveness. NbS are not stand-alone interventions but systemic approaches that operate across ecological, social, and economic domains. By linking them to SDGs, it becomes possible to demonstrate how local interventions generate global sustainability benefits, while ecosystem services act as the tangible mechanisms through which those benefits flow, and KPIs provide the tools to measure and compare impacts in a standardized way. For example, green roofs directly mitigate the urban heat island effect and manage pluvial flooding, thus addressing SDGs 11 (Sustainable Cities), 13 (Climate Action), and 3 (Good Health). Their associated KPIs include rooftop temperature reduction, stormwater retention capacity, and biodiversity indices, which can be monitored to show contributions to regulating and supporting services. Urban forests provide shading, improve air quality, sequester carbon, and enhance biodiversity, aligning with SDGs 3, 11, 13, and 15 (Life on Land). Their ecosystem services include climate regulation and air purification, measurable through KPIs such as canopy cover percentage, pollutant removal rates, and CO_2 sequestration per hectare. Bioswales reduce flooding risk and filter pollutants, supporting SDG 6 (Clean Water) and SDG 15 through regulating and supporting services, with KPIs such as infiltration rates, nutrient removal efficiency, and groundwater recharge volume. Wetlands, peatlands, and coastal ecosystems serve as large-scale carbon sinks while buffering against floods and storms, contributing to SDGs 13, 14 (Life Below Water), and 15. Their KPIs include soil organic carbon stocks, flood damage avoided (economic valuation), and fishery productivity, linking regulating and provisioning services to global targets. Regenerative agriculture integrates food production with ecosystem restoration, improving soil health, conserving biodiversity, and reducing emissions, thereby advancing SDGs 2 (Zero Hunger), 13, and 15. KPIs here include increases in soil organic carbon, crop yield stability under climatic stress, and reductions in fertilizer use. By explicitly mapping NbS types to ES categories—provisioning, regulating, cultural, and supporting—while aligning with SDGs and assigning quantifiable KPIs, policymakers and practitioners gain an evidence-based decision-making framework. This nexus allows trade-offs and synergies to be identified, ensures accountability and comparability across contexts, and highlights the multifunctionality of NbS. Ultimately, the NbS-SDG-ES-KPI nexus strengthens the case for mainstreaming NbS into urban planning, national climate policies, and global sustainability strategies, ensuring that investments in natural systems translate into measurable resilience, equity, and long-term ecological integrity.

6 Challenges and Limitations

NbS offer promising approaches to address environmental challenges, particularly climate change, by utilizing natural systems to enhance biodiversity, mitigate carbon emissions, and support ecosystem resilience. However, the implementation of NbS is often hindered by several challenges and limitations. One significant obstacle is land use conflicts, which arise when NbS interventions, such as afforestation, wetland restoration, or sustainable land management practices, conflict with existing land uses like agriculture, urbanization, or industrial development (Seddon et al., 2020). Landholders, agricultural producers, and local communities may resist NbS initiatives that reduce the availability of land for their livelihoods or limit access to natural resources. This conflict can be exacerbated by the differing priorities of various stakeholders, including government agencies, environmental groups, and private sector actors, each with competing interests in land management.

Governance issues also play a critical role in the challenges facing NbS. The successful implementation of NbS requires effective coordination among multiple sectors, levels of government, and local communities (Snep et al., 2020). However, governance structures often fail to adequately address the complex, multidimensional nature of NbS, leading to fragmented decision-making, insufficient policy integration, and delays in implementation. Moreover, the lack of clear and inclusive governance frameworks that involve local knowledge and community participation can undermine the long-term sustainability of NbS projects (Chong, 2014). In some cases, top-down governance approaches may overlook the needs and aspirations of vulnerable populations, further complicating the adoption of NbS.

Monitoring and measuring the effectiveness of carbon sequestration in NbS is another major challenge. The ability to accurately assess the carbon storage capacity of ecosystems and track long-term carbon sequestration is limited by the variability in natural systems and the complexity of ecosystem processes (Chazdon & Uriarte, 2016). Different ecosystems—such as forests, wetlands, and grasslands—sequester carbon at varying rates, and factors like soil type, climate, and land management practices influence these rates. Furthermore, current carbon accounting methodologies lack standardization, making it difficult to compare the carbon sequestration potential of different NbS approaches across regions. This lack of consistency in monitoring and reporting can undermine the credibility of NbS as a viable climate mitigation strategy and limit their effectiveness in carbon markets.

Beyond governance, funding, and monitoring issues, two additional challenges merit deeper attention: cost of maintenance and economic viability, and long-term performance under climate change. While NbS are often promoted as cost-effective compared to gray infrastructure, their long-term maintenance costs can be substantial. For example, green roofs and bioswales require periodic upkeep, irrigation, and skilled labor, while urban forests require pest control and pruning. Insufficient funding or community disengagement can lead to performance decline over time, undermining climate resilience objectives. Also, the effectiveness of NbS may decrease as climatic conditions shift. For instance, prolonged droughts may reduce wetland water retention, pests may damage urban forests, and sea-level rise may compromise coastal mangroves. Ensuring resilience of NbS under uncertain future scenarios requires adaptive management strategies, long-term monitoring, and incorporation of climate projections into design.

Finally, funding and policy barriers pose significant constraints to the widespread adoption of NbS. Despite the growing recognition of NbS as a cost-effective and sustainable alternative to conventional engineering solutions, financial support for these initiatives remains limited. Government policies often fail to provide sufficient incentives or funding mechanisms to scale up NbS, especially in developing countries where budgetary constraints and competing priorities exist (Stange et al., 2022). Additionally, the financial benefits of NbS, such as improved ecosystem services or carbon credits, are not always adequately valued or incentivized within existing policy frameworks. This leads to underinvestment in NbS projects, even though they can provide significant long-term benefits for climate change mitigation and adaptation (Chong, 2014). While NbS offer great potential for addressing environmental and climate challenges, their successful implementation is fraught with challenges related to land use conflicts, governance issues, monitoring and measuring carbon sequestration, and funding and policy barriers. Overcoming these limitations requires coordinated efforts across sectors, improved methodologies for carbon accounting, and a supportive policy environment that values the full range of benefits provided by NbS.

7 Conclusion and Future Directions

This chapter demonstrates that NbS are not only critical for climate change mitigation and adaptation but also central to delivering co-benefits for biodiversity, human health, and sustainable urban development. Through a structured review, the analysis identified how different NbS types—such as green roofs, urban forests, wetlands, bioswales, mangroves, and regenerative agriculture—address a spectrum of climate hazards including urban heat islands, flooding, storm surges, droughts, and air pollution. When mapped against SDGs and ES, these solutions reveal their multifunctionality: for example, urban forests reduce particulate pollution and heat stress (SDG 3: Good Health), while wetlands improve water regulation and carbon storage (SDG 6: Clean Water, SDG 13: Climate Action). Key Performance Indicators (KPIs) such as surface temperature reduction, infiltration rates, and carbon sequestration metrics provide measurable evidence of these contributions and enable policymakers to evaluate NbS effectiveness in a standardized way.

However, while NbS offer cost-effective and sustainable alternatives to gray infrastructure, their implementation is constrained by multiple challenges. Land-use conflicts often arise between NbS initiatives and competing urban, agricultural, or industrial demands. Governance gaps and fragmented decision-making limit coordination across sectors and scales. Economic barriers—particularly the high upfront investment and ongoing maintenance costs of interventions like green roofs and bioswales—threaten long-term viability, especially in resource-limited contexts. Technical challenges persist in monitoring and verifying carbon sequestration and other ecosystem services, where variability among ecosystems complicates standardization. Moreover, the long-term performance of NbS remains uncertain under changing climate conditions, as prolonged droughts, pest outbreaks, and sea-level rise may reduce their resilience. Without addressing these structural and ecological limitations, NbS risk underperforming relative to their potential.

Looking forward, several pathways can strengthen the role of NbS in global climate strategies. First, the development of standardized KPI frameworks—aligned with SDGs and ecosystem services—would allow comparability across projects and regions, providing an evidence base for scaling up investment. Second, integration of technological innovations such as satellite monitoring, remote sensing, AI-based modeling, and citizen science can significantly enhance the tracking of NbS performance in real time. Third, financing mechanisms including green bonds, carbon markets, and payment for ecosystem services must be expanded to ensure economic sustainability. Fourth, adaptive governance frameworks that emphasize participatory planning, equity, and the inclusion of local and Indigenous knowledge can increase community ownership and long-term success. Finally, embedding NbS into mainstream climate policy and urban planning is essential to ensure they are not treated as isolated projects but as core strategies for resilient and sustainable development. In conclusion, NbS should be viewed not merely as complementary alternatives to gray infrastructure but as transformative pathways that integrate ecological processes with human well-being. Their capacity to mitigate climate hazards, enhance biodiversity, and deliver socioeconomic co-benefits positions them as vital instruments for building climate-resilient societies. Realizing this potential requires overcoming current barriers through coordinated action, innovative financing, robust science, and strong policy commitment. By placing nature at the heart of climate solutions, NbS can drive the transition toward sustainable, equitable, and resilient futures.

References

Alongi, D. M. (2012). Carbon sequestration in mangrove forests. *Carbon Management, 3*(3), 313–322. https://doi.org/10.4155/cmt.12.20

Angelsen, A., Martius, C., De Sy, V., Duchelle, A. E., Larson, A. M., & Pham, T. T. (2018). *Transforming REDD+: Lessons and new directions*. CIFOR.

Arnfield, A. J. (2003). Two decades of urban climate research: A review of turbulence, exchanges of energy and water, and the urban heat Island. *International Journal of Climatology: A Journal of the Royal Meteorological Society, 23*(1), 1–26. https://doi.org/10.1002/joc.859

Barron, O. V., Barr, A. D., & Donn, M. J. (2013). Effect of urbanisation on the water balance of a catchment with shallow groundwater. *Journal of Hydrology, 485*, 162–176. https://doi.org/10.1016/j.jhydrol.2012.04.027

Beck, M. W., Brumbaugh, R. D., Airoldi, L., Carranza, A., Coen, L. D., Crawford, C., et al. (2011). Oyster reefs at risk and recommendations for conservation, restoration, and management. *Bioscience, 61*(2), 107–116. https://doi.org/10.1525/bio.2011.61.2.5

Bosserelle, A. L., Morgan, L. K., & Hughes, M. W. (2022). Groundwater rise and associated flooding in coastal settlements due to sea-level rise: A review of processes and methods. *Earth's Future, 10*(7), e2021EF002580. https://doi.org/10.1029/2021EF002580

Brandt, L., Lewis, A. D., Fahey, R., Scott, L., Darling, L., & Swanston, C. (2016). A framework for adapting urban forests to climate change. *Environmental Science & Policy, 66*, 393–402. https://doi.org/10.1016/j.envsci.2016.06.005

Busch, J., & Engelmann, J. (2017). Cost-effectiveness of reducing emissions from tropical deforestation, 2016–2050. *Environmental Research Letters, 13*(1), 015001. https://doi.org/10.1088/1748-9326/aa907c

Chausson, A., Turner, B., Seddon, D., Chabaneix, N., Girardin, C. A., Kapos, V., et al. (2020). Mapping the effectiveness of nature-based solutions for climate change adaptation. *Global Change Biology, 26*(11), 6134–6155. https://doi.org/10.1111/gcb.15310

Chazdon, R. L., & Uriarte, M. (2016). The role of natural regeneration in large-scale forest and landscape restoration: Challenge and opportunity. *Biotropica, 48*(6), 716–730. https://doi.org/10.1111/btp.12381

Chester, M. V., Markolf, S., & Allenby, B. (2019). Infrastructure and the environment in the Anthropocene. *Journal of Industrial Ecology, 23*(5), 1006–1015. https://doi.org/10.1111/jiec.12848

Chong, J. (2014). Ecosystem-based approaches to climate change adaptation: Progress and challenges. *International Environmental Agreements: Politics, Law and Economics, 14*, 391–405. https://doi.org/10.1007/s10784-014-9242-9

Cohen-Shacham, E., Walters, G., Janzen, C., & Maginnis, S. (2016). Nature-based solutions to address global societal challenges. IUCN: , 97(2016), 2036.

Curry-Lindahl, K. (2019). United Nations Environment Programme. In Earthcare: Global Protection of Natural Areas (pp. 740–753). Routledge.

Czajkowski, J., Engel, V., Martinez, C., Mirchi, A., Watkins, D., Sukop, M. C., & Hughes, J. D. (2018). Economic impacts of urban flooding in South Florida: Potential consequences of managing groundwater to prevent salt water intrusion. *Science of the Total Environment, 621*, 465–478. https://doi.org/10.1016/j.scitotenv.2017.10.251

Davidson, N. C. (2014). How much wetland has the world lost? Long-term and recent trends in global wetland area. *Marine and Freshwater Research, 65*(10), 934–941. https://doi.org/10.1071/MF14173

Donato, D. C., Kauffman, J. B., Murdiyarso, D., Kurnianto, S., Stidham, M., & Kanninen, M. (2011). Mangroves among the most carbon-rich forests in the tropics. *Nature Geoscience, 4*(5), 293–297. https://doi.org/10.1038/ngeo1123

Duarte, C. M., Losada, I. J., Hendriks, I. E., Mazarrasa, I., & Marbà, N. (2013). The role of coastal plant communities for climate change mitigation and adaptation. *Nature Climate Change, 3*(11), 961–968. https://doi.org/10.1038/nclimate1970

Epstein, Y., & Moran, D. S. (2006). Thermal comfort and the heat stress indices. *Industrial Health, 44*(3), 388–398.

Fargione, J. E., Bassett, S., Boucher, T., Bridgham, S. D., Conant, R. T., Cook-Patton, S. C., et al. (2018). Natural climate solutions for the United States. *Science Advances, 4*(11), eaat1869. https://doi.org/10.1126/sciadv.aat1869

Fetting, C. (2020). *The European green deal*. ESDN Report, December, 2(9), 53.

Fourqurean, J. W., Duarte, C. M., Kennedy, H., Marbà, N., Holmer, M., Mateo, M. A., et al. (2012). Seagrass ecosystems as a globally significant carbon stock. *Nature Geoscience, 5*(7), 505–509. https://doi.org/10.1038/ngeo1477

Gill, S. E., Handley, J. F., Ennos, A. R., & Pauleit, S. (2007). Adapting cities for climate change: The role of the green infrastructure. *Built Environment, 33*(1), 115–133. https://doi.org/10.2148/benv.33.1.115

Gilroy, K. L., & McCuen, R. H. (2012). A nonstationary flood frequency analysis method to adjust for future climate change and urbanization. *Journal of Hydrology, 414*, 40–48. https://doi.org/10.1016/j.jhydrol.2011.10.009

Giridharan, R., & Emmanuel, R. (2018). The impact of urban compactness, comfort strategies and energy consumption on tropical urban heat Island intensity: A review. *Sustainable Cities and Society, 40*, 677–687. https://doi.org/10.1016/j.scs.2018.01.024

Grimm, V., Ayllón, D., & Railsback, S. F. (2017). Next-generation individual-based models integrate biodiversity and ecosystems: Yes we can, and yes we must. *Ecosystems, 20*, 229–236. https://doi.org/10.1007/s10021-016-0071-2

Griscom, B. W., Adams, J., Ellis, P. W., Houghton, R. A., Lomax, G., Miteva, D. A., et al. (2017). Natural climate solutions. *Proceedings of the National Academy of Sciences, 114*(44), 11645–11650. https://doi.org/10.1073/pnas.1710465114

Hall, C., Dawson, T. P., Macdiarmid, J. I., Matthews, R. B., & Smith, P. (2017). The impact of population growth and climate change on food security in Africa: Looking ahead to 2050. *International Journal of Agricultural Sustainability, 15*(2), 124–135. https://doi.org/10.1080/14735903.2017.1293929

Herr, D., & Landis, E. (2016) Coastal blue carbon ecosystems. Opportunities for Nationally Determined Contributions. Policy Brief.

Hobbie, S. E., & Grimm, N. B. (2020). Nature-based approaches to managing climate change impacts in cities. *Philosophical Transactions of the Royal Society B, 375*(1794), 20190124. https://doi.org/10.1098/rstb.2019.0124

Honey, C. J., & Sporns, O. (2008). Dynamical consequences of lesions in cortical networks. Human brain mapping, 29(7), 802–809.

Holden, P. B., Rebelo, A. J., Wolski, P., Odoulami, R. C., Lawal, K. A., Kimutai, J., et al. (2022). Nature-based solutions in mountain catchments reduce impact of anthropogenic climate change on drought streamflow. *Communications Earth & Environment, 3*(1), 51. https://doi.org/10.1038/s43247-022-00379-9

Hondula, D. M., Davis, R. E., Saha, M. V., Wegner, C. R., & Veazey, L. M. (2015). Geographic dimensions of heat-related mortality in seven US cities. *Environmental Research, 138*, 439–452. https://doi.org/10.1016/j.envres.2015.02.033

IPCC (2019). *Climate change and land: An IPCC special report on climate change, desertification, land degradation, sustainable land management, food security, and greenhouse gas fluxes in terrestrial ecosystems*.

Joosten, H., Tapio-Biström, M. L., & Tol, S. (2012). *Peatlands: Guidance for climate change mitigation through conservation, rehabilitation and sustainable use*. Food and Agriculture Organization of the United Nations.

Jose, S. (2009). *Agroforestry for ecosystem services and environmental benefits: An overview* (pp. 1–10). Springer.

Kim, J. M., Son, S., Lee, S., & Son, K. (2020). Cost of climate change: Risk of building loss from typhoon in South Korea. *Sustainability, 12*(17), 7107. https://doi.org/10.3390/su12177107

Kjellstrom, T., Briggs, D., Freyberg, C., Lemke, B., Otto, M., & Hyatt, O. (2016). Heat, human performance, and occupational health: A key issue for the assessment of global climate change impacts. *Annual Review of Public Health, 37*(1), 97–112. https://doi.org/10.1146/annurev-publhealth-032315-021740

Ladwig, R., Furusato, E., Kirillin, G., Hinkelmann, R., & Hupfer, M. (2018). Climate change demands adaptive management of urban lakes: Model-based assessment of management scenarios for Lake Tegel (Berlin, Germany). *Water, 10*(2), 186. https://doi.org/10.3390/w10020186

Lal, R. (2020). Regenerative agriculture for food and climate. *Journal of Soil and Water Conservation, 75*(5), 123A–124A. https://doi.org/10.2489/jswc.2020.0620A

Leifeld, J., & Menichetti, L. (2018). The underappreciated potential of peatlands in global climate change mitigation strategies. *Nature Communications, 9*(1), 1071. https://doi.org/10.1038/s41467-018-03406-6

Lele, S., Springate-Baginski, O., Lakerveld, R., Deb, D., & Dash, P. (2013). Ecosystem services: Origins, contributions, pitfalls, and alternatives. *Conservation and Society, 11*(4), 343–358.

Macreadie, P. I., Anton, A., Raven, J. A., Beaumont, N., Connolly, R. M., Friess, D. A., et al. (2019). The future of blue carbon science. *Nature Communications, 10*(1), 3998. https://doi.org/10.1038/s41467-019-11693-w

Markolf, S. A., Chester, M. V., Eisenberg, D. A., Iwaniec, D. M., Davidson, C. I., Zimmerman, R., et al. (2018). Interdependent infrastructure as linked social, ecological, and technological systems (SETSs) to address lock-in and enhance resilience. *Earth's Future, 6*(12), 1638–1659. https://doi.org/10.1029/2018EF000926

McMichael, A., McMichael, C., Berry, H., & Bowen, K. (2010). Climate-related displacement: Health risks and responses. In *Climate change and population displacement: Multidisciplinary perspectives*. Hart Publishing Ltd.

Millennium Ecosystem Assessment, M. E. A. (2005). *Ecosystems and human Well-being* (Vol. 5, p. 563). Island press.

Mitsch, W. J., & Gosselink, J. G. (2015). Wetlands. John wiley & sons.

Mora, C., Dousset, B., Caldwell, I. R., Powell, F. E., Geronimo, R. C., Bielecki, C. R., et al. (2017). Global risk of deadly heat. *Nature Climate Change, 7*(7), 501–506. https://doi.org/10.1038/nclimate3322

Nelson, E. J., Kareiva, P., Ruckelshaus, M., Arkema, K., Geller, G., Girvetz, E., et al. (2013). Climate change's impact on key ecosystem services and the human Well-being they support in the US. *Frontiers in Ecology and the Environment, 11*(9), 483–893. https://doi.org/10.1890/120312

Nepstad, D., McGrath, D., Stickler, C., Alencar, A., Azevedo, A., Swette, B., et al. (2014). Slowing Amazon deforestation through public policy and interventions in beef and soy supply chains. *Science, 344*(6188), 1118–1123. https://doi.org/10.1126/science.1248525

Nesshöver, C., Assmuth, T., Irvine, K. N., Rusch, G. M., Waylen, K. A., Delbaere, B., et al. (2017). The science, policy and practice of nature-based solutions: An interdisciplinary perspective. *Science of the Total Environment, 579*, 1215–1227. https://doi.org/10.1016/j.scitotenv.2016.11.106

Newman, P., & Kenworthy, J. (2015). *The end of automobile dependence* (pp. 105–140). Island Press/Center for Resource Economics.

Nowak, D. J., Greenfield, E. J., Hoehn, R. E., & Lapoint, E. (2013). Carbon storage and sequestration by trees in urban and community areas of the United States. *Environmental Pollution, 178*, 229–236. https://doi.org/10.1016/j.envpol.2013.03.019

Oberndorfer, E., Lundholm, J., Bass, B., Coffman, R. R., Doshi, H., Dunnett, N., et al. (2007). Green roofs as urban ecosystems: Ecological structures, functions, and services. *Bioscience, 57*(10), 823–833. https://doi.org/10.1641/B571005

Oleson, K. W., Monaghan, A., Wilhelmi, O., Barlage, M., Brunsell, N., Feddema, J., et al. (2015). Interactions between urbanization, heat stress, and climate change. *Climatic Change, 129*, 525–541. https://doi.org/10.1007/s10584-013-0936-8

Ordóñez, C., & Duinker, P. N. (2015). Climate change vulnerability assessment of the urban forest in three Canadian cities. *Climatic Change, 131*, 531–543. https://doi.org/10.1007/s10584-015-1394-2

Osaka, S., Bellamy, R., & Castree, N. (2021). Framing "nature-based" solutions to climate change. *Wiley Interdisciplinary Reviews: Climate Change, 12*(5), e729. https://doi.org/10.1002/wcc.729

Paerl, H. W., Gardner, W. S., Havens, K. E., Joyner, A. R., McCarthy, M. J., Newell, S. E., et al. (2016). Mitigating cyanobacterial harmful algal blooms in aquatic ecosystems impacted by climate change and anthropogenic nutrients. *Harmful Algae, 54*, 213–222. https://doi.org/10.1016/j.hal.2015.09.009

Pandit, A., Jain, A., Singha, R., Suting, A., Jamir, S., Pradhan, N. S., & Choudhury, D. (2016). Community perceptions and responses to climate variability: Insights from the himalayas. In *Climate change adaptation strategies–an upstream-downstream perspective* (pp. 179–194). https://doi.org/10.1007/978-3-319-40773-9_10

Paustian, K., Lehmann, J., Ogle, S., Reay, D., Robertson, G. P., & Smith, P. (2016). Climate-smart soils. *Nature, 532*(7597), 49–57. https://doi.org/10.1038/nature17174

Pretty, J. (2006). Agroecological approaches to agricultural development (pp. 2795087-1191427986785). Washington, DC: World Bank.

Radeloff, V. C., Helmers, D. P., Kramer, H. A., Mockrin, M. H., Alexandre, P. M., Bar-Massada, A., et al. (2018). Rapid growth of the US wildland-urban interface raises wildfire risk. *Proceedings of the National Academy of Sciences, 115*(13), 3314–3319. https://doi.org/10.1073/pnas.1718850115

Rosenzweig, C., & Solecki, W. (2014). Hurricane Sandy and adaptation pathways in New York: Lessons from a first-responder city. *Global Environmental Change, 28*, 395–408. https://doi.org/10.1016/j.gloenvcha.2014.05.003

Seddon, N., Chausson, A., Berry, P., Girardin, C. A., Smith, A., & Turner, B. (2020). Understanding the value and limits of nature-based solutions to climate change and other global challenges. *Philosophical Transactions of the Royal Society B, 375*(1794), 20190120. https://doi.org/10.1098/rstb.2019.0120

Seto, K. C., Güneralp, B., & Hutyra, L. R. (2012). Global forecasts of urban expansion to 2030 and direct impacts on biodiversity and carbon pools. *Proceedings of the National Academy of Sciences, 109*(40), 16083–16088. https://doi.org/10.1073/pnas.1211658109

Seymour, F., & Busch, J. (2016). *Why forests? Why now?: The science, economics, and politics of tropical forests and climate change*. Brookings Institution Press.

Shaposhnikov, D., Revich, B., Bellander, T., Bedada, G. B., Bottai, M., Kharkova, T., et al. (2014). Mortality related to air pollution with the Moscow heat wave and wildfire of 2010. *Epidemiology, 25*(3), 359–364. https://doi.org/10.1097/EDE.0000000000000090

Snep, R. P., Voeten, J. G., Mol, G., & Van Hattum, T. (2020). Nature based solutions for urban resilience: A distinction between no-tech, low-tech and high-tech solutions. *Frontiers in Environmental Science, 8*, 599060. https://doi.org/10.3389/fenvs.2020.599060

Stange, E. E., Barton, D. N., Andersson, E., & Haase, D. (2022). Comparing the implicit valuation of ecosystem services from nature-based solutions in performance-based green area indicators across three European cities. *Landscape and Urban Planning, 219*, 104310. https://doi.org/10.1016/j.landurbplan.2021.104310

Sundermann, A., Gerhardt, M., Kappes, H., & Haase, P. (2013). Stressor prioritisation in riverine ecosystems: Which environmental factors shape benthic invertebrate assemblage metrics? *Ecological Indicators, 27*, 83–96. https://doi.org/10.1016/j.ecolind.2012.12.003

Verhoeven, J. T., Arheimer, B., Yin, C., & Hefting, M. M. (2006). Regional and global concerns over wetlands and water quality. Trends in ecology & evolution, 21(2), 96–103.

Woodhead, A. J., Hicks, C. C., Norström, A. V., Williams, G. J., & Graham, N. A. (2019). Coral reef ecosystem services in the Anthropocene. *Functional Ecology, 33*(6), 1023–1034. https://doi.org/10.1111/1365-2435.13331

Zhou, D., Zhao, S., Liu, S., Zhang, L., & Zhu, C. (2014). Surface urban heat Island in China's 32 major cities: Spatial patterns and drivers. *Remote Sensing of Environment, 152*, 51–61. https://doi.org/10.1016/j.rse.2014.05.017

Adaptation and Mitigation: Integrating Strategies for Climate Resilience in the Himalayan Region

Ramesh Kumar, Prity S. Pippal, Alok Kumar, Rupendra Singh, Atar Singh, Payal Sharma, and Rajesh Kumar

Abstract The Himalayan region is facing serious vulnerabilities due to climate change. However, the changing climate threatens the environment, agriculture, and socio-economic conditions of the region. Thus, this region requires a holistic approach to climate resilience through adaptation and mitigation policy. Therefore, this chapter presents the integration of adaptation and mitigation strategies for developing climate resilience in the Himalayan region. The importance of traditional ecological knowledge with scientific strategies is underscored by its potential to enhance the effectiveness of such measures. The strengthening of ecosystems' natural resilience has the potential to reduce vulnerability to climate disasters, enhancing sustainable livelihoods for the inhabitants. Moreover, integrating nature-based solutions with scientific research, policy interventions, and people's participation is required to conceptualize a robust framework for climate resilience. Furthermore, the climate change issues in the Himalayas highlight the potential of nature-based solutions as a viable and sustainable approach to addressing these challenges. It proposes practical strategies, including enhancing local governance, promoting cross-sectoral partnerships, and deriving strength from traditional and ecological knowledge. The study seeks to balance environmental conservation and

R. Kumar (✉)
Department of Civil Engineering, Marwadi University Research Centre, Marwadi University, Rajkot, Gujarat, India

P. S. Pippal · P. Sharma · R. Kumar
Department of Environmental Science, Central University of Rajasthan, Ajmer, Rajasthan, India

A. Kumar
Department of Geology, DST's Centre of Excellence, Sikkim University, Gangtok, Sikkim, India

R. Singh
CPS, School of Social Sciences, Jawaharlal Nehru University, Delhi, India

A. Singh
Center for Cryosphere and Climate Change Studies, National Institute of Hydrology, Roorkee, India

S. Siddiqui et al. (eds.), *Frontiers in Sustainability*,
https://doi.org/10.1007/978-3-032-19740-5_9

socio-economic development by focusing on a strategy for a sustainable Himalayan ecosystem. The findings emphasize the need for inclusive adaptation and mitigation measures tailored to the vulnerabilities of the Himalayan region. Also, promoting nature-based solutions in developing climate resilience and strengthening sustainable development.

Keywords Adaptation and mitigation · Climate resilience · Himalayan region · Climate-induced disasters · Sustainable livelihoods

1 Introduction

The Himalayan region is known as the "Third Pole," which rivals the polar regions, making it an essential environmental and ecological zone of global importance (Singh et al., 2016; Kumar et al., 2016). The Himalayas, home to the world's largest glaciers, are experiencing accelerated glacial retreat due to rising global temperatures (Kumar et al., 2023a, b). This significantly threatens the water supplies of major river systems like the Ganges, Indus, and Brahmaputra, which sustain millions in Afghanistan, India, Nepal, and Pakistan (Pandey et al., 2018; Muccione et al., 2024). Himalayan glaciers act as natural reservoirs, releasing meltwater during the dry months that sustains river systems. This process helps ease water scarcity, reduces the risk of drought, and provides vital support for agriculture and hydropower. In addition, the Himalayas influence the South Asian monsoon, serving as a climatic regulator that balances seasonal rainfall across the region. Because these ecological functions are essential for environmental and social stability, the Himalayas have become a focal point in global discussions on water security and climate resilience. Apart from this, the socio-economic consequences of climate change in the Himalayan region are stark, with vulnerable communities facing disrupted livelihoods and increased disaster risks (Namgyal et al., 2025a). The region is characterized by rugged terrains, steep slopes, and stark seasonal contrasts, contributing to high exposure to climate hazards such as floods, landslides, and glacial lake outbursts (Mishra et al., 2019; Matta et al., 2025). Climate change exacerbates these issues through permafrost thawing, which destabilizes slopes and increases the risk of flash floods (Meetei et al., 2023). Moreover, shrinking glaciers and changes in land use have led to a loss of biodiversity and essential ecosystem services, further complicating adaptation efforts (Dhakal & Kattel, 2019). As traditional knowledge systems increasingly intersect with scientific approaches, there is a growing recognition of the need for effective adaptation strategies that empower local populations and incorporate indigenous knowledge for sustainable resource management. To combat these challenges, various adaptation and mitigation strategies are being explored. Adaptation and mitigation, integrating strategies for climate resilience in the Himalayan region, are crucial for addressing climate change's profound impacts on the environment and local communities (Barua et al., 2014; Darjee et al., 2023). These include nature-based solutions to enhance resilience and renewable energy initiatives that reduce greenhouse gas emissions while supporting community livelihoods.

Implementing adaptation and mitigation strategies for climate resilience in the Himalayan region presents various challenges, primarily influenced by environmental, socio-economic, and governance factors (Patel et al., 2025). The need for regional cooperation is another critical challenge, as many climate-related issues in the Himalayan region are transboundary. Collaborative efforts among neighboring countries are essential for addressing shared vulnerabilities and fostering comprehensive adaptation strategies. However, the existing governance frameworks often fall short in integrating climate adaptation into broader development plans, which leads to incremental and poorly coordinated responses. Furthermore, there is a pressing need for better data and information systems to monitor climate impacts and adaptation measures. Limited access to information and insufficient services hamper communities' ability to make informed decisions and adapt effectively to the changing climate. There is also a call for the development of transformative adaptation knowledge and strategies, as current approaches often do not align well with the realities faced by local populations. Although, the primary challenges in implementing adaptation and mitigation strategies for climate resilience in the Himalayan region include the region's vulnerable geography, socio-economic disparities, the need for regional cooperation, gaps in governance and planning, and inadequate data and information systems. Addressing these challenges is critical for enhancing the region's resilience to climate change. Considering these complexities, future strategies must prioritize collaborative, equity-centered approaches that integrate local needs with scientific insights, ensuring that the diverse socio-economic contexts of Himalayan communities are addressed.

2 Climate Change in the Himalayan Region

The impacts of climate change on the Himalayan region are profound and multifaceted, affecting both the environment and the communities that rely on its resources (Singh et al., 2018). However, the changing climate is characterized by increased temperatures and altered precipitation patterns, leading to shifts in hydrology throughout the region. Rising temperatures have resulted in less snowfall and changed seasonal runoff dynamics, altering river flows. Himalayan glaciers are retreating at alarming rates due to global warming, with significant consequences for millions across the region. This retreat threatens the water supplies of major rivers, including the Ganges, Indus, and Brahmaputra, which are vital for agriculture, drinking water, and hydropower generation in downstream areas (Singh et al., 2022). Additionally, climate change has increased the variability of the South Asian monsoon, contributing to erratic weather patterns that heighten the risks of flooding and drought. One of the most visible impacts of climate change in the Himalayas is the retreat of glaciers. Studies indicate that Himalayan glaciers have been shrinking alarmingly, with some regions experiencing a retreat of over 30 m yearly. A study using remote sensing and field observations highlighted that glacier mass loss has nearly doubled in the twenty-first century compared to the previous decades. Glacial mass loss reduces long-term water storage, which has profound implications for downstream river basins. Temperatures in the Himalayas are rising faster than the global average; some studies suggest a rate of 0.6 °C per decade in parts of the region, compared to the worldwide 0.2 °C (Negi et al., 2021). This accelerated warming is driving glacier melt, estimating that Himalayan glaciers could lose up to a third of their volume by 2100, even if global warming is limited to 1.5 °C (Dimri et al., 2021). Previous research shows that since the 1970s, many glaciers have lost 20–30% of their mass, with smaller, lower-altitude ones being hit hardest. The rate of retreat varies; some areas in the Karakoram range are stable or even growing due to unique weather patterns (the "Karakoram anomaly"), but most, especially in the eastern Himalayas, are shrinking (Hewitt & Hewitt, 2014). A study (Mani, 2021) found that Himalayan glacier melt has doubled since 2000 compared to the 1975–2000 period, driven by rising temperatures and, in some areas, black carbon deposits (soot from burning fossil fuels and biomass) that darken ice and accelerate melting. Such vulnerabilities underscore the importance of developing integrated approaches that combine both climate adaptation and mitigation measures. Figure 1 presents the adverse consequences of climate change and serves as a visual call to action, advocating for a holistic and inclusive framework where adaptation and mitigation are treated as complementary pathways. Together, these strategies pave the way for a future where ecological integrity, community resilience, and sustainable development go hand in hand in the face of ongoing and future climate challenges in the Himalayas. Integrating these approaches makes it possible to reduce the exposure and sensitivity of communities and ecosystems while enhancing their adaptive capacity.

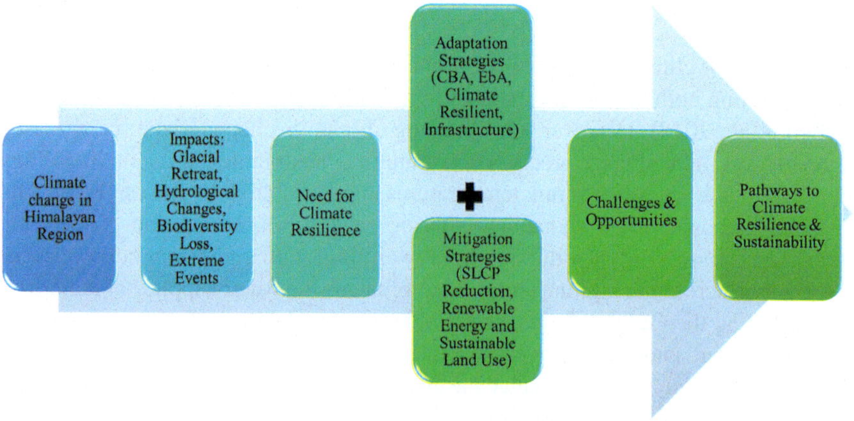

Fig. 1 Flow chart showing the climate change impacts in the Himalayan region and the need for climate resilience by integrating adaptation and mitigation strategies that lead to a sustainable and climate-resilient ecosystem

3 Hydrological Changes and Water Resources

Hydrological models incorporating remote sensing and in situ data have shown that changes in precipitation patterns further complicate the region's water security. Table 1 illustrates the role of hydrological changes and their impacts on water resources with adaptation and mitigation strategies for climate resilience. The Indian monsoon has exhibited increased variability, which is crucial in replenishing the Himalayan rivers (Chandel & Ghosh, 2021). Some regions experience excessive rainfall, leading to floods and landslides, while others suffer from prolonged dry spells, exacerbating water scarcity. Additionally, an increase in rain-on-snow events has been observed, accelerating glacier melt and contributing to flash floods and glacial lake outbursts (GLOFs) (Wang et al., 2024). The warming climate has led to the formation and expansion of glacial lakes in the Himalayas. These lakes are often dammed by unstable moraine walls, making them highly susceptible to sudden outbursts. These events threaten human settlements, infrastructure, and ecosystems, especially in states like Himachal Pradesh, Uttarakhand, Sikkim, and Arunachal Pradesh, where numerous glacial lakes are situated. Studies using satellite imagery and field surveys, have reported a significant increase in the number and size of glacial lakes, particularly in the eastern Himalayas (Chauhan et al., 2023). Glacial lake outburst floods (GLOFs) severely threaten downstream communities and infrastructure. Past events, such as the 2013 Kedarnath disaster in Uttarakhand, India, have demonstrated the catastrophic consequences of such floods. An important example is Nepal's 1985 Dig Tsho GLOF, which had catastrophic impacts on local villages and hydropower plants. The growing risk from expanding high-altitude

Table 1 Role of hydrological changes and their impacts on water resources with adaptation and mitigation strategies for climate resilience

Hydrological change	Impact	Adaptation strategy	Mitigation strategy
Altered snowfall and rainfall patterns	Seasonal water availability uncertainty	Improved water harvesting, storage, and irrigation systems	Promoting green infrastructure to manage runoff
Irregular snowmelt and glacier retreat	Unpredictable river flows, risk of water scarcity and floods	Glacier monitoring, GLOF early warning systems, climate-resilient crops	Reducing black carbon emissions to slow glacier retreat
Decline in groundwater recharge	Water stress for agriculture and domestic use	Recharge wells, rainwater harvesting, watershed management	Sustainable land use practices reduce runoff and enhance infiltration
Reduced hydropower reliability	Threat to clean energy access	Energy diversification (micro-hydro, solar)	Promoting decentralized renewable energy over large dams
Increased water demand in warming climate	Pressure on limited water resources	Demand-side management, water-use efficiency in agriculture	Efficient water-use technologies reduce energy-related emissions

lakes, such as Tsho Rolpa, necessitates immediate planning for risk mitigation, including robust early warning systems and effective glacial lake monitoring. More recently, in October 2023, a GLOF event in Sikkim caused severe damage to the Chungthang dam and downstream infrastructure, highlighting the growing threat of such occurrences in the Eastern Himalayas (Kaushik et al., 2024; Sattar et al., 2025). The Indian Space Research Organisation (ISRO) and other research institutions have identified over 500 glacial lakes in the Indian Himalayas, with many classified as potentially hazardous. The Teesta River basin, the Alaknanda and Bhagirathi basins, and parts of Arunachal Pradesh are particularly vulnerable to GLOFs (Saha et al., 2024). The growing risk has prompted the Indian government and scientific community to use satellite imagery, hydrological modeling, and early warning systems to enhance monitoring efforts. To mitigate GLOF risks, India has undertaken several initiatives, including the National Disaster Management Authority's (NDMA) guidelines for reducing the vulnerability of high-risk regions. Structural measures such as artificial drainage systems, reinforced embankments, and controlled lake lowering have been proposed to prevent sudden breaches. Non-structural measures include community-based disaster preparedness programs, improved land-use planning, and integrating traditional knowledge with modern scientific approaches. However, challenges such as limited accessibility to remote glacial lakes, lack of real-time monitoring infrastructure, and coordination gaps between various government agencies hinder effective risk reduction. The need for a robust early warning system, international collaboration, and climate-resilient development planning cannot be overstated as India navigates the growing risks posed by glacial lake outburst floods. With rising temperatures, the probability of GLOFs is expected to increase, necessitating urgent monitoring and early warning systems. Advanced hydrological models integrating satellite and drone-based observations assess potential risks and improve disaster preparedness.

4 Policy Responses and Adaptation Strategies

Governments and research institutions are increasingly recognizing the urgency of climate adaptation and mitigation efforts in the Himalayas (Rana et al., 2022). Various climate action plans and regional collaborations, such as the Hindu Kush Himalayan Monitoring and Assessment Programme (HIMAP), have been launched to address these challenges. Adaptation measures include improved water management by constructing artificial glaciers, rainwater harvesting, and better irrigation techniques (Wagle et al., 2021). In disaster-prone areas, early warning systems for GLOFs and landslides are being developed using satellite-based monitoring and real-time data collection. Community-based initiatives, such as afforestation programs and climate-resilient agriculture, are also gaining traction. Research has highlighted the effectiveness of integrating indigenous knowledge with modern climate science to enhance resilience at the local level (Makondo & Thomas, 2018). However, challenges such as inadequate funding, lack of coordination between stakeholders, and political constraints remain obstacles to effective climate adaptation. In response to the ongoing challenges posed by climate change, communities in the Himalayan region are exploring various adaptation and mitigation strategies. These include nature-based solutions to enhance resilience against climate change's adverse effects. Effective adaptation involves integrating scientific knowledge with local practices to develop sustainable approaches to safeguard the environment and the livelihoods of the people who depend on it. Table 2 highlights region-specific climate change adaptation strategies across ecologically sensitive and mountainous regions of India. In the Indian Himalayan Region, large-scale national initiatives such as the National Mission for Sustaining the Himalayan Ecosystem (NMSHE) and the Climate Change Action Programme (CCAP) have been implemented to address climate vulnerabilities through scientific research, capacity building, and policy integration (Meetei et al., 2023). In Ladakh, an arid high-altitude region, communities have adopted innovative water storage methods like ice reservoirs (artificial glaciers) to address seasonal water scarcity. These are complemented by community-based adaptation measures that leverage local knowledge and collective action (Srivastava et al., 2024). The Garhwal Himalayas have witnessed agricultural diversification and livelihood transitions, along with the use of traditional ecological knowledge to strengthen resilience among mountain communities (Kumar et al., 2024). In Kullu, Himachal Pradesh, rapid urbanization and climate-induced hazards have focused on eco-sensitive urban planning and disaster-resilient infrastructure development to reduce risks and ensure sustainable growth (Islam et al., 2024). Lastly, in the Similipal Biosphere Reserve, tribal home-gardens are key in enhancing food security, while biodiversity conservation efforts provide a nature-based adaptation framework (Mallick et al., 2024).

Table 2 Community and ecosystem-based adaptation strategies being performed in the different Himalaya regions

Region/Case study	Key adaptation strategies	References
Indian Himalayan Region	National Mission for Sustaining the Himalayan Ecosystem (NMSHE), Climate Change Action Programme (CCAP)	Meetei et al. (2023)
Ladakh, India	Ice reservoirs for water storage, community-based adaptation	Srivastava et al. (2024)
Garhwal Himalayas	Agricultural diversification, livelihood shifts, and traditional knowledge	Kumar et al. (2024)
Kullu, Himachal Pradesh	Eco-sensitive urban planning, disaster-resilient infrastructure	Islam et al. (2024)
Similipal Biosphere Reserve	Tribal home-gardens, biodiversity conservation	Mallick et al. (2024)

5 Government Policies and Institutional Frameworks

Governments in the Himalayan region have recognized the urgency of climate change and have introduced various policies and programs to address its impacts. In India, the National Action Plan on Climate Change (NAPCC) and the National Mission for Sustaining the Himalayan Ecosystem (NMSHE) are key initiatives to reduce greenhouse gas emissions and promote sustainable development in the Himalayas. Similarly, the Indian Himalayan Region (IHR) has specific policies like the Indian Himalayas Climate Adaptation Programme (IHCAP) and the Climate Change Action Programme (CCAP), which focus on climate-resilient infrastructure and ecosystem management (Meetei et al., 2023). In the broader Hindu Kush Himalaya (HKH) region, countries are increasingly adopting climate change adaptation policies, though gaps remain in their implementation and effectiveness. Despite the existence of these policies, there is a significant gap between policy design and implementation. A gap analysis of climate change adaptation policies in the HKH region reveals that while policies address key risks, institutional and governance challenges often hinder operationalisation. The lack of coordination between upstream and downstream water management systems in South Asia's glacier-fed river basins underscores the need for adaptive governance frameworks (Sud et al., 2015). Also, there is a recognized need for improved governance and integrated decision-making in the region, as historical political inequities and a lack of knowledge regarding upstream-downstream linkages hinder effective resource management. The literature suggests that without better coordination and support for local-level hybrid knowledge, the interactive drivers of change will likely undermine ecosystem resilience and social stability, highlighting the importance of engaging local communities in the adaptation process (Xu & Grumbine, 2014). The previous study identifies governance mechanisms related to climate adaptation, highlighting the need for integrated assessments that consider social and ecological factors and the importance of stakeholder participation in policymaking. It also notes the common emphasis on strategic knowledge generation among institutions across the riparian countries, while pointing out the ambiguities in climate policy objectives in several countries, which may hinder the operationalization of adaptive governance principles (Sud et al., 2015).

6 Community Adaptation and Local Resilience

Local communities in the Himalayas have developed innovative adaptation strategies based on their indigenous knowledge. In Ladakh, India, communities have adopted ice reservoirs to solve water scarcity, demonstrating the effectiveness of community-based adaptation initiatives (Saizen, 2023). Similarly, in the Garhwal Himalayas, farmers have shifted to climate-resilient agricultural practices, such as adjusting crop calendars and using high-yielding varieties, to cope with changing climatic conditions (Sharma & Pandey, 2024; Kumar et al., 2024). These challenges seriously threaten their livelihoods, particularly in agriculture, water security, and infrastructure. Local communities have developed and continue to refine adaptive strategies rooted in traditional knowledge, social capital, and evolving scientific understanding. Indigenous knowledge systems often shape community adaptation in the Himalayan region. Populations manage water resources, cultivate crops suited to altitudinal gradients, and predict weather changes through natural indicators. Traditional farming practices, like terracing and agroforestry, have proven resilient to harsh mountain conditions (Romeo et al., 2021). In recent years, communities have had to combine these practices with modern techniques to address new challenges. In many villages, local committees are formed to conduct hazard mapping, build early warning systems, and coordinate emergency responses. These efforts are supported by NGOs, academic institutions, and government agencies working to build local capacity. However, there is a need for more inclusive and region-specific policy frameworks that integrate scientific data with local knowledge systems. Many national adaptation policies still overlook the unique needs of mountain communities, leading to gaps in resource allocation and institutional support. Enhancing community adaptation and resilience in the Himalayan region requires a multi-scalar approach that blends local initiatives with national and global support systems (Ojha et al., 2019). Sustainable development projects, community-led ecosystem restoration, and participatory governance models must be scaled up to safeguard these fragile environments. Strengthening local resilience is essential for the survival of Himalayan communities and the downstream populations who depend on these mountains for water and ecological services. The Himalayan region can build a resilient future amid climatic uncertainties through the synergy of traditional wisdom, local innovation, and external support.

7 Livelihood Diversification and Social-Ecological Vulnerability

Livelihood diversification has become a critical adaptation strategy in the Himalayas (Namgyal et al., 2025a). The livelihoods of Himalayan communities traditionally rely on subsistence agriculture, livestock rearing, forestry, and seasonal migration (Saxena et al., 2005). However, these traditional systems are becoming less viable due to increasing environmental pressures such as glacial retreat, erratic rainfall patterns, land degradation, and biodiversity loss. In the Kaligandaki Basin of Nepal, households have shifted to non-farm livelihoods to reduce their dependence on climate-sensitive agricultural systems. However, social-ecological vulnerability remains high, particularly for marginalized groups like women, who often have limited access to resources and decision-making power. Tribal communities in protected areas like the Similipal Biosphere Reserve in India have demonstrated the importance of home gardens as socio-ecological solutions for climate change adaptation. These gardens enhance biodiversity, provide livelihood security, and contribute to sustainable development goals (SDGs) (Mallick et al., 2024). From a social-ecological systems perspective, the relationship between livelihood diversification and vulnerability is dynamic and context-dependent. Adaptive strategies are more effective when informed by local knowledge, supported by strong institutions, and integrated with conservation efforts. Community-based resource management, improved access to education and markets, and the development of infrastructure such as roads, communication networks, and storage facilities can significantly enhance the capacity of Himalayan communities to pursue sustainable diversification. Government policies and development interventions must thus be designed to support equitable access to diversification opportunities while maintaining environmental integrity.

8 Mitigation Strategies Enhancing Climate Change Resilience

The Himalayan region, being ecologically fragile and socioeconomically diverse, requires integrated mitigation strategies to enhance climate resilience (Sultan et al., 2022; Namgyal et al., 2025b). Mitigation strategies in the Himalayan region aim to reduce the root causes of climate change, particularly the emissions of greenhouse gases and short-lived climate pollutants (SLCPs) like black carbon. One critical approach is the reduction of black carbon and other SLCPS through cleaner cooking technologies, improved transport systems, and modernized brick kilns, which help minimize the deposition of light-absorbing aerosols that accelerate glacier melt. Promoting renewable energy sources such as solar, wind, and micro-hydropower is vital in reducing dependence on fossil fuels in remote mountain communities (Umar & Hussain, 2015). These cleaner energy alternatives contribute to lower carbon emissions while improving local energy access. Sustainable land use practices, including forest conservation, afforestation, and agroforestry, also serve as effective carbon sinks while supporting biodiversity and local livelihoods. Mitigation strategies in the Himalayan region span multiple sectors to reduce emissions and enhance climate resilience (Sethi et al., 2021). Decentralized renewable energy, such as solar and micro-hydropower, is promoted to cut fossil fuel use in remote areas. In forestry, afforestation and reforestation using native species support carbon sequestration and slope stabilization. Climate-smart agriculture, sustainable water management, and glacier monitoring help reduce emissions and build resilience (Sikka et al., 2018). The transport sector is shifting toward electric vehicles and efficient infrastructure to curb black carbon, while eco-tourism and improved waste management aim to minimize environmental impacts. Urban areas are adopting green infrastructure, and disaster risk reduction is being strengthened through early warning systems and nature-based solutions. Robust policy frameworks and regional cooperation for coordinated climate action support these efforts. Major mitigation strategies in different sectors are mentioned in the table (Table 3). Together, these mitigation strategies form a critical pillar of climate resilience by addressing emissions at the source and supporting sustainable development pathways in the ecologically fragile Himalayan ecosystem.

Table 3 Mitigation strategies for enhancing climate change resilience in the Himalayan region, categorized by sector, strategy, purpose, and implementing agencies

Sector	Mitigation strategy	Purpose	Implementing agencies/ Stakeholders
Energy	Promotion of micro-hydro and solar energy	Reduce dependency on fossil fuels	State renewable energy boards, MNRE, and local NGOs
Forestry	Afforestation and reforestation using native species	Enhance carbon sinks, prevent soil erosion	Forest departments, MoEFCC, and community forest user groups
Agriculture	Climate-resilient and low-emission farming practices	Reduce GHG emissions, improve food security	ICAR, state agri departments, Krishi Vigyan Kendras (KVKs)
Water resources	Glacier and watershed monitoring and sustainable water management	Conserve water, monitor climate impact on hydrology	Central Water Commission, State Irrigation Dept., ICIMOD
Transport	Promote electric mobility and improve road efficiency in hilly terrain	Reduce black carbon and CO_2 emissions	MoRTH, State Transport Depts., NITI Aayog
Tourism	Promote eco-tourism and regulate carbon-intensive activities	Lower tourism-related emissions, and conserve fragile ecosystems	Tourism Dept., NGOs, local governing bodies
Waste Management	Strengthening solid waste and sewage treatment infrastructure	Minimize methane & black carbon emissions	Urban local bodies, State Pollution Control Boards
Urban Planning	Green infrastructure and energy-efficient buildings	Reduce urban heat and emissions	Smart Cities Mission, ULBs, Urban Affairs Ministry
Disaster Management	Early warning systems and community-based adaptation	Reduce climate-induced disaster risks	NDMA, SDRF, NGOs, Panchayats
Policy & Governance	Cross-border Himalayan climate action framework	Transboundary coordination on emissions and adaptation	ICIMOD, SAARC, and the national governments of the Himalayan nations

9 Current Challenges and Opportunities

The Himalayan region faces numerous challenges in implementing integrated climate adaptation and mitigation strategies. One of the primary barriers is limited and inconsistent funding, especially for long-term, community-based and ecosystem-focused initiatives. Many remote mountainous communities lack access to financial resources and the infrastructure necessary to adopt clean energy technologies or build climate-resilient infrastructure. Another major hurdle is the scarcity of high-resolution, long-term climate and environmental data across the Himalayan arc. This data gap limits the accuracy of climate projections and hampers effective planning and risk assessments. Additionally, fragmented institutional frameworks and weak coordination among various government departments, NGOs, and regional stakeholders often lead to duplication of efforts and ineffective implementation of adaptation and mitigation programs.

However, significant opportunities are emerging that can accelerate climate resilience. Frameworks like the Hindu Kush Himalaya Call to Action advocate for a shared vision and coordinated response among Himalayan countries, promoting regional cooperation in data sharing, joint research, and transboundary ecosystem management. Advances in remote sensing and Earth observation systems offer potential to bridge data gaps and support evidence-based decision-making. Community-based Forest and water resource management practices present scalable models that integrate traditional knowledge with scientific insights. Moreover, increasing awareness, political will, and international climate finance mechanisms such as the Green Climate Fund (GCF) present avenues to channel resources into local mitigation and adaptation projects. The Himalayan region can transform its vulnerabilities into sustainable development and climate resilience pathways by addressing these challenges through collaborative, data-driven, and inclusive approaches.

10 Conclusion

The Himalayan region, often called the "Third Pole," is experiencing accelerated climate change impacts, including glacial retreat, altered hydrological cycles, biodiversity loss, and heightened risks of floods and landslides. Scientific evidence shows that glaciers in the Himalayas are retreating annually, with mass loss nearly doubling in the twenty-first century. This directly threatens the water security of millions dependent on the Ganges, Indus, and Brahmaputra river systems. Temperatures in the Himalayas are rising at nearly 0.6 °C per decade, significantly higher than the global average of 0.2 °C, placing ecosystems and livelihoods under severe strain. Despite these challenges, opportunities for resilience are emerging. Community-led adaptation strategies such as ice reservoirs in Ladakh, agroforestry in the Garhwal Himalayas, and eco-sensitive urban planning in Himachal Pradesh demonstrate the power of integrating traditional ecological knowledge with modern science. On the mitigation front, reducing short-lived climate pollutants (SLCPs), promoting decentralized renewable energy, strengthening sustainable land-use practices, and enhancing waste management can lower greenhouse gas emissions while improving health and livelihoods. However, significant barriers remain, including fragmented governance, inadequate climate data, and limited financial resources. Addressing these requires regional cooperation, cross-sectoral governance, and scaling up nature-based solutions. The path forward must balance environmental conservation with socio-economic development by strengthening local capacities, advancing scientific monitoring, and ensuring inclusive policies. Also, the Himalayas demand a policy–practice convergence framework where adaptation and mitigation strategies are not isolated programs but interconnected pathways that link research, governance, and community action. A resilient future for the Himalayas will only be possible if policies are co-created with local stakeholders, backed by reliable climate data, and implemented through transparent, multi-scalar governance mechanisms. In doing so, the region can transform its climate vulnerabilities into opportunities for sustainable development, ecological stability, and human security.

Acknowledgements The authors would like to thank the Marwadi University Research Centre for providing the necessary facilities to conduct this research.

References

Barua, A., Katyaini, S., Mili, B., & Gooch, P. (2014). Climate change and poverty: Building resilience of rural mountain communities in South Sikkim, Eastern Himalaya, India. *Regional Environmental Change, 14*, 267–280.

Chandel, V. S., & Ghosh, S. (2021). Components of Himalayan River flows in a changing climate. *Water Resources Research, 57*(2), e2020WR027589.

Chauhan, H. K., Gallacher, D., Bhatt, A., & Bisht, A. K. (2023). The Himalayas: A climate change laboratory. *Environmental Development, 45*, 100814.

Darjee, K. B., Neupane, P. R., & Köhl, M. (2023). Proactive adaptation responses by vulnerable communities to climate change impacts. *Sustainability, 15*(14), 10952.

Dhakal, B., & Kattel, R. R. (2019). Effects of global changes on ecosystems services of multiple natural resources in mountain agricultural landscapes. *Science of the Total Environment, 676*, 665–682.

Dimri, A. P., Allen, S., Huggel, C., Mal, S., Ballesteros-Cánovas, J. A., Rohrer, M., et al. (2021). Climate change, cryosphere and impacts in the Indian Himalayan Region. *Current Science, 120*(5), 774–790.

Hewitt, K., & Hewitt, K. (2014). Karakoram glaciers and climate change. *Glaciers of the Karakoram Himalaya: Glacial Environments, Processes, Hazards and Resources*, 291–326.

Islam, K. M. B., Mitra, A., Khan, A., & Das, S. (2024). Toward sustainable climate governance: A case study of risk assessment and Management in Kullu, Himachal Pradesh. *IIMS Journal of Management Science*. https://doi.org/10.1177/0976030x241226520

Kaushik, S., Rafiq, M., Dharpure, J. K., Howat, I., Moortgat, J., Joshi, P. K., et al. (2024). Increasing risk of glacial lake outburst flood in Sikkim, eastern Himalaya under climate warming. *Remote Sensing Applications: Society and Environment, 36*, 101286.

Kumar, R., Singh, S., Kumar, R., Singh, A., Bhardwaj, A., Sam, L., et al. (2016). Development of a glacio-hydrological model for discharge and mass balance reconstruction. *Water Resources Management, 30*, 3475–3492.

Kumar, R., Kumar, R., Bhardwaj, A., Singh, A., Singh, S., Kumari, A., & Sinha, R. K. (2023a). Multivariate statistical analysis and geospatial approach for evaluating hydro-geochemical characteristics of meltwater from Shaune Garang Glacier, Himachal Pradesh, India. *Acta Geophysica, 71*(1), 323–339.

Kumar, R., Pippal, P. S., Kumar, R., Kumar, P., Singh, A., & Sharma, P. (2023b). The global scenario of hydrogeochemical research on glacier meltwater: A bibliometric and visualization analysis. *Environmental Science and Pollution Research, 30*(30), 74612–74627.

Kumar, R., Sharma, A., Rathore, J., Negi, A., Sharma, K. K., & Patel, S. (2024). *Climate-induced vulnerability, adaptation, and mitigation strategies: A case study of Chamoli District, Garhwal Himalayas* (pp. 103–113). Springer International Publishing. https://doi.org/10.1007/978-3-031-55821-4_7

Makondo, C. C., & Thomas, D. S. (2018). Climate change adaptation: Linking indigenous knowledge with western science for effective adaptation. *Environmental Science & Policy, 88*, 83–91.

Mallick, M., Singh, P. K., & Pandey, R. (2024). Harvesting resilience: Tribal home-gardens as socio-ecological solutions for climate change adaptation and sustainable development in a protected area. *Journal of Cleaner Production, 445*, 141174. https://doi.org/10.1016/j.jclepro.2024.141174

Mani, M. (2021). *Glaciers of the Himalayas: Climate change, black carbon, and regional resilience*. World Bank Publications.

Matta, G., Kumar, A., Singh Tomar, D., & Kumar, R. (2025). Climate change in the Himalayan region: Susceptible impacts on environment and human settlements. *Frontiers in Environmental Science, 13*, 1550843.

Meetei, K. B., Tsopoe, M., Giri, K., Mishra, G., Verma, P. K., & Rohatgi, D. (2023). Climate-resilient pathways and nature-based solutions to reduce vulnerabilities to climate change in the Indian Himalayan Region. In *Climate change in the Himalayas* (pp. 89–119). Academic Press.

Mishra, A., Appadurai, A. N., Choudhury, D., Regmi, B. R., Kelkar, U., Alam, M., ... Sharma, U. (2019). Adaptation to climate change in the Hindu Kush Himalaya: Stronger action urgently needed. *The Hindu Kush Himalaya assessment: Mountains, climate change, sustainability and people*, 457–490.

Muccione, V., Aguilera Rodriguez, J., Scolobig, A., Witton, R., Zwahlen, J., Mackey, A., et al. (2024). Trends in climate adaptation solutions for mountain regions. *Mitigation and Adaptation Strategies for Global Change, 29*(7), 74.

Namgyal, P., Sarkar, S., & Kumar, R. (2025a). Vulnerability assessment of rural households to climate change using livelihood vulnerability framework approach in the trans-Himalayan region of Ladakh, India. *Anthropocene, 49*, 100467.

Namgyal, P., Sarkar, S., Kumar, R., Dolkar, T., & Rawat, V. (2025b). Assessment of livelihood vulnerability across different altitudinal zones under the influence of a changing climate in the trans-Himalayan region. *Natural Hazards*, 1–21.

Negi, V. S., Thakur, S., Dhyani, R., Bhatt, I. D., & Rawal, R. S. (2021). Climate change observations of indigenous communities in the Indian Himalaya. *Weather, Climate, and Society, 13*(2), 245–257.

Ojha, H. R., Ghate, R., Dorji, L., Shrestha, A., Paudel, D., Nightingale, A., ... Kotru, R. (2019). Governance: Key for environmental sustainability in the Hindu Kush Himalaya. *The Hindu Kush Himalaya assessment: Mountains, climate change, sustainability and people*, 545–578.

Pandey, R., Kumar, P., Archie, K. M., Gupta, A. K., Joshi, P. K., Valente, D., & Petrosillo, I. (2018). Climate change adaptation in the western-Himalayas: Household level perspectives on impacts and barriers. *Ecological Indicators, 84*, 27–37.

Patel, S. K., Sharma, A., Barla, A., Tiwari, A. K., Singh, R., Kumar, S., & Singh, G. S. (2025). Socio-ecological challenges and adaptation strategies of farmers towards changing climate in Vindhyan highlands, India. *Environmental Management, 75*(1), 46–62.

Rana, S. K., Dangwal, B., Negi, V. S., & Bhatt, I. D. (2022). Scientific research in the Himalaya: Current state of knowledge, funding paradigm and policy implications. *Environmental Science & Policy, 136*, 685–695.

Romeo, R., Manuelli, S., Geringer, M., & Barchiesi, V. (2021). *Mountain farming systems–seeds for the future: Sustainable agricultural practices for resilient mountain livelihoods*. Food & Agriculture Org.

Saha, S., Bera, B., Bhattacharjee, S., Ghosh, D., Tamang, L., Shit, P. K., & Sengupta, N. (2024). Identification of the multiple causes of recent series of landslides and related damage by extreme rainfall and GLOF in Sikkim Himalaya, India, during October 2023. *Landslides, 21*(12), 2993–3009.

Saizen, I. (2023). Social innovation perspective of community-based climate change adaptation: A framework-based study of Ladakh, India. *Water, 15*(7), 1424. https://doi.org/10.3390/w15071424

Sattar, A., Cook, K. L., Rai, S. K., Berthier, E., Allen, S., Rinzin, S., et al. (2025). The Sikkim flood of October 2023: Drivers, causes, and impacts of a multihazard cascade. *Science, 387*(6740), eads2659.

Saxena, K. G., Maikhuri, R. K., & Rao, K. S. (2005). Changes in agricultural biodiversity: Implications for sustainable livelihood in the Himalaya. *Journal of Mountain Science, 2*, 23–31.

Sethi, M., Sharma, R., Mohapatra, S., & Mittal, S. (2021). How to tackle complexity in urban climate resilience? Negotiating climate science, adaptation and multi-level governance in India. *PLoS One, 16*(7), e0253904.

Sharma, K., & Pandey, V. (2024). Impact of climate change on agricultural activities in Garhwal Himalayas: A comprehensive analysis on shifts, changes and adaptation strategies. *Journal of Geography, Environment and Earth Science International*. https://doi.org/10.9734/jgeesi/2024/v28i2746

Sikka, A. K., Islam, A., & Rao, K. V. (2018). Climate-smart land and water management for sustainable agriculture. *Irrigation and Drainage, 67*(1), 72–81.

Singh, S., Kumar, R., Bhardwaj, A., Sam, L., Shekhar, M., Singh, A., et al. (2016). Changing climate and glacio-hydrology in Indian Himalayan Region: A review. *Wiley Interdisciplinary Reviews: Climate Change, 7*(3), 393–410.

Singh, S., Kumar, R., Bhardwaj, A., Kumar, R., & Singh, A. (2018). Changing climate and glacio-hydrology: A case study of Shaune Garang basin, Himachal Pradesh. *International Journal of Hydrology Science and Technology, 8*(3), 258–272.

Singh, A., Kumar, R., Kumar, R., Singh, K. K., & Randhawa, S. S. (2022). Quantification of volume loss and snout retreat from 1980 to 2019 of baspa basin glaciers, western Himalaya. *Materials Today: Proceedings, 49*, 3331–3339.

Srivastava, S., Pandey, B. W., & Negi, V. S. (2024). Community-based approaches to climate change adaptation and livelihood security in the high-altitude Bhagirathi Valley of Garhwal Himalaya. *Journal of Resources and Ecology*. https://doi.org/10.5814/j.issn.1674-764x.2024.01.017

Sud, R., Mishra, A., Varma, N., & Bhadwal, S. (2015). Adaptation policy and practice in densely populated glacier-fed river basins of South Asia: A systematic review. *Regional Environmental Change, 15*(5), 825–836. https://doi.org/10.1007/S10113-014-0711-Z

Sultan, H., Zhan, J., Rashid, W., Chu, X., & Bohnett, E. (2022). Systematic review of multi-dimensional vulnerabilities in the Himalayas. *International Journal of Environmental Research and Public Health, 19*(19), 12177.

Umar, M., & Hussain, A. (2015). Micro hydro power: A source of sustainable energy in rural communities: Economic and environmental perspectives. *The Pakistan Development Review*, 487–504.

Wagle, N., Dhakal, M. P., & Shrestha, A. B. (2021). Adaptation strategies to address challenges of traditional agricultural water management in the Upper Indus Basin. *Mountain Research and Development, 41*(3), R24–R31.

Wang, H., Bin-Bin, W., Cui, P., Yao-Ming, M., Wang, Y., Jian-Sheng, H., et al. (2024). *Disaster effects of climate change in high-Mountain Asia: State of art and scientific challenges*. Advances in Climate Change Research.

Xu, J., & Grumbine, R. E. (2014). Building ecosystem resilience for climate change adaptation in the Asian highlands. *Wiley Interdisciplinary Reviews: Climate Change, 5*(6), 709–718. https://doi.org/10.1002/WCC.302

Sustainable Land Use and Biodiversity Conservation

Mohsin Khan, Nizakat Bibi, Zahoor ul Haq, Hafsa Ijaz, Muqaddas Bano, Maria Kalsoom, Fazal ur Rehman, and Junqiang Zheng

Abstract Overcoming global ecological challenges requires both biodiversity conservation and sustainable land use. This study demonstrates the significance of forging an interaction between ecological health and human growth by investigating the relationship between safeguarding biodiversity and sustainable land management techniques. Massive biodiversity loss has resulted from ecological collapse driven on by urbanization, agriculture, deforestation, and industrial activity, endangering ecosystem services vital to human survival. Such impacts can be mitigated through the utilization of sustainable land use techniques such as agroecology, forestry, habitat restoration, and urban planning with green infrastructure. Such impacts are boosted when biodiversity protection is incorporated into economic planning, legal frameworks, and engagement among communities. This chapter analyzes valuable case studies, evaluates the efficacy of policies, and provides a comprehensive

M. Khan · Z. ul Haq · H. Ijaz
Taohuayu Yellow River Floodplain Ecosystem Observation and Research Station of Henan Provincearch Station, School of Life Sciences, Henan University, Xingyang, Henan, China

School of Life Sciences, Henan University, Kaifeng, Henan, China

N. Bibi
School of Life Sciences, Henan University, Kaifeng, Henan, China

International Joint Research Laboratory for Global Change Ecology, School of Life Sciences, Henan University, Kaifeng, Henan, China

M. Bano
Key Laboratory of Plant Germplasm Enhancement and Specialty Agriculture, Wuhan Botanical Garden, Chinese Academy of Sciences, Wuhan, China

M. Kalsoom
Institute of Food Science and Nutrition, University of Sargodha, Sargodha, Pakistan

F. ur Rehman
Quaid i Azam, University, ISLAMABAD, Pakistan

J. Zheng (✉)
International Joint Research Laboratory for Global Change Ecology, School of Life Sciences, Henan University, Kaifeng, Henan, China
e-mail: jqzheng@henu.edu.cn

S. Siddiqui et al. (eds.), *Frontiers in Sustainability*,
https://doi.org/10.1007/978-3-032-19740-5_10

method of land use planning that emphasizes ecological integrity first while attending to socioeconomic demands. Effective land use can be a key component of accomplishing biodiversity conservation and global sustainability objectives by promoting cross-sector collaboration and integrating indigenous knowledge.

Keywords Biodiversity Conservation · Sustainable Land Use · Ecological Health · Ecosystem Services

1 Introduction

Sustainable land use is the use and management of land resources in a manner that satisfies the present needs, negotiating the capacity of future generations to meet their own (Worlanyo and Jiangfeng 2021). Land should be more resilient and more productive by means of balancing environmental health, social well-being, and economic growth (Bilderback, 2024). Sustainable land use includes reforestation, sustainable agriculture, and urban design helps to stop long-term ecological stability as well as land deterioration. Implementing sustainable land use plans will help to offset the climate changes, improve food security, and safeguard significant natural resources (Deng et al., 2024). These methods slow down deforestation, preserve water quality, assist us to maintain soil fertility, and provide effective continuity of the ecosystem (Farooq, 2023). Maintaining the equilibrium of the surroundings and ensuring the lives of the many species of plants and animals rely fundamentally on the protection of biodiversity. Very vital for human existence, biodiversity supports ecological functions like pollination, soil building, and water purification. It also strengthens ecosystems by letting nature adapt to environmental fluctuations and recover from other human activities like natural catastrophes (Verma 2025). Conservation of biodiversity helps to preserve genetic variation; this is crucial for the development of agricultural sustainability and pharmaceutical discovery. Environmental pollution, climate change, overexploitation, and habitat destruction all contribute to the loss of biodiversity endangering human welfare and ecosystems (Smith et al., 2022). Reducing biodiversity causes agricultural output to drop, increases disease vulnerability, and reduces natural defenses against climate change. Protecting species from extension depends much on conservation techniques like the management of protected areas, wildlife corridors, and sustainable lands (Pettorelli et al., 2021). Habitat and species loss follow from anthropogenic activities like urbanization, deforestation, and intensive agriculture. Unmanageable land use methods influence reproduction, food location, species migration ability, and degradation of ecosystems. This disturbance reduces biodiversity and increases the vulnerability of the surroundings to outside factors such as shifting temperatures and invading species. Furthermore, resulting from land mismanagement could be habitat degradation, water pollution, soil erosion, and impact on animal population (Feng et al., 2022).

Likewise, sustainable land use practices can help to improve biodiversity by means of the maintenance of biological corridors, conservation of natural ecosystems, and adoption of regenerative agricultural methods. To live with nature, agroforestry, ethical urban planning, and conservation techniques help to lower environmental damage and support human growth (Kumar et al. 2022). One may generate well managed land that balances nature with human civilizations by regulating the preservation of biodiversity. These strategies provide flexibility, enhance ecosystem services, and help to preserve natural resources for following generations. The degree of ecosystem robustness and the facilities offered depend on the direct use of land (Arneth et al., 2021). By means of responsible land use, the practice of conservation helps to mitigate the negative consequences of habitat degradation, deforestation, and climate change. Sustainable land methods include urban green design, reforestation, and agroforestry allow one to embrace environmental flexibility, social advantages, and financial ones. Policies protecting biodiversity and ensuring that land resources remain sustainable and productive for next generations need for a combined approach encompassing governments, businesses, and communities (Ahirwal & Pandey, 2021).

The chapter aims to assess these activities in terms of their ecological effects and sustainability possibilities, therefore revealing paths promoting human development as well as preservation of biodiversity. From intensive agriculture and urban growth to forestry and conservation, this chapter seeks to provide a targeted framework for the discussion on several land use patterns that impact biodiversity and the supply of significant ecosystem services.

2 Methodology

This chapter is based on a comprehensive literature review of peer-reviewed articles, global assessment reports and relevant policy documents. The analysis involves a comparative assessment of different land use practices including agriculture, urban development, forestry, and conservation, in terms of their impacts on biodiversity and ecosystem services. Selected case studies and existing empirical data are used to illustrate key points and support visual representations. This approach ensures a broad yet evidence-based understanding of the ecological consequences and sustainability potential of various land use strategies.

3 Principles of Sustainable Land Management

Sustainable land management (SLM) is based on four principles (FAO, 2021):

(a) Targeted policy and institutional assistance including the creation of local income generating incentive systems for SLM acceptance.
(b) Participative and land user-driven strategies.
(c) The coordinated application of natural resources on farms and at the level of an ecosystem; and.
(d) Multilevel, multi stakeholder engagement and partnerships at all levels land users, technical specialists and legislators.

4 Ecosystem-Based Approach

Maintaining the condition of the environment used for human activities requires ecosystems-based methods. This approach reveals that, for human survival, ecosystems provide vital amenities such as temperature control, water purification, soil fertility, and others (Hernández & Camerin, 2024). Policies of land management may boost biodiversity and lower the danger of environmental damage. Wetland restoration, forestry, and sustainable agriculture assist to preserve the balance of nature and ensure that the ecosystem stays robust to outside pressure like human population increase and climate change (Roy et al., 2024). Using ecosystem-based strategy demands cooperation among politicians, local communities, and scientists to reduce the negative environmental impacts. Natural basis solutions included into land use plans serve to boost sustainability by means of green infrastructure and agricultural conservation. This strategy largely depends on adaptive management strategies, which consist in constant observation and modification based on environmental input. Furthermore, improving sustainability is inspiring land use plans involving green infrastructure and nature-based solutions like conservation farming. This method guarantees long-term environmental stability and resource availability for next generations by means of a balance between human growth and natural processes (Wamsler et al. 2020).

Many land uses incorporate urban development, forestation, and agricultural integrated land management strategies maximize output while conserving environmental quality by way of approaches of coordinated land management. Examining social, economic, and environmental factors of decisions guarantees that this all--encompassing approach guarantees that land resources are managed morally and profitably (Fischer, 2023). To lower erosion and promote biodiversity, agroforestry systems, for instance, mix tree planting with farming using soil fertility. In keeping with this, intentional urban design incorporates green patches and buffer zones to guard natural surroundings in addition to human needs (Fahad et al., 2022).

Good integrated land management calls for local, national, and international cross-sector cooperation as well as policy consistency. Among other technologies for land use planning, remote sensing and geographic information systems (GIS) assist to monitor environmental conditions and advise direct sustainable decisions (Bourbonnais, 2022). Encouragement of participatory ways whereby local communities and stakeholders help to coordinate land use guarantees that policies are inclusive and efficient. Combining multiple points of view and land use needs helps this approach to minimize resource conflicts and environmental damage and so promote sustainability, resilience, and long-term productivity (Kanojia et al., 2024). Reaching sustainable growth without sacrificing Earth's natural systems requires balancing ecological integrity with human demands. Among human activities often stressing ecosystems contributing to habitat loss and pollution as well as deforestation include industry, infrastructure development, and agriculture. First order resource efficiency comes from sustainable land use; it also helps to lower waste and protect important ecosystems. Low-impact farming, circular economy ideas, and sustainable forestry are among other ways that meet human needs while maintaining environmental services and biodiversity (Chen et al., 2024).

Encouragement of prudent land use among governments, companies, and people depends heavily on awareness-raising education. Policies supporting sustainable land management including fines for environmental damage and incentives for conservation agriculture support more ecologically friendly behavior (Liang et al., 2021). Investing in renewable energy, water saving, and sustainable building will assist to lower human effect on natural environments even more. Knowing that environmental health is strongly related to human prosperity, society can create rules and procedures guaranteeing a harmonic coexistence between nature and development (Meyfroidt et al., 2022).

5 Major Threats to Biodiversity from Land Use Changes

Major threats to the biodiversity at global level have been depicted in Fig. 1. Deforestation and habitat loss are among the most significant threats to biodiversity since they ruin ecosystems sustaining many different plant and animal species.

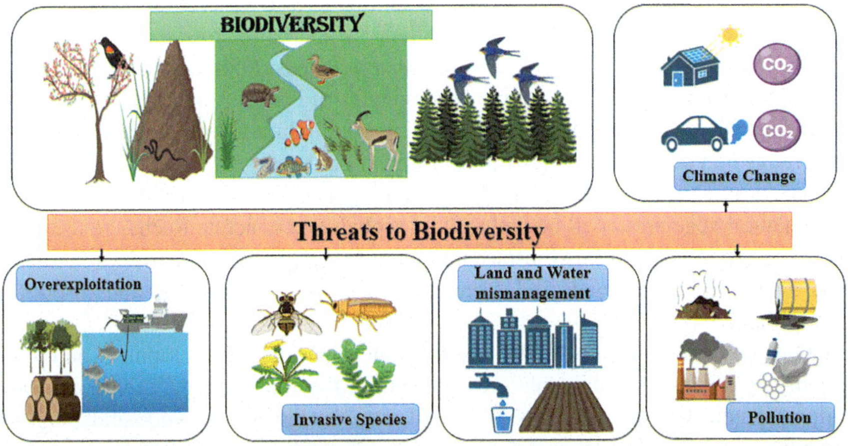

Fig. 1 Biodiversity and its major threats

Large-scale logging, agricultural land clearance, and infrastructure development fragment habitats, therefore compromising species survival and reproduction (Bodo et al., 2021). Key habitat for biodiversity, forests, wetlands, and grasslands are swiftly being destroyed by human activities. Apart from threatening species, its absence disrupts ecological processes including carbon sequestration, water management, and soil fertility (Shah et al., 2022). Beyond only reducing biodiversity, deforestation fuels environmental instability and climatic change. Greenhouse gas emissions increase without trees absorbing carbon dioxide, therefore hastening global warming. Furthermore, aggravating human-wildlife conflicts is habitat degradation as displaced animals are compelled to visit human communities in quest of food and cover (Kumar et al., 2022). Reducing these effects depends on efforts at conservation including reforestation, afforestation, and the creation of protected areas. Governments and companies have to use sustainable forestry techniques, land policies, policy emphasis on habitat preservation (Kumar et al., 2022).

Major sources of biodiversity loss include monoculture and agricultural development as farms built to meet the growing food and resource demands transform marshes, grasslands, and woodlands (Emmerson et al., 2016). Many times, substituting single-crop farms for different ecosystems, extensive agricultural operations decrease habitat complexity and eliminate food sources for many different species. When one kind of crop is grown across large regions, monoculture farming depletes soil nutrients, increases susceptibility to pests and diseases, and mostly relies on chemical fertilizers and pesticides (Buzhdygan & Petermann, 2023). First then this impacts pollinators, soil creatures, and aquatic life by means of runoff pollution. Sustainable farming methods cut down the loss of biodiversity even if food production is stable. Organic farming, agroforestry, and crop rotation help to increase these features by virtue of their effects on soil fertility, water retention, and preservation of beneficial insects and animals. Rotation of crops on rotation Encouragement of small-scale, varied farming systems might also assist to preserve genetic variability in cattle and crops, therefore defending agricultural ecosystems against disease outbreaks and climate change. Governments and businesses as well as conscientious land use rules and research funding must help standards for sustainable farming to develop environmental-friendly agricultural approaches (Dhiman, 2020).

Urbanization and infrastructure development drastically alter the natural surroundings; they harm habitat, alter the ecology, and induce animal migration. Rising dwellings and infrastructure for transportation split ecosystems and hence separate groups of plants and animals, thereby reducing genetic diversity (Tong et al. 2023). The natural vegetation will be replaced by asphalt and concrete to change the local temperature trends and raise surface temperature. Even more undermines biodiversity is building dams, roads, industrial zones often generating water and air pollutants (Mehra et al., 2022).

One may help to minimize the negative consequences of development on biodiversity by means of sustainable urban planning. Rooftop gardens, urban parks are among green infrastructure initiatives that may develop urban woodlands, animal corridors, and ecosystems in cities and boost ecological connectedness. Emphasizing compact building, efficient public transportation, and preservation of green spaces, smart development laws help to balance environmental conservation with urban growth. Policies supporting environmental impact assessments (EIAs) ensure that land use planning guarantees conservation of biodiversity even before major constructions (Ashinze et al., 2024). Mining companies and extractive industry Mining and extractive businesses gravely damage biodiversity by disrupting ecosystems, contaminating water sources, and changing landscapes. Open-pit mining to oil drilling to mineral extraction leads to deforestation, soil erosion, and toxic mercury and arsenic chemical contamination of rivers and lakes. Apart from the loss of species of plants and animals, these activities affect the means of life for indigenous people depending on natural resources. Usually resulting in habitat fragmentation, mining operations limit animal migration and adaptation to environmental change (Seki, 2023). Every stricter regulation implemented, environmental protection laws followed, and the others decided how mining lessens its consequences. Companies have to use moral mining techniques like land protection, trash reduction, and reduced water consumption. Moreover, reducing the need for freshly acquired minerals are recycling initiatives and replacement materials (Dehkordi et al., 2024). Working together, governments and environmental groups may identify protected areas where mining is prohibited, therefore preserving rich areas of biodiversity for next generations free from disturbance. Investments in renewable energy sources and sustainable extraction techniques help even more to reduce dependency on harmful mining technologies (Sharifi et al., 2023).

6 Sustainable Agricultural Practices for Biodiversity Conservation

Combining animals, crops, and trees in agroforestry and permaculture produces a balanced and self-sustaining environment. Agroforestry gives habitat for species, lowers erosion, and increases soil fertility by way of tree cover. It also accelerates carbon sequestration, hence slowing down world temperature (Biswas et al., 2023). Conversely, permaculture stresses developing agricultural systems modeled by natural ecosystems, therefore fostering variety, resilience, and sustainability. These strategies guarantee long-term production by including native plants, natural water management, and restricted outside inputs, which therefore preserve biodiversity and ecological services by means of which one may insure long-term sustainability (Singh & Singh, 2024).

To lower pesticide usage hence reducing environmental harm and encouraging biodiversity, organic farming stresses natural inputs, crop variety, and soil health. Eliminating synthetic fertilizers and pesticides in organic farming lowers chemical runoff, therefore safeguarding pollinators, helpful insects, and aquatic life from dangerous contact (Kharel & Sahoo, 2023). Instead, using composting, biological pest control, and natural fertilizers, it preserves soil fertility and ecological balance. Less pesticides mean more birds, bees, and other pollinators required for food production can live. Adoption of organic farming generally would promote biodiversity, assist to restore damaged land, and develop improved food systems (Saikanth et al., 2023).

Maintaining soil health and avoiding land degradation depend critically on crop rotation and soil conservation methods (Liang et al., 2023). Seasonally rotating crops helps to minimize the need for chemical treatments by reducing the accumulation of pests and diseases and by nitrogen-fixing plants, therefore improving the soil fertility. Cover cropping, mulching, and no-till farming help to prevent erosion of the soil, increase moisture retention, and maintain microbial life, all vital for plant development. These techniques guarantee long-term output while maintaining biodiversity in and around farms, hence building a more resilient agricultural system (Shah et al., 2021).

By use of sensor-based irrigation systems, drones, and GPS mapping among other technologies, precision farming optimally uses resources and reduces environmental effect. Water, fertilizers, and pesticides used just, when necessary, helps to lower too strong runoff that can endanger biodiversity (Getahun et al., 2024). Through soil conditions, weather patterns, and crop conditions, data-driven agricultural methods enable farmers to make informed decisions for sustainable development. By means of water saving, habitat protection, and maintenance of healthy soil ecosystems, smart farming technologies assist in maintaining biodiversity (Sishodia et al., 2020).

7 Urban Land Use Planning for Biodiversity

Urban land use planning for biodiversity comprises the managing and designing urban space to sustain and increase biodiversity guaranteeing that cities and natural ecosystems can co-occur. This method is very important because the process of urbanization leads to degradation, fragmentation, and habitat loss, which impact biodiversity (Reid et al., 2019). Below are some important approaches and reflections for the planning for integrating biodiversity into urban land use planning.

7.1 Ecological Corridor and Green Infrastructure

Ecosystem, wildlife and its habitat are in danger due to the urbanization process but this process can be recovered by ecological corridor as genes are drifted and biodiversity moves, and ecosystems turn out to be stable by it (Peng et al., 2017). Sustainability of wildlife and balance of environment is dependent on ecological corridors (Gregory et al., 2021). Incorporation of natural solutions such as green infrastructure in urban development is useful for our ecosystem (Beaugeard et al., 2021). In contrast to the grey infrastructure, green infrastructure meets the goal of sustainability as it provides vegetation and other fundamentals to support wildlife

(Ahern, 2007; Dhakal & Chevalier, 2017). On the sideways of rivers and other wetlands, riparian buffers, green roofs as well as forests in urban areas are some of the examples of green infrastructure as they decrease temperature in urban areas and quality of air and water gets improved (Cole et al., 2017). A lot of countries have included green infrastructure in their cities such as Singapore's Park Connector Network, and have natural conservation areas with it (Tanuwidjaja, 2011). Another example is the natural reserve for biodiversity in the Green Belt Plan of Europe which encourages wildlife in it (Lafortezza et al., 2013). But there are several challenges such as lack of policies, disagreement of land consumption and economic restrictions in implementing the idea of green infrastructure and ecological corridors. Improvement in the funding system, strict rules and urban policies are required to address these problems (Dhakal & Chevalier, 2017). One effective way is to use geographic information systems for ecological corridor monitoring and mapping (Popescu et al., 2022). Implementing the idea of supportable urban development requires awareness of the community by starting awareness programs and encouraging people to participate in the journey toward sustainable living.

7.2 *Sustainable Urban Development Models*

The purpose of sustainable urban development models is to make much safer and comfortable infrastructure for cities. Some examples of the sustainable urban development models are as follows:

7.2.1 Sustainability Indicator Model

In order to build sustainable urban areas, this model emphasizes on the ecofriendly planning in the development of cities (de Roo & Miller, 2017).

7.2.2 Low Carbon Eco-City Planning Model

The purpose of this model is to promote the development of low-carbon, resource-efficient, and energy-efficient urban systems (Lambrechts & Sinha, 2017).

7.2.3 Garden City Model

This model combines the natural and modern characteristics. In this model, cities have green areas among modern infrastructure to produce an ecofriendly atmosphere (Evalina & Sawab, 2011).

7.2.4 Resilient City Model

The focus of the resilient city model is to make an environment of cities which adapt and recover after disaster via sustainable infrastructure, nature-based solutions and community involvement. Various strategies like green corridors, infrastructure which is flood resilient and decentralized energy systems boost urban resistance in contrast to ecological stress (Hernantes et al., 2019).

7.2.5 Smart Growth Model

By encouraging walkability, public conveyance, and effective use of land, the smart growth model focuses on compact, mixed-use development. This model lowers environmental impact and urban sprawl by giving priority to infill construction, improves community cohesiveness and maintains open areas (Daniels, 2001).

7.3 Urban Woods and Wetlands: Their Functions

Wetlands and woods play an important part to increase biodiversity and provide stability of ecosystems and other economic and social services that are beneficial for human beings. Combination of wetlands and woods in urban areas develop the better living style and well-being of inhabitants as well as for wildlife (Alberti, 2005).

Habitation of a number of animals such as insects, birds, small mammals, and so on is dependent on the wetlands and woods in urban areas. So, the enhancement of biodiversity and conservation of natural environment is the key role of wetland and woods within cities because they offer home to a lot of aquatics as well as terrestrial species (Bolund & Hunhammar, 1999; Sharma & Naik, 2024). Moreover, CO_2 in the environment is dropped by increasing trees in cities as they clean air quality by taking it up and captivate harmful contaminants in air and leads to a clean environment (Nowak & Dwyerm, 2007). Reduction in temperature and cooling of the surrounding environment in cities is seen by woods due to the cooling effect of the transpiration process (Grimm et al., 2008). Wetlands, on the other hand, are indispensable for water management. They filter out pollutants from runoff, improve the quality of water, this spongey action absorbs heavy rain and reduces flood risk too (Mitsch & Gosselink, 2015). Furthermore, health of soil, biogeochemical cycling and decomposition of living beings is also carried out by wetlands (Mustafa et al., 2024). Emotional and physical health of human beings is improved by natural bodies such as green forests and wetlands in urban environments as they offer aesthetic natural views, entertainment, walking, and cycling (Tzoulas et al., 2007). Natural environments boost living standards of humans and decrease anxiety and stress of inhabitants, engage and attract tourists and enhance the overall economic growth of cities (Nowak & Dwyerm, 2007). Conservation of these wetlands and woods is significant because they are beneficial for human beings and biodiversity. For a suitable and habitable ecosystem, management of wetlands and woods is crucial in cities.

8 Protected Areas and Conservation Strategies

National parks, animal reserves, marine protected areas (MPAs), and their defense of ecosystems against human intrusion help to greatly preserve biodiversity. These protected zones give threatened species safe haven by preserving their native habitats and allowing population rebound. By preventing habitat loss, poaching, and deforestation, national parks and animal reserves assist to maintain ecological balance (Rahman et al., 2022). By thus ensuring the long-term survival of oceans and coastal ecosystems, MPAs help protect marine biodiversity by reducing overfishing, pollution, and destructive activities such as coral mining

(Li et al., 2022; Wani et al., 2025).

Beyond protection, protected areas support local businesses, tourists, and scientific study. They provide chances for research on ecosystems, tracking effects of climate change, and creation of sustainable development plans. Many national parks and reserves also encourage ecotourism, which brings money for local people and conservation projects back-off-stream (Forje & Tchamba, 2022). To stop unlawful activity and guarantee that these regions really serve as hotspots for biodiversity, nevertheless, good administration and law enforcement of conservation policies are very vital. Enhancing the efficacy of protected areas by including local people into conservation strategies and strengthening legal safeguards would help projects led by communities. Projects organized by communities enable local people to participate actively in preserving their natural surroundings. Having coexisted peacefully for millennia, many rural and indigenous people have been very good animal caretakers. Including neighbors in environmental initiatives assures rational use of resources and fosters responsibility. Though they provide income, projects include sustainable fishing techniques, wildlife corridors, community-managed forests assist to protect ecosystems even now (Esmail et al., 2023). These initiatives also assist capacity-building and education, therefore empowering communities with the tools needed for environmental conservation. Many effective conservation programs include participatory decision-making, in which local stakeholders work with governments and NGOs to create and carry out simultaneously plans benefitting people and the environment. Particularly helping attempts at preservation are incentives like ecotourism, sustainable agriculture, and payment for ecosystem services (PES). When communities feel that maintaining biodiversity will clearly benefit them, they are more likely to actively support conservation and reject environmentally detrimental activities (Schaal et al., 2023).

Projects in rewilding and restoration ecology aim to bring damaged ecosystems back to their native condition. Reforestation, wetland rehabilitation, and soil rehabilitation among other things help to increase ecosystem services and biodiversity. Restoring natural vegetation, enhancing water quality, and reintroducing important species helps ecosystems to be in a normal state of functioning. This helps not only animals but also slows down erosion, fights climate change, increases carbon sequestration (Yurui et al., 2021). Rewilding is a specific technique used in restoration ecology to restore important Keystone species back into an ecosystem and allow natural processes to shape the topography. For instance, reintroduction of wolves in Yellowstone National Park helped control deer numbers, therefore promoting more biodiversity and healthier flora. Projects involving large-scale rewilding throughout North America and Europe are helping to assist ecosystem recovery by restoring natural habitats, and building animal corridors. By giving ecosystem restoration priority, communities can guarantee the long-term existence of many species and undo the harm resulting from human activity (Merola, 2023).

9 Policies and Legal Frameworks for Sustainable Land Use

Globally agreements greatly help to encourage sustainable land use by way of international standards and requirements for environmental protection (Table 1). The fundamental focus of the Convention on Biological Diversity (CBD) is preserving biodiversity, assuring sustainable use of natural resources, and hence promoting fair profits from genetic resources (Hoban et al., 2021). Particularly Goal 15, Life on Land, and the Sustainable Development Goals (SDGs) provide great importance land restoration, avoidance of deforestation, and conservation of biodiversity. By means of sustainable land management techniques to increase carbon sequestration, the Paris Agreement urges countries to reduce carbon emissions, therefore tackling climate change. These agreements promote global collaboration in addressing environmental concerns, support conservation initiatives, and influence national policies (Hoban et al., 2023).

Land use regulations both nationally and locally provide the legal basis for sustainable management of resources while juggling environmental preservation with economic growth. Protected area designations, zoning policies, and environmental impact studies (EIs) assist governments in controlling land development and guaranteeing sustainable practices. Projects for afforestation, incentives for organic farming, and bans on deforestation among policies aimed to help lower habitat loss and soil degradation (Ren et al. 2026). Moreover, regional integrated land use

Table 1 Case studies of sustainable land use and indigenous as well as traditional land management in China and other countries

Examples of sustainable land use worldwide		
Case studies	Description	References
Agroecology in Brazil	Emphasizing crop diversity, reduced pesticide use, and soil health restoration, Brazilian agroecological approaches help to create ecosystem resilience	Stratton (2021)
Forest Conservation in Costa Rica	The payment for environmental services (PES) program of Costa Rica effectively combines local economic activities with forest preservation to help to restore biodiversity and stop deforestation	Cordero (2018)
Habitat Restoration in the Netherlands	Projects involving wetland and peat bog restoration in the Netherlands have helped to restore aquatic life especially as well as improved water quality	Nienhuis et al. (2002)
Ecological Restoration of the Loess Plateau	Reforestation and grassland rehabilitation among large-scale restoration initiatives in China's Loess Plateau have improved soil quality, raised biodiversity, and helped to slow down desertification	Kong et al. (2024)
Sustainable Rice Intensification in Southern China	By lowering water use, limiting chemical fertilizers, and so improving soil health, sustainable rice intensification (SRI) methods boost yields while maintaining biodiversity in southern China	Dahlgreen & Parr (2024)
Green Belt and Ecological Corridors in Beijing, China	Beijing's development of green belts and ecological corridors has helped to rebuild split ecosystems by means of animal corridors and so improve urban biodiversity	Yang et al. (2024)
Green Urban Planning in Singapore	In order to solve urbanization issues such as heat islands, Singapore combines green infrastructure including vertical gardens, urban woods, and green rooftops, therefore preserving biodiversity	Priya & Senthil (2024)
Lessons from indigenous and traditional land management		
Indigenous Fire Management in Australia	Controlled burns and other Aboriginal fire management methods help to preserve ecosystems, rebuild plant species, and stop major wildfires	Ota (2024)
Sustainable Agriculture Practices in the Andes	By means of terrace farming, crop rotation, and organic pest control—Which guarantee soil fertility and water retention— Indigenous people in the Andes preserve biodiversity	Vikas & Ranjan (2024)
Community-led Mangrove Conservation in the Philippines	Mangroves are protected and restored by local Philippine communities, therefore enhancing coastal biodiversity, water filtering, and storm protection	Grimm et al. (2024)
Pastoralism in the Sahel Region	Properly executed traditional nomadic pastoralism improves rotational grazing and aids in the restoration of degraded land, therefore promoting biodiversity	Herrera Calvo (2024)
Traditional Agroforestry Practices in the Tibetan Plateau	Agroforestry methods combining cattle grazing with tree planting have long been used by Tibetan herders to increase land fertility and stop soil erosion	Li (2024)

(continued)

Table 1 (continued)

Examples of sustainable land use worldwide		
Case studies	Description	References
Sustainable Forestry in the Daliang Mountains	For millennia, indigenous people living in the Daliang Mountains have managed forests and sustainably logged, therefore balancing local lives with environmental preservation	Ha (2024)
Traditional Water Management in the Loess Plateau	Terracing and water collecting among other traditional water management techniques used on the Loess Plateau have served to preserve biodiversity while advancing sustainable agriculture.	Li et al. (2024)

planning guarantees that urban, industrial, and agricultural development takes place free from compromising ecological integrity. Effective policy-making execution, transparency, and community engagement are the means by which one may achieve long-term sustainability (Roy et al., 2022). Indigenous land rights and traditional knowledge are fundamental components of sustainable land use as traditionally indigenous people have managed ecosystems in ways that help to preserve biodiversity.

By acknowledging and safeguarding land rights for indigenous people, one may assist to save wetlands, forests, and other important ecosystems from unsustainable use (Dawson et al., 2021). Traditional ecological knowledge provides insightful analysis of sustainable resource management including rotational farming, water conservation methods, and herbal medical practices. Many conservation projects mix indigenous knowledge with scientific research more and more in order to create more effective and culturally appropriate strategies. Strengthening legal recognition of indigenous land ownership and integrating indigenous communities in decision--making would assist to preserve biodiversity even while social justice and cultural heritage are preserved (Estrada et al., 2022).

10 Role of Technology and Innovation

Remote sensing and Geographic Information Systems (GIS) provide ways of land use planning for Remote sensing and GIS with real-time data on land cover, deforestation, soil condition, and water supply define sustainable land use planning most of all. These instruments allow legislators and environmentalists to track human activity, investigate topographical changes, and apply well-informed laws aimed to prevent habitat degradation. By means of integration of satellite photos with GIS mapping, planners may detect areas of great risk, maximize land use, and build projects balancing development with biodiversity preservation (Hussain et al., 2020). Artificial intelligence (AI) and big data analytics have revolutionized biodiversity preservation by means of trend identification, environmental change prediction, and automatically monitoring species monitoring by vast volumes of ecological data (Ayoola et al., 2024).

Machine learning techniques could identify threatened species, locate illicit deforestation, and evaluate how ecosystems are changing in connection to climate change. Simulating several conservation scenarios helps AI-driven models also aid in restoring habitat, enabling researchers to find the most efficient ways to protect biodiversity and hence promote ecological balance (Mishra et al. 2025).

Blockchain technology ensures secure and tamper-proof records of land ownership, resource usage, and conservation activities, therefore offering a clear and distributed way of sustainable land management (Friedman & Ormiston, 2022). Blockchain smart contracts support fair trade in sustainable agriculture, ethical land use practices, and land corruption prevention as well as others. This technology also enhances traceability in supply chains, thus ensuring that items come from legally managed forests and ecologically friendly sources and so fostering sustainable behaviors among businesses and consumers (Polas et al., 2025).

11 Future Directions and Recommendations

Strong policies assuring preservation of biodiversity and sustainable land use have to be developed by governments and interested parties. This includes in national development plans integration of conservation aims with national progress, application of strict environmental rules, and incentives for sustainable living. Policies should give first priority to careful land management and sustainable farming coupled with preservation of significant places of importance. Furthermore, the long-term survival and effective implementation of conservation efforts require the involvement of governments, corporations, and local communities. Understanding their duty as defenders of biodiversity will enable one to additionally enhance conservation projects by increasing land ownership rights especially for indigenous people.

Promoted should be sustainable livelihoods that complement environmental goals, such tourism and organic farming, thereby generating financial incentives for preservation. A society founded on sustainability and preservation depends much on awareness-raising events as well as education. Environmental education should be included into classrooms, businesses, schools, and community projects to enable people to appreciate the need of several land management strategies and values. Public awareness initiatives, seminars, and media exposure help to encourage good environmental behavior. Education could provide people the means to make wise land use decisions by integrating current scientific methods with traditional ecological knowledge. Moreover, promoting early environmental stewardship might result in long-term favorable changes in attitudes and behavior linked with sustainability.

12 Conclusion

Maintaining long-term environmental sustainability primarily rests on allowing the neighbors to choose suitable land use plans. Agroforestry, regenerative farming, and careful water management community-led projects all help to successfully balance environmental preservation with financial needs. Providing technical and financial assistance to small businesses, landowners, farmers, and developers would help increase the acceptance of environmentally friendly solutions. Participatory decision-making involving people for land use planning and resource management also promotes accountability and commitment to conservation initiatives.

References

Ahern, J. (2007). Green infrastructure for cities: The spatial dimension. *Cities of the future: Towards integrated sustainable water and landscape management, 13*, 267–283.

Ahirwal, J., & Pandey, V. C. (2021). Restoration of mine degraded land for sustainable environmental development. *Restoration Ecology, 29*, e13268.

Alberti, M. (2005). The effects of urban patterns on ecosystem function. *International Regional Science Review, 28*(2), 168–192.

Arneth, A., Olsson, L., Cowie, A., Erb, K.-H., Hurlbert, M., Kurz, W. A., Mirzabaev, A., & Rounsevell, M. D. (2021). Restoring degraded lands. *Annual Review of Environment and Resources, 46*, 569–599.

Ashinze, U. K., Edeigba, B. A., Umoh, A. A., Biu, P. W., & Daraojimba, A. I. (2024). Urban green infrastructure and its role in sustainable cities: A comprehensive review. *World Journal of Advanced Research and Reviews, 21*, 928–936.

Ayoola, V. B., Idoko, I. P., Eromonsei, S. O., Afolabi, O., Apampa, A., & Oyebanji, O. (2024). The role of big data and AI in enhancing biodiversity conservation and resource management in the USA. *World Journal of Advanced Research and Reviews, 23*, 1851–1873.

Beaugeard, E., Brischoux, F., & Angelier, F. (2021). Green infrastructures and ecological corridors shape avian biodiversity in a small French city. *Urban Ecosystems, 24*(3), 549–560.

Bilderback, S. (2024). Integrating training for organizational sustainability: The application of sustainable development goals globally. *European Journal of Training and Development, 48*, 730–748.

Biswas, S., Pakhira, R., Khan, O., & Bose, S. (2023). *The role of agroforestry-based farming in sustainable agriculture* (p. 99). AkiNik Publ.

Bodo, T., Gimah, B. G., & Seomoni, K. J. (2021). Deforestation and habitat loss: Human causes, consequences and possible solutions. *Journal of Geographical Research, 4*, 22–30.

Bolund, P., & Hunhammar, S. (1999). Ecosystem services in urban areas. *Ecological Economics, 29*(2), 293–301.

Bourbonnais, M. (2022). Applications of geographic information systems, spatial analysis, and remote sensing in environmental impact assessment. In Routledge handbook of environmental impact assessment (pp. 201–220). Routledge.

Buzhdygan, O. Y., & Petermann, J. S. (2023). Multitrophic biodiversity enhances ecosystem functions, services and ecological intensification in agriculture. *Journal of Plant Ecology, 16*, rtad019.

Chen, F., Jiang, F., Ma, J., Alghamdi, M. A., Zhu, Y., & Yong, J. W. H. (2024). Intersecting planetary health: Exploring the impacts of environmental stressors on wildlife and human health. *Ecotoxicology and Environmental Safety, 283*, 116848.

Cole, L. B., McPhearson, T., Herzog, C. P., & Russ, A. (2017). Green infrastructure. *Urban Environmental Education Review*, 261–270.

Cordero, S. (2018). Local regulatory and economic instruments to encourage tropical forestry conservation: an analysis of the policy process in Costa Rica and Mexico.

Dahlgreen, J., & Parr, A. (2024). Exploring the impact of alternate wetting and drying and the system of rice intensification on greenhouse gas emissions: A review of rice cultivation practices. *Agronomy, 14*(2), 378.

Daniels, T. (2001). Smart growth: A new American approach to regional planning. *Planning Practice and Research, 16*(3–4), 271–279.

Dawson, N. M., Coolsaet, B., Sterling, E. J., Loveridge, R., Gross-Camp, N. D., Wongbusarakum, S., Sangha, K. K., Scherl, L. M., Phuong Phan, H., & Zafra-Calvo, N. (2021). *The role of Indigenous peoples and local communities in effective and equitable conservation.*

de Roo, G., & Miller, D. (2017). *Integrating city planning and environmental improvement: Practicable strategies for sustainable urban development.* Routledge.

Dehkordi, M. M., Nodeh, Z. P., Dehkordi, K. S., Khorjestan, R. R., & Ghaffarzadeh, M. (2024). Soil, air, and water pollution from mining and industrial activities: Sources of pollution, environmental impacts, and prevention and control methods. *Results in Engineering, 102729.*

Deng, G., Jiang, H., Zhu, S., Wen, Y., He, C., Wang, X., Sheng, L., Guo, Y., & Cao, Y. (2024). Projecting the response of ecological risk to land use/land cover change in ecologically fragile regions. *Science of the Total Environment, 914*, 169908.

Dhakal, K. P., & Chevalier, L. R. (2017). Managing urban stormwater for urban sustainability: Barriers and policy solutions for green infrastructure application. *Journal of Environmental Management, 203*, 171–181.

Dhiman, V. (2020). Organic farming for sustainable environment: Review of existed policies and suggestions for improvement. *International Journal of Research and Review, 7*, 22–31.

Emmerson, M., Morales, M. B., Oñate, J. J., Batary, P., Berendse, F., Liira, J., Aavik, T., Guerrero, I., Bommarco, R., & Eggers, S. (2016). How agricultural intensification affects biodiversity and ecosystem services. *Advances in Ecological Research, 55*, 43–97.

Esmail, N., McPherson, J. M., Abulu, L., Amend, T., Amit, R., Bhatia, S., Bikaba, D., Brichieri-Colombi, T. A., Brown, J., & Buschman, V. (2023). What's on the horizon for community-based conservation? Emerging threats and opportunities. *Trends in Ecology & Evolution, 38*, 666–680.

Estrada, A., Garber, P. A., Gouveia, S., Fernández-Llamazares, Á., Ascensão, F., Fuentes, A., Garnett, S. T., Shaffer, C., Bicca-Marques, J., & Fa, J. E. (2022). Global importance of indigenous peoples, their lands, and knowledge systems for saving the world's primates from extinction. *Science Advances, 8*, eabn2927.

Evalina, E., & Sawab, H. (2011). Garden City': The suitability of its principles as a model to the contemporary planning. *NALARs, 10*, 1.

Fahad, S., Chavan, S. B., Chichaghare, A. R., Uthappa, A. R., Kumar, M., Kakade, V., Pradhan, A., Jinger, D., Rawale, G., & Yadav, D. K. (2022). Agroforestry systems for soil health improvement and maintenance. *Sustainability, 14*, 14877.

FAO. (2021). *Sustainable land management | land & water | Food and Agriculture Organization of the United Nations.* https://www.fao.org/land-water/land/sustainable-land-management/en/#:~:text=SLM%20is%20based%20on%20four%20principles%3A%201%20targeted,levels%20%E2%80%93%20land%20users%2C%20technical%20experts%20and%20policy-makers.

Farooq, M. (2023). Conservation agriculture and sustainable development goals. *Pakistan Journal of Agricultural Sciences, 60*, 3.

Feng, S., Zhao, W., Zhan, T., Yan, Y., & Pereira, P. (2022). Land degradation neutrality: A review of progress and perspectives. *Ecological Indicators, 144*, 109530.

Fischer, T. B. (2023). *Strategic environmental assessment in transport and land use planning.* Routledge.

Forje, G. W., & Tchamba, M. N. (2022). Ecotourism governance and protected areas sustainability in Cameroon: The case of Campo Ma'an National Park. *Current Research in Environmental Sustainability, 4*, 100172.

Friedman, N., & Ormiston, J. (2022). Blockchain as a sustainability-oriented innovation?: Opportunities for and resistance to Blockchain technology as a driver of sustainability in global food supply chains. *Technological Forecasting and Social Change, 175*, 121403.

Getahun, S., Kefale, H., & Gelaye, Y. (2024). Application of precision agriculture technologies for sustainable crop production and environmental sustainability: A systematic review. *Scientific World Journal, 2024*, 2126734.

Gregory, A., Spence, E., Beier, P., & Garding, E. (2021). Toward best management practices for ecological corridors. *Land, 10*(2), 140.

Grimm, N. B., Faeth, S. H., Golubiewski, N. E., Redman, C. L., Wu, J., Bai, X., & Briggs, J. M. (2008). Global change and the ecology of cities. *Science, 319*(5864), 756–760.

Grimm, K., Spalding, M., Leal, M., Kincaid, K., Aigrette, L., Amoah-Quiminee, P., Amoras, L., Amouine, D., Areki, F., & Arends, W. (2024). Including local ecological knowledge (LEK) in Mangrove conservation & restoration. A *Best-practice guide for practitioners and researchers.*

Ha, J. (2024). *Frontier above the clouds: A trans-Pacific history of mountain engineering in South Korea*. University of California.

Hernández, R. C., & Camerin, F. (2024). The application of ecosystem assessments in land use planning: A case study for supporting decisions toward ecosystem protection. *Futures, 103399.*

Hernantes, J., Maraña, P., Gimenez, R., Sarriegi, J. M., & Labaka, L. (2019). Towards resilient cities: A maturity model for operationalizing resilience. *Cities, 84*, 96–103.

Herrera Calvo, P. M. (2024). *Global land outlook: Thematic report on rangelands and pastoralists.*

Hoban, S., Campbell, C. D., da Silva, J. M., Ekblom, R., Funk, W. C., Garner, B. A., Godoy, J. A., Kershaw, F., MacDonald, A. J., & Mergeay, J. (2021). Genetic diversity is considered important but interpreted narrowly in country reports to the convention on biological diversity: Current actions and indicators are insufficient. *Biological Conservation, 261*, 109233.

Hoban, S., da Silva, J. M., Mastretta-Yanes, A., Grueber, C. E., Heuertz, M., Hunter, M. E., Mergeay, J., Paz-Vinas, I., Fukaya, K., & Ishihama, F. (2023). Monitoring status and trends in genetic diversity for the convention on biological diversity: An ongoing assessment of genetic indicators in nine countries. *Conservation Letters, 16*, e12953.

Hussain, S., Mubeen, M., Ahmad, A., Akram, W., Hammad, H. M., Ali, M., Masood, N., Amin, A., Farid, H. U., & Sultana, S. R. (2020). Using GIS tools to detect the land use/land cover changes during forty years in Lodhran District of Pakistan. *Environmental Science and Pollution Research, 27*, 39676–39692.

Kanojia, M., Kamani, P., Kashyap, G. S., Naz, S., Wazir, S., & Chauhan, A. (2024). Alternative agriculture land-use transformation pathways by partial-equilibrium agricultural sector model: A mathematical approach. *International Journal of Information Technology*, 1–20.

Kharel, P., & Sahoo, S. (2023). Advancing sustainable agriculture: A comprehensive review of organic farming methods and their implications for a resilient future. *Journal Of Food Chemistry And Nanotechnology, 9*, S335–S341.

Kong, C., Huang, J., Du, S., & Li, G. (2024). Exploring the driving forces of vegetation greening on the loess plateau at the county scale. *Forests, 15*(3), 486.

Kumar, R., Kumar, A., & Saikia, P. (2022). Deforestation and forests degradation impacts on the environment. In *Environmental degradation: Challenges and strategies for mitigation* (pp. 19–46). Cham: Springer International Publishing.

Lafortezza, R., Davies, C., Sanesi, G., & Konijnendijk, C. C. (2013). Green infrastructure as a tool to support spatial planning in European urban regions. *iForest-Biogeosciences and Forestry, 6*(3), 102.

Lambrechts, W., & Sinha, S. (2017). Modeling a low-carbon city: Eco-city and eco-planning. In *Carbon footprint and the industrial life cycle: From urban planning to recycling* (pp. 407–443).

Li, T. (2024). *Sustainable development and management of ecosystem in arid and semi-arid areas: A focus on the Qinghai-Tibet plateau and inner Mongolian grasslands.*

Li, Q., Zhou, Y., & Yi, S. (2022). An integrated approach to constructing ecological security patterns and identifying ecological restoration and protection areas: A case study of Jingmen, China. *Ecological Indicators, 137*, 108723.

Li, P., Kou, X., Wang, Y., & Niu, L. (2024). Building a more sustainable Chinese loess plateau. *Journal of Earth Science, 35*(1), 283–287.

Liang, X., Guan, Q., Clarke, K. C., Liu, S., Wang, B., & Yao, Y. (2021). Understanding the drivers of sustainable land expansion using a patch-generating land use simulation (PLUS) model: A case study in Wuhan, China. *Computers, Environment and Urban Systems, 85*, 101569.

Liang, Z., Xu, Z., Cheng, J., Ma, B., Cong, W.-F., Zhang, C., Zhang, F., van der Werf, W., & Groot, J. C. (2023). Designing diversified crop rotations to advance sustainability: A method and an application. *Sustainable Production and Consumption, 40*, 532–544.

Mehra, S., Singh, M., Sharma, G., Kumar, S., & Navishi, C. P. (2022). Impact of construction material on environment. In *Ecological and health effects of building materials* (pp. 427–442).

Merola, E. (2023). The role of rewilding in combating biodiversity loss. *Ukrainian Journal of Ecology, 13*, 8.

Meyfroidt, P., De Bremond, A., Ryan, C. M., Archer, E., Aspinall, R., Chhabra, A., Camara, G., Corbera, E., DeFries, R., & Díaz, S. (2022). Ten facts about land systems for sustainability. *Proceedings of the National Academy of Sciences of the United States of America, 119*, e2109217118.

Mishra, R. K., Mishra, D., & Agarwal, R. (2025). Environmental sustainability and ecological balance. Implementation of Innovative Strategies in Integral Plant Protection, First Edition: January, 81–96.

Mitsch, W. J., & Gosselink, J. G. (2015). *Wetlands* (5th ed.). John Wiley & Sons.

Mustafa, G., Hussain, S., Liu, Y., Ali, I., Liu, J., & Bano, H. (2024). Microbiology of wetlands and the carbon cycle in coastal wetland mediated by microorganisms. *Science of the Total Environment, 175734.*

Nienhuis, P., Bakker, J., Grootjans, A., Gulati, R., & De Jonge, V. (2002). The state of the art of aquatic and semi-aquatic ecological restoration projects in The Netherlands. *Ecological restoration of aquatic and semi-aquatic ecosystems in the Netherlands (NW Europe)*, 219–233.

Nowak, D. J., & Dwyerm, J. F. (2007). Understanding the benefits and costs of urban forest ecosystems. In *Urban and community forestry in the northeast* (pp. 25–46). Springer.

Ota, A. (2024). *A scoping study of collaborative wildfire management: Association between indigenous and non-indigenous peoples.*

Peng, J., Zhao, H., & Liu, Y. (2017). Urban ecological corridors construction: A review. *Acta Ecologica Sinica, 37*(1), 23–30.

Pettorelli, N., Graham, N. A., Seddon, N., Bustamante, M. D. C., Lowton, M. J., Sutherland, W. J., Koldewey, H. J., Prentice, H. C., & Barlow, J. (2021). Time to integrate global climate change and biodiversity science-policy agendas. *Journal of Applied Ecology, 58*, 2384–2393.

Polas, M. R. H., Afshar Jahanshahi, A., Islam, M. E., Kabir, A. I., Sohel-Uz-Zaman, A. S. M., & Fahad, A. A. (2025). A journey from traditional supply chain processes to sustainability-oriented Blockchain supply chain: The critical role of organizational capabilities. *Business Strategy and the Environment.*

Popescu, O. C., Tache, A. V., & Petrişor, A. I. (2022). Methodology for identifying ecological corridors: A spatial planning perspective. *Land, 11*(7), 1013.

Priya, U. K., & Senthil, R. (2024). Framework for enhancing urban living through sustainable plant selection in residential green spaces. *Urban Science, 8*(4), 235.

Rahman, M. K., Masud, M. M., Akhtar, R., & Hossain, M. M. (2022). Impact of community participation on sustainable development of marine protected areas: Assessment of ecotourism development. *International Journal of Tourism Research, 24*(1), 33–43.

Reid, A. J., Carlson, A. K., Creed, I. F., Eliason, E. J., Gell, P. A., Johnson, P. T., Kidd, K. A., MacCormack, T. J., Olden, J. D., Ormerod, S. J., & Smol, J. P. (2019). Emerging threats and persistent conservation challenges for freshwater biodiversity. *Biological Reviews, 94*(3), 849–873.

Ren, Y., Zhang, Y., Shi, Y., Fan, X., & Cao, Y. (2026). Assessing the effectiveness of large-scale ecological restoration programs: a spatiotemporal analysis of ecological efficiency in forest and landscape restoration. Journal of Forestry Research, 37(1), 68.

Roy, P. S., Ramachandran, R. M., Paul, O., Thakur, P. K., Ravan, S., Behera, M. D., Sarangi, C., & Kanawade, V. P. (2022). Anthropogenic land use and land cover changes—A review on its environmental consequences and climate change. *Journal of the Indian Society of Remote Sensing, 50*(8), 1615–1640.

Roy, S. K., Alam, M. T., Mojumder, P., Mondal, I., Kafy, A.-A., Dutta, M., Ferdous, M. N., Al Mamun, M. A., & Mahtab, S. B. (2024). Dynamic assessment and prediction of land use alterations influence on ecosystem service value: A pathway to environmental sustainability. *Environmental and Sustainability Indicators, 21*, 100319.

Saikanth, K., Singh, B. V., Sachan, D. S., & Singh, B. (2023). Advancing sustainable agriculture: A comprehensive review for optimizing food production and environmental conservation. *International Journal of Plant & Soil Science, 35*(16), 417–425.

Schaal, T., Mitchell, M., Scheele, B. C., Ryan, P., & Hanspach, J. (2023). Using the three horizons approach to explore pathways towards positive futures for agricultural landscapes with rich biodiversity. *Sustainability Science, 18*(3), 1271–1289.

Seki, H. (2023). *Environmental impacts of mining on biodiversity and ecosystem services*. Univ York.

Shah, K. K., Modi, B., Pandey, H. P., Subedi, A., Aryal, G., Pandey, M., & Shrestha, J. (2021). Diversified crop rotation: An approach for sustainable agriculture production. *Advances in Agriculture, 1*, 8924087.

Shah, S. A. H., Bilal, A., Ahmad, M. M., & Bukhari, S. S. (2022). Deforestation is causing a great loss in avian diversity in Pakistan. *American Journal of Zoology, 5*(3), 24–29.

Sharifi, S. A., Zaeimdar, M., Jozi, S. A., & Hejazi, R. (2023). Effects of soil, water and air pollution with heavy metal ions around lead and zinc mining and processing factories. *Water, Air, and Soil Pollution, 234*(12), 760.

Sharma, L. K., & Naik, R. (2024). Wetland ecosystems. In *Conservation of saline wetland ecosystems: An initiative towards UN decade of ecological restoration* (pp. 3–32). Springer Nature Singapore.

Singh, S., & Singh, G. (2024). Agroforestry for sustainable development: Assessing frameworks to drive agricultural sector growth. *Environment, Development and Sustainability, 26*(9), 22281–22317.

Sishodia, R. P., Ray, R. L., & Singh, S. K. (2020). Applications of remote sensing in precision agriculture: A review. *Remote Sensing, 12*(19), 3136.

Smith, P., Arneth, A., Barnes, D. K., Ichii, K., Marquet, P. A., Popp, A., Pörtner, H. O., Rogers, A. D., Scholes, R. J., & Strassburg, B. (2022). How do we best synergize climate mitigation actions to co-benefit biodiversity. *Global Change Biology, 28*(8), 2555–2577.

Stratton, A. (2021). *Assessing social, ecological, and nutritional outcomes of crop diversification: Transitions to agroecological management in Southern Brazil*.

Tanuwidjaja, G. (2011). *Park connector network planning in Singapore: integrating the green in the garden city*.

Tong, S., Beggs, P. J., Davies, J. M., Jiang, F., Kinney, P. L., Liu, S., ... & Ebi, K. L. (2023). Compound impacts of climate change, urbanization and biodiversity loss on allergic disease. International Journal of Epidemiology, 52(3), 655–663.

Tzoulas, K., Korpela, K., Venn, S., Yli-Pelkonen, V., Kaźmierczak, A., Niemela, J., & James, P. (2007). Promoting ecosystem and human health in urban areas using Green Infrastructure: A literature review. *Landscape and Urban Planning, 81*(3), 167–178.

Verma, S. K., Dave, A., Israr, J., Singh, D., & Misra, S. (2025). Rhizosphere engineering for agriculture and ecological sustainability. In Plant-microbiome Interactions for Climate-resilient Agriculture (pp. 259–279). Singapore: Springer Nature Singapore.

Vikas, & Ranjan, R. (2024). Agroecological approaches to sustainable development. *Frontiers in Sustainable Food Systems, 8*, 1405409.

Wamsler, C., Wickenberg, B., Hanson, H., Alkan Olsson, J., Stalhammar, S., Bjorn, H., Falck, H., Gerell, D., Oskarsson, T., Simonsson, E., Torffvit, F., Zelmerlow, F. (2020). Environmental and climate policy integration: Targeted strategies for overcoming barriers to nature-based solutions and climate change adaptation. Journal of Cleaner Production, 247(119154):1–10.

Wani, Z. A., Singh, L., Uniyal, S., Rana, S. K., Bhatt, I. D., & Nautiyal, S. (2025). Improving ecosystem vitality in India: Overcoming barriers to meet National and International targets. *Environmental Sustainability*, 1–13.

Worlanyo AS, Jiangfeng L (2021) Evaluating the environmental and economic impact of mining for post-mined land restoration and land-use: a review. J Environ Manag 279:111623

Yang, M., Wang, Z., Zhang, Z., Chen, P., Zhaom, D., Cheng, E., Wang, C., & Yan, Y. (2024). Pathways for ecological restoration of territorial space based on ecosystem integrity: A case study of approach to protecting and restoring mountains, rivers, forests, farmlands, lakes, and grasslands in Beijing, China. *Ecological Frontiers, 44*(6), 1214–1223.

Yurui, L., Xuanchang, Z., Zhi, C., Zhengjia, L., Zhi, L., & Yansui, L. (2021). Towards the progress of ecological restoration and economic development in China's loess plateau and strategy for more sustainable development. *Science of the Total Environment, 756*, 143676.

Challenges to Ecosystem Restoration in the Era of Climate Change

Satendra Kumar Rathaude, Abdul Rahim P.P., Aamir Nazir Lone, Shailesh Kumar, Shairq Irtiqa, and Javid Ahmad Dar

Abstract Ecosystem restoration is a critical strategy in addressing global challenges such as biodiversity loss, climate change, and environmental degradation. It describes the planning and activities aimed at recovering an ecosystem that has changed from its original condition. This chapter focuses on the major problems and new frontiers in restoration science, highlighting gaps in long-term ecological research, interdisciplinary integration, and climate mitigation. Artificial intelligence, GIS, and remote sensing, along with citizen science, are changing the scope of data collection, monitoring, and stewardship for ecosystem restoration by altering the boundaries for possible community engagement. The chapter also discusses restoration strategies that align the needs of the Anthropocene with environmental sustainability in the face of ongoing environmental change. Restorative work can be done to support recovery and human well-being when science, policy, and practice are integrated, which this chapter seeks to explain by using the examples of the Great Green Wall, mangrove restoration in Southeast Asia, and urban Miyawaki forests as accessible, scalable, and community-based restoration projects. Integrated within the scope of the UN Decade on Ecosystem Restoration, this chapter discusses proposed strategies, technologies, collaboration, and action frameworks necessary to achieve the global restoration targets set forth by the Decade.

Keywords Global Biodiversity Framework · Target 30x30 · Nature-based solutions · Restoration challenges

---

S. K. Rathaude (✉) · A. R. P.P. · A. N. Lone · S. Kumar · S. Irtiqa
Terrestrial Ecology and Modeling (TEaM) Lab, Department of Environmental Science and Engineering, SRM University AP, Amaravati, Andhra Pradesh, India
e-mail: satendrakumar_r@srmap.edu.in

J. A. Dar
Terrestrial Ecology and Modeling (TEaM) Lab, Department of Environmental Science and Engineering, SRM University AP, Amaravati, Andhra Pradesh, India

Centre for Geospatial Technology (CGT), SRM University AP, Amaravati, Andhra Pradesh, India

S. Siddiqui et al. (eds.), *Frontiers in Sustainability*,
https://doi.org/10.1007/978-3-032-19740-5_11

1 Introduction

Ecosystem deterioration is one of the major concerns for the sustainable development of planet Earth (Yeremeyev et al., 2021). It is impractical to restore an ecosystem to its pre-existing state (Chen & Zhang, 2024). The regrettable reality is that people are affecting most ecosystems worldwide, even though preventing deterioration in the first place should be a primary goal of environmental management (Barrow, 2006; Polasky et al., 2011). About one-third of the world's ecosystems have been transformed into human settlements, towns, industries, and agricultural land. Most of the remaining ecosystems are also under threat from pollution, habitat fragmentation and loss, land-use change, and invasive species (Suding, 2011). One practical method for undoing the negative consequences of environmental degradation is ecosystem restoration (Dou et al., 2021). This system is aided by numerous global activities ranging from local to global dimensions (Frietsch et al., 2023). Climate change refers to global temperature and long-term changes in weather patterns, and its impact on the natural ecosystem is mainly inspired by human activities such as fossil fuels, deforestation, construction, and industrial emissions (Kabir et al., 2023). These activities increase greenhouse gas concentrations such as carbon dioxide and methane, which trap heat in the Earth's atmosphere and lead to global warming (Omemen & Aldbbah, 2025). The results of climate change are far--reaching, including rising temperatures, melting glaciers, increase in sea level (Mondal & Lee, 2025), and more extreme weather events such as storms, droughts, and forest fires (Rawat et al., 2024; Wong et al., 2025). Subsequently, there are direct or indirect causes of severe weather incidence, loss of biodiversity, and disruption of ecosystems and damage to human livelihoods; in addition, climate change threatens food security and human health (Raza et al., 2024). Due to the increase in temperatures on the planet, extreme forms of weather are being experienced with greater prevalence and severity across different regions while displacing various species (Upadhyay, 2020). In the context of climate change and the Anthropocene, we can protect natural ecosystems from decline through restoration. Scientific, nature-based solutions are a key part of global strategies to combat the impact of climate change (Simonson et al., 2021; Suding, 2011).

Addressing climate change requires immediate global cooperation to reduce emissions, investment in renewable energy, and flexibility as there is failure to risk irreversible damage to the planet and future generations. The quick speed of climate change faces unprecedented challenges for traditional restoration practices marked by the rapid speed of climate change, shifting rain patterns, extreme weather events, and distributed species (Hoegh-Guldberg et al., 2019; Abbass et al., 2022). These challenges require the development of innovative strategies to reevaluate the existing approaches and to be compatible with the changing environmental conditions. The demand for ecological restoration across many biomes and continents has increased due to the startling rate of ecosystem degradation. The future involves increasing flexibility. Increased environmental harm as well as projected changes in the future have raised concerns about the health and stability of natural systems.

This chapter aims to explore the drivers of degradation and review various restoration strategies within the context of global frameworks and national efforts.

2 Major Challenges in Ecological Restoration

Worldwide, many initiatives from local to global scales contribute to ecosystem restoration (Goffner et al., 2019). However, the scope, magnitude, effectiveness, and future potential of ecological restoration are constrained by several adverse environmental factors (Yang et al., 2024) and strongly rely on the specific context and the assessment of the area about reference ecosystems (Bustamante et al., 2019). The United Nations has established the 2021–2030 "Decade of Ecosystem Restoration" to halt, prevent, and reverse environmental degradation (IUCN, 2020). Among the various ecosystems, riparian ecosystems are currently facing significant threats due to diverse developmental activities worldwide, necessitating restoration efforts (Mohan et al., 2022). Riparian ecosystems have become vulnerable due to natural calamities such as landslides and floods. In addition, human interventions like dam construction and expanding road networks disrupt the natural flow of rivers and affect their genetic connectivity. Consequently, the restoration of riparian ecosystems is crucial because they are vital for safeguarding and rehabilitating the habitats that sustain both flora and fauna, as well as land and water biodiversity (Basak et al., 2021). Riparian zones are also associated with wetlands, and restoring these highly productive and biodiverse ecosystems is essential to recover their lost functions and well-being (Bentley et al., 2022).

Transitioning from damaged ecosystems to a restored state, conservation and restoration play a major role in ecosystem services; however, there are several gaps in understanding the valuation of ecosystem services associated with restoration techniques. Ensuring multifunctionality in the medium and long term is a special issue because the restoration of one ecosystem service may come at the expense of another (Suding, 2011). The complexity of ecosystems, with their intricate interdependencies, makes full restoration difficult (Wang et al., 2011), compounded by limited knowledge of historical baselines needed for accurate benchmarks (Young et al., 2005). The constraints on resources, such as financial capability and efforts, limit project scalability and reduce community engagement due to low public awareness (Lengnick-Hall & Lengnick-Hall, 2002). Apart from that, local and regional-level policy and legal barriers, including conflicting regulations and land--use disputes, complicate implementation (Butler et al., 2015; Shaw & Cui, 2025). Invasive species tend to suppress the native flora and fauna, thereby hindering the recovery of native species, altering the plant species assemblage and host-specific fauna (D'Antonio et al., 2016). Climate change alters environmental conditions, making it difficult to replicate past conditions (Weidlich et al., 2020). There is a lack of regular monitoring to improve the performance of restoration programs, which hinders the sustainable management of restored lands by environmental managers and indigenous communities (Aronson et al., 2020). Overcoming these barriers

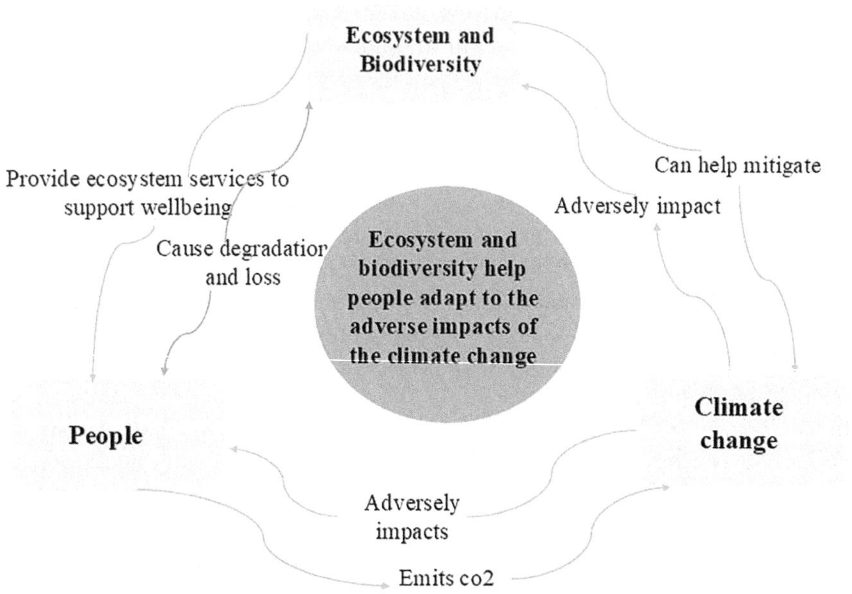

Fig. 1 Ecosystem-based adaptations and climate change mitigation

requires an interdisciplinary approach, policymaking, scientific and adaptive management, and stakeholders' participation with local communities to achieve lasting ecological recovery. Essentially, ecosystem restoration involves interacting with major challenges, including climate change, social-ecological and overexploitation of resources, and political instability.

The theoretical model suggests potential trajectories of degraded ecosystems following various management approaches. Ecosystem-based adaptation harnesses the strength of biodiversity and robust ecosystems to enable people to resist the effects of climate change and reduce emissions, build resilience, and maintain important ecosystem services for people and nature (Fig. 1).

2.1 Climate Change: Soil and Water Resource Limitations

Ecosystem restoration efforts increasingly acknowledge soil health's pivotal role as a foundational element for achieving long-term success (Nolan et al., 2021). Soil degradation has become increasingly severe and is often related to factors such as deforestation and overgrazing, intensifying the tension between population, resources, and the environment, which has caused global concern. Accelerated erosion, depletion of the soil organic carbon (SOC) reserves, and decrease in biodiversity are among the dominant soil degradation processes. Also, reductions in soil fertility, acidification, and salinization contribute to the deterioration of soil quality (Lal, 2015).

Land degradation refers to the progressive deterioration in the productivity of ecosystems over time. Integrating agricultural production and other land uses with ecosystem preservation is crucial to mitigate future land degradation and restore degraded lands (Turner et al., 2016). Over two-thirds of the world's drylands experience moderate-to-severe soil degradation, which is continuing to deteriorate, resulting in an annual economic impact exceeding $300 billion (Jiang et al., 2021). The intensifying drought (Christian et al., 2021), soil erosion (Hossini et al., 2022), desertification (Yan et al., 2024), deforestation, pressure on natural resources, urbanization and agriculture production (Molnárová et al., 2023) increase the severity of land degradation, which reduces land productivity, threatening food security and livelihoods globally. To combat land degradation is a complex process that includes land cover change analysis (Domingo-Marimon et al., 2024), process control, spatial technologies (Di Leginio et al., 2024), and interactions between humans and the natural environment within ecosystems (Fu et al., 2023). Severe land degradation, marked by soil nutrient depletion, reduced vegetation cover, and loss of productivity, necessitates a thorough assessment to protect terrestrial ecosystems (Wu et al., 2024). This global issue affects approximately one-quarter of the Earth's surface and impacts the livelihoods of 250 million individuals in developing nations (Ngabire et al., 2023). Therefore, monitoring and forecasting are essential for tackling environmental and societal issues for environmental restoration by identifying its different driving factors (Saha et al., 2022; Wang et al., 2024a, b).

2.2 *Land-Use Conflicts and Human Activities*

Understanding and assessing the state of regional ecological security and identifying land-use conflicts are crucial for both environmentally sustainable development and regional social stability, but it is still difficult to do so using a scientific methodology (Zhang et al., 2025). Population increase, urbanization, and economic growth have considerably changed the pattern of land use through extensive human activities, enhancing land-use conflict and encouraging environmental risks, thus posing challenges to the cohabitation of human societies and the natural world (Zhang et al., 2020; Chen & Zhang, 2024). The utilization of land for residential and related purposes has been an integral part of the encroachment of human civilization upon natural habitats since the beginning (Liang et al., 2023). By 2050, the anticipated urbanization rate is expected to reach 70%, necessitating an additional land area of 90–150 million hectares for urban development (Grafton et al., 2015). The complex nature of socioeconomic systems has given rise to multifaceted land-use conflicts. These conflicts exhibit various characteristics, such as competitive pressures among different land-use types, ecological risks occurring from disorderly landscape patterns, and functional trade-offs among distinct land categories (Steinhäußer et al., 2015); therefore, they pose a major challenge to sustainable land planning and management and a serious threat to food and ecological security. With residential development, agricultural activities are also examples of mutually exclusive land uses that frequently compete for the same land resources (Gottero, 2019). However, ecosystem restoration can be achieved in agricultural areas where the demand for land-use conversion is reduced, because of population movement from rural regions to urban centers, and shifts in agricultural policies (Waring, 2024).

2.3 Socioeconomic and Policy Challenges

Land is the fundamental basis for socioeconomic activities and critical for human sustenance (Zhou et al., 2025). Rapid socioeconomic development has resulted in more prominent regional competition and conflicts, urban expansion occupying high-quality cropland, the disorderly spread of rural settlements, and other phenomena (Lu et al., 2024). Failure in the planning of land-use results in an imbalance to reconcile economic development, food security, and ecological preservation (Yi et al., 2020). Migration, a socioeconomic burden, tends to damage the ecology of home countries by increasing energy use and encroachment into restricted forest areas by undocumented and illegal migrants (Alola et al., 2019). Meaningful change and better public services rely on strong, forward-thinking policies that are thoughtfully designed and effectively implemented through good governance. Adopting an integrative approach that links social and economic policies with the goals of economic democratization is essential, which brings inclusive growth and promotes good governance from both social and economic perspectives. This approach should consider other policy domains and stakeholders, as well as the linkages between public institutions and the social and solidarity economy sector (Eizaguirre Anglada, 2021). Well-targeted financial measures, such as taxes or subsidies, can incentivize landowners to avoid environmental damage, but must consider regional disparities, economic differences, multiple goals, and budget constraints (Piñeiro et al., 2020). The use of specific subsidies or taxation on landowners will actually minimize environmental degradation, provided the incentives are efficient, effective, and appropriate for local, economic, and budget circumstances (Piñeiro et al., 2020). Also, targeted economic and policy measures at local, national, and international levels can promote sustainable management of regrowth forests on former agricultural lands and strengthen rural communities and economies. However, there is insufficient collaborative learning among forest and conservation planners, inadequate funding to preserve forest habitats with suitable size, quality, and connectivity, and national political agendas that disregard evidence-based knowledge (Angelstam et al., 2023; Wani et al., 2025a, b).

3 Emerging Strategies for Climate-Resilient Restoration

Current and projected climate conditions emphasize the urgent need to update and improve conservation policies to reduce their effects on ecosystems. Developing these strategies requires collaboration among diverse stakeholders to promote scientific knowledge and encourage innovative approaches that support sustainable development. Such cooperation helps us better understand the intricate relationships within nature and find balanced ways to support economic growth and improve livelihoods. Climate change refers to long-term shifts in environmental conditions that can significantly affect life across ecosystems. To predict these changes, scientists have created future climate models called Representative Concentration Pathways (RCPs), which are based on anticipated greenhouse gas emissions and radiative forcing. However, these models originally overlooked socioeconomic factors. To address this, additional elements like population growth, education, urbanization, economic development, and technological progress were later integrated into the models, forming what are known as Shared Socioeconomic Pathways (SSPs). These SSPs outline five distinct scenarios of possible future development. Using these frameworks, researchers have also built habitat distribution models for plants and animals, which help forecast changes in species populations and their geographic ranges. Understanding these predictions is crucial for designing conservation efforts and restoration plans that can withstand the impacts of climate change and safeguard biodiversity (Wani et al., 2025a, b).

3.1 Genetic Diversity: Support for Resilient Ecosystem

Understanding genetic diversity and population genetics is crucial for effective management and conservation in a changing climate (Sharma et al., 2024). Genetic variability can enhance the population's ability to adapt and be resilient to extreme climatic conditions and pest threats. So, integrating evolutionary resilience into management and policy is essential for mitigating ecological damage (Sgrò et al., 2011). There are some strategies for reducing ecosystem degradation using genetic methods.

- Sourcing seeds from different populations to ensure adaptive potential (Breed et al., 2018).
- Ex situ conservation of genetic materials through seed banks (Wambugu et al., 2023).
- Preserving and designing ecological corridors to ensure gene flow between the populations (De et al., 2021).

3.2 *Nature-Based Strategies through Ecosystem Restoration*

Ecosystem restoration is one of the effective components of nature-based solutions (NbS), which play a significant role in conservation, especially for biodiversity, ecosystem services, and mitigation of climate change impacts, even as ecological degradation continues despite progress in conservation efforts (Cheng & Li, 2024).

(a) Mangrove and coastal wetlands:

- These ecosystems are crucial in mitigating climate change by acting as carbon sinks that sequestrate 3 to 5 times more carbon per area than terrestrial forests (Duarte et al., 2013).
- It reduces coastal erosion and damage from storm surges and acts as a natural buffer against critical weather, as mangroves reduce wave energy by 66% (Alongi Daniel, 2015).

(b) Forest and Grassland:

- Forests enhance watershed protection and carbon sequestration. Urban forests can help reduce temperatures during heat waves. Fire-adapted restoration strengthens the resilience of ecosystems prone to fire by reintroducing controlled fire practices, reducing fuel load, and facilitating ecosystem regeneration (Hessburg et al., 2019).
- Grasslands: Reintroduced native species have greater drought resilience than non-native grass-dominated ecosystems, highlighting their importance in climate-adaptive restoration strategies (Kirwan et al., 2021).

(c) Agricultural system:

- Agroecological approaches and polyculture reduce vulnerability to climate change by stabilizing yield and ecological functions. Restoring a sustainable agricultural ecosystem enhances water retention, soil health, and nutrient cycling and increases robustness in extreme climatic events (Lin, 2011).

3.3 *Community Engagement, Local Knowledge, and Indigenous-Led Restoration*

Indigenous communities have been maintaining the landscape for millennia through activities such as prescribed burning, rotational farming, and watershed management (Ansell & Evans, 2020). Indigenous traditional ecological knowledge (TEK) is very valuable in restoration planning, augmenting contemporary scientific knowledge (Berkes et al., 2018). Traditional restoration methods ignore social and cultural aspects of ecological recovery. Indigenous and local people have rich knowledge of their ecosystems and are thus important stakeholders in restoration processes (Fernández et al., 2021). Research indicates that restoration activities by local people are more efficient and climate change resistant because they have a complete comprehension of the dynamics of ecosystems (Kimmerer, 2013). The escalating consequences of climate change have raised the need for ecosystem restoration to promote biodiversity, carbon sinks, and climate resilience.

3.3.1 Case Studies of Indigenous-Led Restoration

3.3.1.1 Australian Indigenous Fire Management

Indigenous fire management has been part of the Australian ecological strategy for thousands of years. The Aboriginal people used a method called cool burning, which is a low-severity controlled burn to reduce big fires and promote vegetation growth and biodiversity (Ansell & Evans, 2020). Unlike high-severity, high--intensity fires that burn large areas, controlled burning is a highly effective method for reducing fuel loads and lowering the risk of high-intensity, large-scale wildfire (McKenna et al., 2023). Restoring Indigenous fire management has brought more biodiversity and climate resilience. Northern Australian data show areas where traditional fire use is sustained have lower carbon emissions and less intense fires (Moura et al., 2019). The West Arnhem Land Fire Abatement Project (WALFA) is an example of this, where Aboriginal fire management is part of national carbon credit schemes and is working environmentally and economically for Indigenous Australians (Whitehead et al., 2009).

3.3.1.2 Stewardship of the Amazon Rainforest

Indigenous communities have been taking care of the Amazon rainforest for thousands of years, using agroforestry, rotational agriculture, and land guardianship. Research has shown that forests managed by Indigenous people have lower deforestation rates and more carbon storage than those managed by governments and corporations (Nepstad et al., 2006; Walker et al., 2020). A well-known documented example of Indigenous conservation is the Xingu Indigenous Park in Brazil, established in 1960 and managed by local communities. However, large-scale deforestation around Xingu Park has increased the risk to its conservation, particularly in terms of forest cover, carbon sink, and its role as a biodiversity hotspot. Indigenous people like the Kayapó use natural resource conservation strategies like controlled fishing, agroforestry, and forest conservation to take care of their environment (Garcia et al., 2024).

3.3.2 Challenges in Ensuring Participation and Sustainability

Although the role of Indigenous and local knowledge in ecological restoration is widely acknowledged, Indigenous people typically experience institutional constraints to their participation on a systemic scale (Wunder, 2015). These constraints arise from an entanglement of legal, economic, institutional, and epistemological obstacles to their ability to participate or lead restoration processes (Coulthard, 2014). These restrictions need to be overcome by undertaking radical policy changes, facilitating funding mechanisms, and shifting to more participatory knowledge systems.

3.3.2.1 Legal and Policy Limitations

The greatest obstacle to Indigenous people's involvement in restoration is the absence of legal recognition of Indigenous land rights. Most national policies fail to give formal recognition to Indigenous land tenure and governance institutions, thus preventing these people from exercising jurisdiction over their ancestral lands (Agrawal & Ostrom, 2001). Despite the United Nations Declaration on the Rights of Indigenous Peoples (UNDRIP) calling for Indigenous Peoples' self-determination over natural resources, most states have fallen behind in enacting its principles into national law (Anaya, 2009). For example, in Canada, the complications of legal regimes governing Indigenous land claims pose bureaucratic obstacles that hinder or slow down Indigenous-led conservation (Borrows, 2017). Most projects initiated by Indigenous communities are dependent on external finance derived from non-governmental organizations (NGOs) or international institutions, making them susceptible to interruption when funds run out (Mistry & Berardi, 2016). The greatest challenge lies in the fact that funding mechanisms are typically made for large-scale conservation organizations and not for community-initiated conservation (Agrawal & Gibson, 1999). In addition, restoration programs funded by government agencies tend to promote technocratic, top-down approaches that rule out Indigenous involvement.

3.3.2.2 Community-Led Restoration Sustainability Challenges

Long-term community-led restoration is a complex and multifaceted process. Local and indigenous communities have the ecological knowledge and commitment to land care; however, outside factors such as modernization, economic pressure, policy loopholes, and financial dependency tend to undermine the initiative's sustainability (McKenna et al., 2023). These restoration and sustainability challenges can be addressed by bringing indigenous and modern conservation techniques together, with calibration of the economic benefit with community value and the long-term support and funding. The knowledge gap about indigenous conservation techniques from one generation to the next poses a major barrier in terms of sustainability. Globalization, urbanization, and formal education systems are driving youths away from native ways of life, thus eroding ancestral land stewardship practices like rotational farming, water harvesting, and fire protection systems (Fernández-Llamazares et al., 2021).

3.3.2.3 Dependence on External Funding

The major challenge of community-led restoration projects is inadequate funding, as they often rely heavily on NGOs and other international donors or government organizations. While this funding is necessary to initiate restoration efforts, its short-term nature is a problem for long-term sustainability (Hoffmann, 2022). When funding cycles end, most projects struggle to continue activities, pay staff, or sustain conservation work, and then projects get disrupted and may even reverse environmental gains (Maffi & Woodley, 2012). Funding mechanisms are mainly in the form of grants where support is offered for a short period of 2 to 5 years, and then funding may not be renewed or may be deflected to other projects. This creates a lot of uncertainty for communities, especially when projects involve long-term ecological restoration (Wunder, 2015). For example, in East Africa, donor-funded reforestation initiatives in Tanzania collapsed after financial supporters withdrew their funding, with local communities being unable to fund or support tree nurseries or afforestation activities (Pfliegner, 2010). In India, international conservation groups initially supported mangrove restoration initiatives, but community participation declined after funding ended, resulting in degradation of previously restored coastal habitats (Mohan, 2019; Fleischman et al., 2025).

3.3.3 Community Engagement and Competing Priorities

Success of restoration projects depends on community engagement, while competing socioeconomic challenges may hinder the progress of work. Promoting inclusive and locally relevant benefits can motivate communities to engage in long-term restoration programs.

The livelihoods of Indigenous and local peoples are based mainly on agriculture, resource extraction, and urbanization, which may induce a trade-off between economic need and conservation objectives (Bharucha & Pretty, 2010). In addition, as climate change impacts increase, vulnerable communities are likely to focus on short-term coping mechanisms like disaster preparedness at the expense of long-term restoration processes (Adger et al., 2005). These competing interests need to be resolved by integrative solutions that balance conservation objectives with economic incentives and ensure that restoration processes address both ecological and social requirements (Hoffmann, 2022).

3.3.3.1 Agricultural Expansion Vs Reforestation

Agricultural practices and food security in most areas are at odds with reforestation. Land restoration for afforestation or rewilding is seen as an obstacle to agricultural production, especially in smallholder agriculture and subsistence farming (FAO, 2021). In Brazil, for example, there have been conflicts between soybean farmers and reforestation; the farmers see afforestation as shrinking land for crop growth (Garnett et al., 2018). In India, tree-planting operations have displaced smallholder farmers, resulting in land tenure conflicts between conservationists and the local community (Fleischman et al., 2025). Encouraging agroforestry practices that support the co-cultivation of crops and trees can support biodiversity without compromising food production. Economic incentives, for instance, Payments for Ecosystem Services (PES), can motivate farmers to embrace sustainable land management (Wunder, 2015).

3.3.3.2 Climate Change Adaptation Vs Ecological Restoration

Climate change poses immediate survival issues, especially for the vulnerable. In the majority of situations, short-term adaptive measures like the building of seawalls, emergency drought relief, and water diversion programs tend to come at the expense of long-term ecological recovery (Adger et al., 2005). In Bangladesh, sea-level rise communities give top priority to embankment and flood protection construction, which demolishes natural wetlands required for long-term ecosystem resilience (Rahman, 2019). In California, huge investments in wildfire prevention measures (e.g., firebreaks) have, at times, compromised forest restoration initiatives (Schumann III et al., 2025). Moreover, in the African Sahel, extreme drought has forced communities to increase irrigation and groundwater abstraction, with a bias toward draining the natural ecosystems required for long-term climate resilience (Chagnaud et al., 2022).

4 Future Prospectus and Requirements

Several ecosystem restoration practices have already been implemented across different regions of the world, while many others are currently ongoing. These efforts aim to address environmental degradation, restore biodiversity, and enhance ecosystem services. In Table 1, a selection of such restoration practices is presented, highlighting key projects from diverse ecosystems, including tropical forests, wetlands, grasslands, coral reefs, and peatlands. The table outlines the restoration goals, funding sources, implementation strategies, and notable outcomes, providing insight into the practical application of ecological restoration in varying socioecological contexts.

Table 1 Overview of global ecosystem restoration initiatives: Concepts, goals, and outcomes

Ecological theory/concept	Agreement/funded by	Country/ Region	Ecosystem type	Restoration Goals and conceptual	Timeframe	Examples of practice	Outcomes and achievements	References
Atlantic forest restoration	International Climate Initiative (IKI)	Brazil	Tropical Forest	Restore biodiversity, increase carbon sequestration, and support sustainable livelihoods	2017–2022	Native species reforestation, agroforestry practices	Over 2000 hectares restored; increased wildlife presence; community engagement in conservation	Bizuti et al. (2022)
Loess plateau watershed rehabilitation	World Bank	China	Arid/Upland Agricultural Land	Combat soil erosion, improve productivity, reduce poverty	1994–2005	Terracing, reforestation, grass planting, sediment control	Significant vegetation recovery, reduced erosion, improved agricultural yields	Qu et al. (2020)
Aral Sea basin restoration	International Fund for Saving the Aral Sea	Uzbekistan	Desertified Land	Reverse desertification, restore flora and fauna	Ongoing since early 2000s	Sand stabilization, tree planting (Saxaul species)	Over 350,000 ha of desert land stabilized; biodiversity slowly returning	Micklin (2010)
Mangrove restoration	Wetlands International	Senegal	Coastal Mangroves	Protect coastlines, improve fisheries, adapt to climate change	2006–Present	Community-led planting of mangrove saplings, nursery development	10,000+ hectares restored; increased fish stocks; community resilience	Feka & Manzano (2008)
Coral reef restoration	Reef Check Foundation	Indonesia	Coral Reefs	Rebuild reef ecosystems damaged by blast fishing and bleaching	2010–Present	Artificial reef structures, coral transplanting	Coral cover increased by up to 60%; rise in marine biodiversity	Lange et al. (2020)

(continued)

Table 1 (continued)

Ecological theory/concept	Agreement/funded by	Country/ Region	Ecosystem type	Restoration Goals and conceptual	Timeframe	Examples of practice	Outcomes and achievements	References
Wetland restoration	Government of Israel	Hula valley, Israel	Wetlands	Reverse wetland drainage, support bird migration	1990s–Present	Water level regulation, habitat reconstruction	Return of migratory birds; rise in ecotourism and biodiversity	Hambright and Zohary (1999)
Rangeland restoration	UNDP and Local Government	Mongolia	Grasslands	Prevent overgrazing, restore productivity	2008–2018	Rotational grazing, reseeding native grasses	Enhanced forage quality; improved livelihoods of herders	Wang et al. (2013)
Peatland restoration	UNEP and Central African Forest Initiative (CAFI)	Congo Basin	Peat Swamps	Carbon storage, biodiversity conservation	2016–Present	Hydrology restoration, conservation agreements	Improved water table levels; carbon emissions mitigation	Dargie et al. (2017)
Forest landscape restoration	World Resources Institute	Rwanda	Montane Forest	Increase forest cover, protect water sources	2011–2025	Agroforestry, tree planting on farms, natural regeneration	30% forest cover target on track; improved rural incomes	Nsabimana et al. (2008)
River thames restoration	Environment Agency (UK)	United Kingdom	Freshwater River	Improve water quality, restore habitats	1990s–Present	Riverbed reengineering, pollution control	Fish species returned; recreational activities revived	Everard and Moggridge (2012)
Shan-Shui initiative	Shan-Shui initiative in China	China	Mountains to coastal estuaries	Addressing issues such as soil, water degradation and promoting biodiversity and sustainable livelihood	2016–2030	Species and ecosystem conservation	Provided 3.7 million jobs, Oujiang River Headwaters project helped to boost biodiversity	Jiang et al. (2025)

Dry corridor in Central America	Food and Agriculture Organization (FAO), United Nations Environment Programme (UNEP), World Food Programme (WFP)	Central America	Farmlands, forests	Mitigate climate change and to support livelihoods	2015–2030	Reforestation, agroforestry, soil management, climate services	Afforestation, enhanced soil protection systems and land restoration	Gotlieb et al. (2019); Huber et al. (2023)
Sehal integrated resiliences programms (SIRP)	World Food Programme (WFP)	Africa	Farmlands, forests, freshwaters, grasslands, urban settlements etc		2018–2030	Land degradation, food insecurity, climate change	Restored 420,000 ha of degraded land, improved food security, and safeguarded biodiversity	Grist et al. (2014)
Life peat restore Poland	Klub Przyrodników	Poland	Peatlands	Protect freshwaters, biodiversity and climate change mitigations	2016–2022	Protect biodiversity and improve water retention and flood control	1350 ha land restored, improved water regulation and habitat of rare species	Pakalne et al. (2021)

(continued)

Table 1 (continued)

Ecological theory/concept	Agreement/funded by	Country/Region	Ecosystem type	Restoration Goals and conceptual	Timeframe	Examples of practice	Outcomes and achievements	References
Forest landscape restoration (FLR) hub	International Climate Initiative (IKI)	Global	Farmlands, forests, freshwaters, grasslands		2021–2030	Safeguard biodiversity, mitigate climate change and support livelihood	200,000 ha land restored, improved livelihood of local communities through sustainable management	Stanturf et al. (2015); Chazdon et al. (2021)
Great Green Wall for Restoration and Peace	UNCCD, FAO, Global landscape forum, World food program (WFP), UNEP	Africa	Forests and land	To restore 100 million hectares of land, sequester 250 million tonnes of carbon and create ten million jobs	2007–2030	Land restoration, climate resilience and peacebuilding	Created jobs, carbon sequestration, land restored and helped conflict prevention and community stability	Mirzabaev et al. (2022)

4.1 Importance of Long-Term Ecological Monitoring

As an evaluation of restoration efforts, long-term ecological monitoring, tracking the recovery of biodiversity, is fundamental. To improve the attention received by lesser-studied areas, the ability to recover ecosystems (Cuvelier et al., 2018) under different conditions needs to be standardized, as soil recovery indicators have yet to be developed. There exists a variety of features and indicators that are monitored and evaluated over long periods of time based on the ecosystem method as part of the international long-term ecological research program in many countries around the world (Suding et al., 2015). One major objective is, by gathering and statistically analyzing information, to defend, prove, or bolster some theoretical or workable managerial concepts (Müller et al., 2016). The aim for restoration is to attain complementation of ecosystems with their environment: self-sustaining, resilient systems whose surroundings and topography are natural features of the region. When it comes to restoring local tradition or replicating lost processes, such as fire or invasive species management, actions taken should be minimal and have no anthropogenic influences in the long term (Suding et al., 2015). Globally, decision-makers and organizations are increasingly engaging citizen science and volunteers through standardized protocols and community science initiatives to enhance monitoring, manage natural resources, and support conservation across diverse landscapes (Conrad & Hilchey, 2011; Fraisl et al., 2022). Long-term ecological studies are vital for understanding gradual ecological changes and rare, episodic events such as succession, soil development, and natural disturbances like wildfires or hurricanes. Recognizing this, the United States National Science Foundation established the Long-Term Ecological Research (LTER) Network in 1980, which has since played a key role in advancing ecological theory and informing adaptive management through sustained data collection and collaboration among scientists (Franklin et al., 1990; Hobbie et al., 2003).

4.2 Integrating Restoration with Carbon Sequestration and Climate Mitigation

Restoration of ecosystems and global climate goals synergistically work toward improving biodiversity and carbon results. "Carbon hotspots" such as peatlands, mangroves, and even old-growth forests should be prioritized as they have the highest carbon sequestration potential (Wood, 2023; Pichancourt et al., 2014). Taking monoculture reforestation and the enrichment of soil organic matter, Indigenous practices, native species, and mixed-species planting serve as a more appealing option toward fulfilling climate obligations. von Unger & Emmer (2018) and Chazdon et al. (2020) argue that carbon credits and payments for ecosystem services (PES) financing schemes provide flexible funding opportunities while upholding conservation standards. The African Great Green Wall Initiative is one of the bold endeavors which attempts to reclaim 100 million hectares of land, capture 250 million tons of carbon, and strengthen the local economy (UNCCD, 2015; Wall, 2024).

4.3 Strengthening Interdisciplinary Collaboration in Restoration Science

Restoration requires an interdisciplinary integration of ecology, hydrology, social sciences, and economics alongside Indigenous Knowledge Systems (Uprety et al., 2012). Interdisciplinary research with policymakers and practitioners creates culturally relevant and scientifically informed approaches (Sievanen et al., 2012). These collaborations are instrumental in resolving complex, multifaceted restoration challenges such as invasive species management, land-use conflicts, and dependence on natural resources (Islam et al., 2024). Programs like Landscape Restoration for People, Nature and Climate by WWF and IUCN illustrate the value of interdisciplinary approaches in achieving large-scale restoration that is highly inclusive and ecologically and socially beneficial (Winsor, 2021).

4.4 Rethinking Restoration Goals in the Anthropocene

Human-driven changes to the Earth's systems have irreversibly transformed our planet. In the contemporary context, restoration or rehabilitation efforts outdatedly centered around recreating historical baselines are becoming less applicable. It is becoming increasingly important to construct adaptable, robust, and functioning ecosystems that can endure extreme shifts, both in climate and land use (Cooke et al., 2023). This involves the addition of novel ecosystems, climate-smart plant species, and even assisted migration for enhanced ecological flexibility (Hussain et al., 2024). Furthermore, these plans should also incorporate water regulation, flood control systems, biodiversity corridors, and other relevant primary or secondary socioecological systems, thereby fulfilling ecological and social requirements (Gregory et al., 2021). Projects like Oostvaardersplassen in the Netherlands are stepping away from static historical recreations and moving toward the acceptance of more dynamic rewilded ecosystems (Lorimer & Driessen, 2014).

4.5 The Way Forward: Integrating Science, Policy, and Practice

Sustaining ecosystems to be restored and maintained in the future requires the following: appropriate scientific research, well- thought-out policies, and practical groundwork. It is important for policies to make sense within a political framework that is realistic and captures the local context as well as deliberation being informed by science (Abhilash, 2021). Policies for Ecosystem Restoration are strategically placed at the end with their placement having the intent that ad contractual compliance, usable funds, and stimulation from the public do work together toward goals for ecosystem restoration (Pokorny et al., 2019; Aronson et al., 2020).

4.6 The Role of Restoration in Building Climate-Resilient Ecosystems

Restoration of ecosystems is crucial for climate change mitigation as it attempts to lessen the impact of extremes such as droughts and floods. The process of restoring wetlands, forests, and mangroves also contributes to the support of biodiversity and livelihood by bolstering natural defenses (Seddon et al., 2020). Acts like stitching together the various elements of a landscape through habitat corridors enable the movement of broader spatial ranges of species, which is important for genetic variation and resilience (Christie & Knowles, 2015). This is also demonstrated in the case of the Mangrove Restoration Projects in the Philippines, which not only restored necessary coastal ecosystems but also reduced the community's vulnerability to typhoons and sea level rises (Primavera & Esteban, 2008).

4.7 Climate-Resilient Urban Planning

Climate-resilient urban planning is a crucial pathway toward building sustainable and adaptive cities capable of withstanding the impacts of climate change (World Bank, 2010). By integrating green infrastructure, smart technologies, and inclusive design, cities can enhance both environmental performance and community resilience (Johnson & Kastner, 2018). While challenges remain—such as securing funding and fostering political commitment—the examples set by global cities demonstrate that with innovation and collaboration, a climate-resilient urban future is both achievable and essential (Praveen et al., 2025).

4.8 Use of Artificial Intelligence, Remote Sensing, and Geospatial Information System in Monitoring Restoration Progress

The latest technologies, such as the Internet of Things (IoT), remote sensing, and artificial intelligence (AI), augment practices in sustainable forest management (Ali et al., 2025). Forests can be monitored continuously with the aid of IoT devices such as drones, which assist in real-time temperature, humidity, soil moisture, and tree growth data collection. Remote sensing technologies provide satellite and aerial imagery, which offers high-resolution data for large-scale assessments of forest cover changes, biomass estimation, and early detection of illegal logging (Wang et al., 2024a, b). Global Forest Watch is a leading example that employs satellite data and AI to oversee worldwide forest alteration and subsequently make informed decisions in restoration (Deng et al., 2024). Such technologies together with AI improve predictive modeling, data analysis, and decision-making, which fosters effective forest management strategies. The affordability and high-resolution monitoring provided by satellites and drones make machine learning algorithms notice fine details in ecological data. All these technologies enhance real-time monitoring systems, resource allocation, and data-driven decisions, making them crucial in preserving forest ecosystems and mitigating climate change impacts (Ali et al., 2025).

4.9 Miyawaki Forest Restoration Approach

To stop global crises and carbon dioxide emissions, and restore the ecosystem, Miyawaki pioneered a new type of reforestation method in Japan that aimed to do just that in the 1980s (Schirone et al., 2011). The Miyawaki method allows for ecosystems to recuperate without human intervention, thus enabling effortless recovery of biodiversity. This method makes it possible to restore ecosystems in cities and other heavily polluted areas at a much faster rate. The Miyawaki method is also termed "virtual forcing," where gradual coaxing from the soil to forest accelerates the movement through successional stages of natural vegetation. The ecotechnology and its vegetation-ecological approach enable the development of self-sustaining forests in two to three decades, emulating natural succession through the intensive planting of indigenous species (Miyawaki, 1998). It has been especially useful for community-driven urban afforestation, with remarkable success from India's many nonprofit and private organizations, has built over 150 miniature forests referred to as 'Tiny Forests', transforming micro-urban climates. Scotland's 2021 initiative 'Wee Forests' showcased how community participation could be incorporated and—along with showing ecological restoration—paved the way for integrating urban planning, illustrating the potential reach (Charkow, 2022).

5 Conclusion

Ecosystem Restoration: Substantial evidence has shown that ecological restoration programs have made significant contributions to the improvement of ecosystem services. Ecosystem restoration has become a vital approach in tackling the dual challenges of biodiversity decline and climate change. As ecosystems around the globe experience unparalleled degradation, restoration provides a pathway to regain ecological integrity, bolster resilience, and promote sustainable livelihoods. This chapter highlights the significance of long-term ecological monitoring, interdisciplinary collaboration, and the fusion of scientific knowledge with local and traditional practices. Long-term research, such as that conducted by the LTER network, offers crucial insights into ecosystem dynamics and the often-delayed responses seen in restoration efforts. Effective restoration encompasses not only ecological aspects but also socioeconomic and political dimensions. It necessitates inclusive planning that involves local communities, values traditional knowledge, and guarantees fair outcomes. Contemporary methods, including the application of smart materials, adaptive technologies, and data-driven decision-making tools, are increasingly utilized to monitor, evaluate, and steer restoration efforts. Concurrently, innovative strategies like the Miyawaki method illustrate the potential for swift, community-driven forest restoration, especially in urban environments.

Restoration initiatives must also be in harmony with broader climate adaptation and mitigation objectives. Restored ecosystems play a role in carbon sequestration, regulate hydrological cycles, and provide communities with protection against natural disasters. The incorporation of green infrastructure in urban planning—backed by public policy and local participation—further strengthens resilience and sustainability. Looking forward, the effectiveness of global restoration efforts under the UN Decade on Ecosystem Restoration will hinge on continuous funding, political commitment, and cooperative action. By restoring ecosystems, we revive life-supporting functions that are crucial for human survival and well-being. Ultimately, ecosystem restoration transcends mere healing of nature; it aims to reestablish a balanced relationship between humanity and the environment.

References

Abbass, K., Qasim, M. Z., Song, H., Murshed, M., Mahmood, H., & Younis, I. (2022). A review of the global climate change impacts, adaptation, and sustainable mitigation measures. *Environmental Science and Pollution Research, 29*(28), 42539–42559. https://doi.org/10.1007/s11356-022-19718-6

Abhilash, P. C. (2021). Restoring the unrestored: Strategies for restoring global land during the UN decade on ecosystem restoration (UN-DER). *Land, 10*(2), 201.

Adger, W. N., Arnell, N. W., & Tompkins, E. L. (2005). Successful adaptation to climate change across scales. *Global Environmental Change, 15*(2), 77–86.

Agrawal, A., & Gibson, C. C. (1999). Enchantment and disenchantment: The role of community in natural resource conservation. *World Development, 27*(4), 629–649.

Agrawal, A., & Ostrom, E. (2001). Collective action, property rights, and decentralization in resource use in India and Nepal. *Politics and Society, 29*(4), 485–514.

Ali, G., Mijwil, M. M., Adamopoulos, I., & Ayad, J. (2025). Leveraging the internet of things, remote sensing, and artificial intelligence for sustainable forest management. *Babylonian Journal of Internet of Things*, 1–65. https://doi.org/10.58496/BJIoT/2025/001

Alola, A. A., Yalçiner, K., Alola, U. V., & Saint Akadiri, S. (2019). The role of renewable energy, immigration and real income in environmental sustainability targets. Evidence from Europe's largest states. *Science of the Total Environment, 674*, 307–315.

Alongi Daniel, M. (2015). The impact of climate change on mangrove forests. *Current Climate Change Reports, 1*(1), 30–39. https://doi.org/10.1007/s40641-015-0002-x

Anaya, S. J. (2009). *Indigenous peoples in international law*. Oxford University Press.

Angelstam, P., Bush, T., & Manton, M. (2023). Challenges and solutions for forest biodiversity conservation in Sweden: Assessment of policy, implementation outputs, and consequences. *Land, 12*(5), 1098.

Ansell, J., & Evans, J. (2020). *Indigenous fire management and biodiversity conservation in Australia*. Springer.

Aronson, J., Goodwin, N., Orlando, L., Eisenberg, C., & Cross, A. T. (2020). A world of possibilities: Six restoration strategies to support the United Nation's Decade on Ecosystem Restoration. *Restoration Ecology, 28*(4), 730–736. https://doi.org/10.1111/rec.13170

Barrow, C. (2006). *Environmental management for sustainable development*. Routledge. https://doi.org/10.4324/9780203016671

Basak, S. M., Hossain, M. S., Tusznio, J., & Grodzińska-Jurczak, M. (2021). Social benefits of river restoration from ecosystem services perspective: A systematic review. *Environmental Science & Policy, 124*, 90–100.

Bentley, S. B., Tomscha, S. A., & Deslippe, J. R. (2022). Indicators of wetland health improve following small-scale ecological restoration on private land. *Science of the Total Environment, 837*, 155760.

Berkes, F., Colding, J., & Folke, C. (2018). *Navigating social-ecological systems: Building resilience for complexity and change*. Cambridge University Press.

Bharucha, Z., & Pretty, J. (2010). The roles and values of wild foods in agricultural systems. *Philosophical Transactions of the Royal Society B: Biological Sciences, 365*(1554), 2913–2926.

Bizuti, D. T. G., Robin, A., Soares, T. M., Moreno, V. S., Almeida, D. R. A., Andreote, F. D., Casagrande, J. C., Guillemot, J., Herrmann, L., van Melis, J., Perim, J. E. L., de Medeiros, S. D. S., Sorrini, T. B., & Brancalion, P. H. S. (2022). Multifunctional soil recovery during the restoration of Brazil's Atlantic Forest after bauxite mining. *Journal of Applied Ecology, 59*(9), 2262–2273.

Borrows, J. (2017). *Recovering Canada: The resurgence of indigenous law*. University of Toronto Press.

Breed, M. F., Harrison, P. A., Bischoff, A., Durruty, P., Gellie, N. J. C., Gonzales, E. K., Havens, K., et al. (2018). Priority actions to improve provenance decision-making. *Bioscience, 68*(7), 510–516. https://doi.org/10.1093/biosci/biy050

Bustamante, M. M., Silva, J. S., Scariot, A., Sampaio, A. B., Mascia, D. L., Garcia, E., et al. (2019). Ecological restoration as a strategy for mitigating and adapting to climate change: Lessons and challenges from Brazil. *Mitigation and Adaptation Strategies for Global Change, 24*, 1249–1270.

Butler, W. H., Monroe, A., & McCaffrey, S. (2015). Collaborative implementation for ecological restoration on US public lands: Implications for legal context, accountability, and adaptive management. *Environmental Management, 55*, 564–577. https://doi.org/10.1007/s00267-014-0430-8

Chagnaud, G., Panthou, G., Vischel, T., & Lebel, T. (2022). A synthetic view of rainfall intensification in the West African Sahel. *Environmental Research Letters, 17*(4), 044005.

Charkow, M. (2022). Forests for the future. *Science and Opinion. Avtree*. [online], 29–81.

Chazdon, R. L., Lindenmayer, D., Guariguata, M. R., Crouzeilles, R., Benayas, J. M. R., & Chavero, E. L. (2020). Fostering natural forest regeneration on former agricultural land through economic and policy interventions. *Environmental Research Letters, 15*(4), 043002.

Chazdon, R. L., Wilson, S. J., Brondizio, E., Guariguata, M. R., & Herbohn, J. (2021). Key challenges for governing forest and landscape restoration across different contexts. *Land Use Policy, 104*, 104854.

Chen, L., & Zhang, A. (2024). Identification of land use conflicts and dynamic response analysis of Natural-Social factors in rapidly urbanizing areas—A case study of urban agglomeration in the middle reaches of Yangtze River. *Ecological Indicators, 161*, 112009.

Cheng, C., & Li, F. (2024). Ecosystem restoration and management based on nature-based solutions in China: Research progress and representative practices. *Nature-Based Solutions*, 100176. https://doi.org/10.1016/j.nbsj.2024.100176

Christian, J. I., Basara, J. B., Hunt, E. D., Otkin, J. A., Furtado, J. C., Mishra, V., et al. (2021). Global distribution, trends, and drivers of flash drought occurrence. *Nature Communications, 12*(1), 6330.

Christie, M. R., & Knowles, L. L. (2015). Habitat corridors facilitate genetic resilience irrespective of species dispersal abilities or population sizes. *Evolutionary Applications, 8*(5), 454–463.

Conrad, C. C., & Hilchey, K. G. (2011). A review of citizen science and community-based environmental monitoring: Issues and opportunities. *Environmental Monitoring and Assessment, 176*, 273–291. https://doi.org/10.1007/s10661-010-1582-5

Cooke, S. J., Heger, T., Murphy, S. D., Shackelford, N., Febria, C. M., Rochefort, L., & Higgs, E. S. (2023). Ecological restoration in support of sustainability transitions: Repairing the planet in the Anthropocene. In *Introduction to designing environments: Paradigms & approaches* (pp. 93–112). Springer International Publishing.

Coulthard, G. S. (2014). *Red skin, white masks: Rejecting the colonial politics of recognition*. University of Minnesota Press.

Cuvelier, D., Gollner, S., Jones, D. O., Kaiser, S., Arbizu, P. M., Menzel, L., & Colaço, A. (2018). Potential mitigation and restoration actions in ecosystems impacted by seabed mining. *Frontiers in Marine Science, 5*, 467.

D'Antonio, C. M., August-Schmidt, E., & Fernandez-Going, B. (2016). Invasive species and restoration challenges. In *Foundations of restoration ecology* (pp. 216–244). https://doi.org/10.5822/978-1-61091-698-1_8

Dargie, G. C., Lewis, S. L., Lawson, I. T., Mitchard, E. T. A., Page, S. E., Bocko, Y. E., & Ifo, S. A. (2017). Age, extent and carbon storage of the Central Congo Basin peatland complex. *Nature, 542*, 86–90.

De, R., Sharma, R., Davidar, P., Arumugam, N., Sedhupathy, A., Puyravaud, J. P., Selvan, K. M., et al. (2021). Pan-India population genetics signifies the importance of habitat connectivity for wild Asian elephant conservation. *Global Ecology and Conservation, 32*, e01895. https://doi.org/10.1016/j.gecco.2021.e01888

Deng, Y., Zhang, Y., Pan, D., Yang, S. X., & Gharabaghi, B. (2024). Review of recent advances in remote sensing and machine learning methods for lake water quality management. *Remote Sensing, 16*, 4196.

Di Leginio, M., Agrillo, A., Congedo, L., Munafò, M., Riitano, N., Terribile, F., & Manna, P. (2024). Analysis of trends in productivity metrics in assessing land degradation: A case study in the Campania region of southern Italy. *Ecological Indicators, 161*, 111962.

Domingo-Marimon, C., Jenerowicz-Sanikowska, M., Pesquer, L., Ruciński, M., Krupiński, M., Woźniak, E., & Quader, M. A. (2024). Developing an early warning land degradation indicator based on geostatistical analysis of ecosystem functional types dynamics. *Ecological Indicators, 169*, 112815.

Dou, Y., Yu, X., & Liu, Y. (2021). Rethinking non-material links between people and drylands from a cultural ecosystem services perspective. *Current Opinion in Environmental Sustainability, 48*, 110–114. https://doi.org/10.1016/j.cosust.2020.12.007

Duarte, C. M., Losada, I. J., Hendriks, I. E., Mazarrasa, I., & Marbà, N. (2013). The role of coastal plant communities for climate change mitigation and adaptation. *Nature Climate Change, 3*, 961–968. https://doi.org/10.1038/nclimate1970

Eizaguirre Anglada, S. (2021). Challenges in strategies for socioeconomic democratization. Assessing solidarity economy policies in Barcelona. *European Planning Studies, 29*, 2145–2163.

Everard, M., & Moggridge, H. L. (2012). Rediscovering the value of urban rivers. *Urban Ecosystems, 15*, 293–314.

FAO. (2021). *The state of the world's forests 2021: Pathways for recovery and resilience*. Food and Agriculture Organization of the United Nations.

Feka, N. Z., & Manzano, M. G. (2008). The implications of wood exploitation for fish smoking on mangrove ecosystem conservation in the south West Province, Cameroon. *Tropical Conservation Science, 1*, 222–241.

Fernández-Llamazares, Á., Lepofsky, D., Lertzman, K., Armstrong, C. G., Brondizio, E. S., Gavin, M. C., & Vaughan, M. B. (2021). Scientists' warning to humanity on threats to indigenous and local knowledge systems. *Journal of Ethnobiology, 41*, 144–169.

Fleischman, F., Rana, P., Fischer, H., Guleria, V., Rana, R., & Ramprasad, V. (2025). Planting and replanting: Continuity and change over four decades of forest restoration in Himachal Pradesh, India. *Plants, People, Planet.* https://doi.org/10.1002/ppp3.10639

Fraisl, D., Hager, G., Bedessem, B., Gold, M., Hsing, P. Y., Danielsen, F., & Haklay, M. (2022). Citizen science in environmental and ecological sciences. *Nature Reviews Methods Primers, 2*, 64. https://doi.org/10.1038/s43586-022-00144-4

Franklin, J. F., Bledsoe, C. S., & Callahan, J. T. (1990). Contributions of the long-term ecological research program. *Bioscience, 40*, 509–523.

Frietsch, M., Loos, J., Löhr, K., Sieber, S., & Fischer, J. (2023). Future-proofing ecosystem restoration through enhancing adaptive capacity. *Communications Biology, 6*, 377. https://doi.org/10.1038/s42003-023-04736-y

Fu, B., Liu, Y., & Meadows, M. E. (2023). Ecological restoration for sustainable development in China. *National Science Review, 10*, nwad033.

Garcia, B., Morgan, E. A., Aruch, M., Ferreira, I. R., Jerozolimski, A., Mackey, B., & Hugh, S. (2024). Large-scale forest protection: The successful case of the Kayapo people in the Brazilian Amazon. *Regional Environmental Change, 24*, 148.

Garnett, S. T., Burgess, N. D., Fa, J. E., Fernández-Llamazares, Á., Molnár, Z., Robinson, C. J., & Watson, J. E. (2018). A spatial overview of the global importance of indigenous lands for conservation. *Nature Sustainability, 1*, 369–374.

Goffner, D., Sinare, H., & Gordon, L. J. (2019). The Great Green Wall for the Sahara and the Sahel initiative as an opportunity to enhance resilience in Sahelian landscapes and livelihoods. *Regional Environmental Change, 19*, 1417–1428.

Gotlieb, Y., Pérez-Briceño, P. M., Hidalgo, H., & Alfaro, E. (2019). The Central American Dry Corridor: A consensus statement and its background. *Revista Yu'am, 3*, 42–51.

Gottero, E. (2019). Identifying vulnerable farmland: An index to capture high urbanization risk areas. *Ecological Indicators, 98*, 61–67.

Grafton, R. Q., Daugbjerg, C., & Qureshi, M. E. (2015). Towards food security by 2050. *Food Security, 7*, 179–183.

Gregory, A., Spence, E., Beier, P., & Garding, E. (2021). Toward best management practices for ecological corridors. *Land, 10*, 140.

Grist, N., Mosello, B., Roberts, R., & Hilker, L. M. (2014). Resilience in the Sahel. *Building better lives with children, Plan International and ODI.*

Hambright, K. D., & Zohary, T. (1999). The Hula Valley (Northern Israel) wetlands rehabilitation project. In W. Streever (Ed.), *An international perspective on wetland rehabilitation* (pp. 173–180). Springer.

Hessburg, P. F., Miller, C. L., Parks, S. A., Povak, N. A., Taylor, A. H., Higuera, P. E., Prichard, S. J., et al. (2019). Climate, environment, and disturbance history govern resilience of Western North American forests. *Frontiers in Ecology and Evolution, 7*, 239. https://doi.org/10.3389/fevo.2019.00239

Hobbie, J. E., Carpenter, S. R., Grimm, N. B., Gosz, J. R., & Seastedt, T. R. (2003). The US long term ecological research program. *Bioscience, 53*, 21–32.

Hoegh-Guldberg, O., Jacob, D., Taylor, M., Guillén Bolaños, T., Bindi, M., Brown, S., & Zhou, G. (2019). The human imperative of stabilizing global climate change at 1.5 C. *Science, 365*, eaaw6974. https://doi.org/10.1126/science.aaw6974

Hoffmann, S. (2022). Challenges and opportunities of area-based conservation in reaching biodiversity and sustainability goals. *Biodiversity and Conservation, 31*, 325–352.

Hossini, H., Karimi, H., Mustafa, Y. T., & Al-Quraishi, A. M. F. (2022). Role of effective factors on soil erosion and land degradation: A review. In *Environmental degradation in Asia: Land degradation, environmental contamination, and human activities* (pp. 221–235).

Huber, J., Madurga-Lopez, I., Murray, U., McKeown, P. C., Pacillo, G., Laderach, P., & Spillane, C. (2023). Climate-related migration and the climate-security-migration nexus in the Central American Dry Corridor. *Climatic Change, 176*, 79.

Hussain, S., Sharma, S., Bhatti, R. C., & Singh, A. N. (2024). Sustainability in the Indian Himalayan region: Understanding the ecosystem services, climate change impacts, land use shifts and their threats. In *The Himalayas in the Anthropocene: Environment and development* (pp. 33–57). Springer Nature Switzerland.

Islam, M. M., Pranto, M. A., Shabab, M. R., Rone, M. R. I., Al Miraj, M. A., Hossen, M. M., & Shoumi, S. (2024). Revitalizing the Land: Ecosystem Restoration in Post-Mining Areas. N. Am. Acad. Res, 7, 11-19.

IUCN. (2020). *The Bonn challenge*. Retrieved June 19, 2021, from www.bonnchallenge.org

Jiang, C., Guo, H., Wei, Y., Yang, Z., Wang, X., Wen, M., & Zhou, P. (2021). Ecological restoration is not sufficient for reconciling the trade-off between soil retention and water yield: A contrasting study from catchment governance perspective. *Science of the Total Environment, 754*, 142139.

Jiang, Y., Shan, X., Liu, Q., Feng, Y., Lai, Y., Xu, K., & Feng, X. (2025). Evaluating the effectiveness of the Shan-Shui initiatives in China. *Geography and Sustainability*, 100271.

Johnson, C., & Kastner, P. (2018). Building urban resilience: Sustainable planning and development in the face of climate change. In *Urban sustainability* (pp. 123–140).

Kabir, M., Habiba, U. E., Khan, W., Shah, A., Rahim, S., De los Rios-Escalante, P. R., & Shafiq, M. (2023). Climate change due to increasing concentration of carbon dioxide and its impacts on environment in 21st century; a mini review. *Journal of King Saud University-Science, 35*, 102693. https://doi.org/10.1016/j.jksus.2023.102693

Kimmerer, R. W. (2013). *Braiding Sweetgrass: Indigenous wisdom, scientific knowledge, and the teachings of plants*. Milkweed Editions.

Kirwan, W., Smart, A. J., Trooien, T., Clay, D. E., & Hatfield, G. (2021). Drought resistance and resilience of non-native versus invaded-native grassland in the Northern Tallgrass Prairie. *Rangeland Ecology & Management, 79*, 100–109. https://doi.org/10.1016/j.rama.2021.08.004

Lal, R. (2015). Restoring soil quality to mitigate soil degradation. *Sustainability, 7*, 5875–5895. https://doi.org/10.3390/su7055875

Lange, I. D., Perry, C. T., & Alvarez-Filip, L. (2020). Carbonate budgets as indicators of functional reef "health": A critical review of data underpinning the current paradigm. *Ecological Indicators, 110*, 105857.

Lengnick-Hall, M., & Lengnick-Hall, C. (2002). *Human resource management in the knowledge economy: New challenges, new roles, new capabilities*. Berrett-Koehler Publishers.

Liang, C., Zhang, R. C., & Zeng, J. (2023). Optimizing ecological and economic benefits in areas with complex land-use evolution based on spatial subdivisions. *Landscape and Urban Planning, 236*, 104782.

Lin, B. B. (2011). Resilience in agriculture through crop diversification: Adaptive management for environmental change. *Bioscience, 61*, 183–193. https://doi.org/10.1525/bio.2011.61.3.4

Lorimer, J., & Driessen, C. (2014). Wild experiments at the Oostvaardersplassen: Rethinking environmentalism in the Anthropocene. *Transactions of the Institute of British Geographers, 39*, 169–181.

Lu, Q., Sun, J., Xiao, Y., Zhu, S., Zhu, G., Liu, J., & Liu, H. (2024). The progression traits of spatial conflicts within the production–living–ecological space among varying geomorphological types of mountain–basin areas in karst regions, China. *Ecological Indicators, 166*, 112324.

Maffi, L., & Woodley, E. (2012). *Biocultural diversity conservation: A global sourcebook*. Routledge.

McKenna, P. B., Ufer, N., Glenn, V., Doley, D., Phinn, S., & Erskine, P. D. (2023). Old Man Saltbush mortality following fire challenges the resilience of post-mine rehabilitation in Central Queensland, Australia. *Ecological Management & Restoration, 24*, 36–46.

Micklin, P. (2010). The past, present, and future Aral Sea. *Lake and Reservoir Management, 25*, 123–137.

Mirzabaev, A., Sacande, M., Motlagh, F., Shyrokaya, A., & Martucci, A. (2022). Economic efficiency and targeting of the African Great Green Wall. Nature Sustainability, 5(1), 17-25.

Mistry, J., & Berardi, A. (2016). Bridging indigenous and scientific knowledge. *Science, 352*, 1274–1275.

Miyawaki, A. (1998). Restoration of urban green environments based on the theories of vegetation ecology. *Ecological Engineering, 11*, 157–165.

Mohan, A. T. (2019). *Working from the middle: A multidirectional approach to mangrove conservation*. https://scholarworks.calstate.edu/downloads/rv042v775

Mohan, M., Chacko, A., Rameshan, M., Gopikrishna, V. G., Kannan, V. M., Vishnu, N. G.,... & Baiju, K. R. (2022). Restoring riparian ecosystems during the UN-decade on ecosystem restoration: a global perspective. Anthropocene Science, 1(1), 42-61.

Molnárová, K. J., Sklenička, P., Bohnet, I. C., Lowther-Harris, F., van den Brink, A., Moghaddam, S. M., & Azadi, H. (2023). Impacts of land consolidation on land degradation: A systematic review. *Journal of Environmental Management, 329*, 117026.

Mondal, S., & Lee, M. A. (2025). Impact of global warming on fisheries. In *Food security, nutrition and sustainability through aquaculture technologies* (pp. 227–253). Springer Nature Switzerland.

Moura, L. C., Scariot, A. O., Schmidt, I. B., Beatty, R., & Russell-Smith, J. (2019). The legacy of colonial fire management policies on traditional livelihoods and ecological sustainability in savannas: Impacts, consequences, new directions. *Journal of Environmental Management, 232*, 600–606.

Müller, F., Bergmann, M., Dannowski, R., Dippner, J. W., Gnauck, A., Haase, P., & Theuerkauf, M. (2016). Assessing resilience in long-term ecological data sets. *Ecological Indicators, 65*, 10–43. https://doi.org/10.1016/j.ecolind.2015.10.066

Nepstad, D., Schwartzman, S., Bamberger, B., Santilli, M., Ray, D., Schlesinger, P., & Rolla, A. (2006). Inhibition of Amazon deforestation and fire by parks and indigenous lands. *Conservation Biology, 20*, 65–73.

Ngabire, M., Wang, T., Liao, J., & Sahbeni, G. (2023). Quantitative analysis of desertification-driving mechanisms in the Shiyang River basin: Examining interactive effects of key factors through the Geographic detector model. *Remote Sensing, 15*, 2960.

Nolan, M., Stanton, K. J., Evans, K., Pym, L., Kaufman, B., & Duley, E. (2021). From the ground up: Prioritizing soil at the forefront of ecological restoration. *Restoration Ecology, 29*, e13453.

Nsabimana, D., Klemedtsson, L., & Kaplin, B. A. (2008). Soil CO_2 flux in six monospecific forest plantations in southern Rwanda. *Soil Biology and Biochemistry, 40*, 829–835.

Omemen, K. M., & Aldbbah, M. O. (2025). Climate change: Key contributors and sustainable solutions. *International Journal of Electrical Engineering and Sustainability*, 10–27.

Pakalne, M., Etzold, J., Ilomets, M., Jarašius, L., Pawlaczyk, P., Bociag, K., & Kirschey, T. (2021). *Best practice book for peatland restoration and climate change mitigation*. Experiences from LIFE Peat Restore Project.

Pfliegner, K. (2010). *The impacts of joint forest management on forest condition, livelihoods and governance: Case studies from Morogoro region in Tanzania* (Doctoral dissertation, University of East Anglia).

Pichancourt, J. B., Firn, J., Chadès, I., & Martin, T. G. (2014). Growing biodiverse carbon-rich forests. *Global Change Biology, 20*, 382–393.

Piñeiro, V., Arias, J., Dürr, J., Elverdin, P., Ibáñez, A. M., Kinengyere, A., & Torero, M. (2020). A scoping review on incentives for adoption of sustainable agricultural practices and their outcomes. *Nature Sustainability, 3*, 809–820.

Pokorny, B., Sotirov, M., Kleinschmit, D., & Kanowski, P. (2019). Forests as a global commons: International governance and the role of Germany. Report to the Science Platform Sustainability 2030. Freiburg: Universität Freiburg. 10.2312/iass.2019.037.

Polasky, S., Carpenter, S. R., Folke, C., & Keeler, B. (2011). Decision-making under great uncertainty: Environmental management in an era of global change. *Trends in Ecology & Evolution, 26*, 398–404.

Praveen, G., Kashi Vishwanath, N., Sree, C. C., & Rohith, D. (2025). Climate-resilient urban planning. *International Journal of Advances in Agricultural Science and Technology, 12*, 226–234.

Primavera, J. H., & Esteban, J. M. A. (2008). A review of mangrove rehabilitation in The Philippines: Successes, failures and future prospects. *Wetlands Ecology and Management, 16*, 345–358.

Qu, L., Huang, Y., Yang, L., & Li, Y. (2020). Vegetation restoration in response to climatic and anthropogenic changes in the loess plateau, China. *Chinese Geographical Science, 30*, 89–100.

Rahman, M. A. (2019). *Impact of structural development projects on vulnerability of coastal communities to disaster* (Doctoral dissertation, Queensland University of Technology).

Raza, A., Safdar, M., Adnan Shahid, M., Shabir, G., Khil, A., Hussain, S.,... & Akram, H. M. B. (2024). Climate Change Impacts on Crop Productivity and Food Security: An Overview. Transforming Agricultural Management for a Sustainable Future: Climate Change and Machine Learning Perspectives, 163-186.

Rawat, A., Kumar, D., & Khati, B. S. (2024). A review on climate change impacts, models, and its consequences on different sectors: A systematic approach. *Journal of Water and Climate Change, 15*, 104–126.

Saha, A., Pal, S. C., Chowdhuri, I., Islam, A. R. M. T., Roy, P., & Chakrabortty, R. (2022). Land degradation risk dynamics assessment in red and lateritic zones of eastern plateau, India: A combined approach of K-fold CV, data mining and field validation. *Ecological Informatics, 69*, 101653.

Schirone, B., Salis, A., & Vessella, F. (2011). Effectiveness of the Miyawaki method in Mediterranean forest restoration programs. *Landscape and Ecological Engineering, 7*, 81–92.

Schumann, R. L., III, Mockrin, M. H., Balachandran, B., Binder, S. B., & Greer, A. (2025). Flashpoints between wildfire recovery & mitigation in northern California. *International Journal of Disaster Risk Reduction, 119*, 105270.

Seddon, N., Chausson, A., Berry, P., Girardin, C. A., Smith, A., & Turner, B. (2020). Understanding the value and limits of nature-based solutions to climate change and other global challenges. *Philosophical Transactions of the Royal Society B, 375*, 20190120.

Sgrò, C. M., Lowe, A. J., & Hoffmann, A. A. (2011). Building evolutionary resilience for conserving biodiversity under climate change. *Evolutionary Applications, 4*, 326–337. https://doi.org/10.1111/j.1752-4571.2010.00157.x

Sharma, R., De, R., Puyravaud, J. P., Parida, J., Sedhupathy, A., Kalam, T., Rahim, A., et al. (2024). Patterns of genetic diversity, gene flow and genetic structure of three Peninsular Indian elephant populations indicate population connectivity. *Conservation Genetics*. https://doi.org/10.1007/s10592-024-01630-w

Shaw, R., & Cui, Y. (2025). Disaster law: An overview on its history, potentials and challenges. In *Disaster law: Implications to governance and implementation* (pp. 1–14). https://doi.org/10.1007/978-981-97-7374-9_1

Sievanen, L., Campbell, L. M., & Leslie, H. M. (2012). Challenges to interdisciplinary research in ecosystem-based management. *Conservation Biology, 26*, 315–323.

Simonson, W. D., Miller, E., Jones, A., García-Rangel, S., Thornton, H., & McOwen, C. (2021). Enhancing climate change resilience of ecological restoration—A framework for action. *Perspectives in Ecology and Conservation, 19*, 300–310. https://doi.org/10.1016/j.pecon.2021.05.002

Stanturf, J. A., Kant, P., Lillesø, J. P. B., Mansourian, S., Kleine, M., Graudal, L., & Madsen, P. (2015). Forest landscape restoration as a key component of climate change mitigation and adaptation. *Forest Ecology and Management, 34*, 34.

Steinhäußer, R., Siebert, R., Steinführer, A., & Hellmich, M. (2015). National and regional land-use conflicts in Germany from the perspective of stakeholders. *Land Use Policy, 49*, 183–194.
Suding, K. N. (2011). Toward an era of restoration in ecology: Successes, failures, and opportunities ahead. *Annual Review of Ecology, Evolution, and Systematics, 42*, 465–487. https://doi.org/10.1146/annurev-ecolsys-102710-145115
Suding, K., Higgs, E., Palmer, M., Callicott, J. B., Anderson, C. B., Baker, M., & Schwartz, K. Z. (2015). Committing to ecological restoration. *Science, 348*, 638–640. https://doi.org/10.1126/science.aaa4216
Turner, K. G., Anderson, S., Gonzales-Chang, M., Costanza, R., Courville, S., Dalgaard, T., & Wratten, S. (2016). A review of methods, data, and models to assess changes in the value of ecosystem services from land degradation and restoration. *Ecological Modelling, 319*, 190–207.
UNCCD. (2015). *Decision 3/COP.12*. https://www.unccd.int/sites/default/files/inline-files/dec3-COP.12eng.pdf
Upadhyay, R. K. (2020). Markers for global climate change and its impact on social, biological and ecological systems: A review. *American Journal of Climate Change, 9*, 159. https://doi.org/10.4236/ajcc.2020.93012
Uprety, Y., Asselin, H., Bergeron, Y., Doyon, F., & Boucher, J. F. (2012). Contribution of traditional knowledge to ecological restoration: Practices and applications. *Ecoscience, 19*, 225–237.
von Unger, M., & Emmer, I. (2018). *Carbon market incentives to conserve, restore and enhance soil carbon*. Silvestrum and The Nature Conservancy.
Wall, G. G. (2024). The Great Green Wall. ECHOES of the GGW. http://www.grandemuraille-verte.org/images/docs/Publications/Echoes_GGW_EN_web_compressed.pdf
Walker, W. S., Gorelik, S. R., Baccini, A., Aragon-Osejo, J. L., Josse, C., Meyer, C., & Schwartzman, S. (2020). The role of forest conversion, degradation, and disturbance in the carbon dynamics of Amazon indigenous territories and protected areas. *Proceedings of the National Academy of Sciences, 117*, 3015–3025.
Wambugu, P. W., Nyamongo, D. O., & Kirwa, E. C. (2023). Role of seed banks in supporting ecosystem and biodiversity conservation and restoration. *Diversity, 15*, 896. https://doi.org/10.3390/d15080896
Wang, R., Li, F., Hu, D., & Li, B. L. (2011). Understanding eco-complexity: Social-economic-natural complex ecosystem approach. *Ecological Complexity, 8*, 15–29. https://doi.org/10.1016/j.ecocom.2010.11.001
Wang, Z., Jiao, J., & Li, X. (2013). Towards sustainable integrated watershed ecosystem management: A case study in Dingxi on the Loess Plateau, China. *Environmental Management, 51*, 126–137.
Wang, R., Sun, Y., Zong, J., Wang, Y., Cao, X., Wang, Y., & Zhang, W. (2024a). Remote sensing application in ecological restoration monitoring: A systematic review. *Remote Sensing, 16*, 2204. https://doi.org/10.3390/rs16122204
Wang, J., Wang, Y., & Xu, D. (2024b). Desertification in northern China from 2000 to 2020: The spatial–temporal processes and driving mechanisms. *Ecological Informatics, 82*, 102769.
Wani, Z. A., Abdul Rahim, P. P., Dar, J. A., Lone, A. N., & Siddiqui, S. (2025a). Predicting the potential distribution of *Podophyllum hexandrum* Royle in the Himalaya under CMIP6 climate projections. *Scientific Reports, 15*, 25374. https://doi.org/10.1038/s41598-025-10862-w
Wani, Z. A., Singh, L., Uniyal, S., Rana, S. K., Bhatt, I. D., & Nautiyal, S. (2025b). Improving ecosystem vitality in India: Overcoming barriers to meet national and international targets. *Environmental Sustainability, 1*, 1–13. https://doi.org/10.1007/s42398-025-00339-x
Waring, B. G. (2024). Grand challenges in ecosystem restoration. *Frontiers in Environmental Science, 11*, 1353829.
Weidlich, E. W., Flórido, F. G., Sorrini, T. B., & Brancalion, P. H. (2020). Controlling invasive plant species in ecological restoration: A global review. *Journal of Applied Ecology, 57*, 1806–1817. https://doi.org/10.1111/1365-2664.13656

Whitehead, P., Purdon, P., Cooke, P., Russell-Smith, J., & Sutton, S. (2009). The West Arnhem Land Fire Abatement (WALFA) project. In *Culture, ecology, and economy of fire management in North Australian Savannas: Rekindling the Wurrk tradition* (pp. 287–312). CSIRO.

Winsor, P. (2021). Observer Review Report-WWF-2021.

Wong, R., Zhang, L., & Zhang, K. (2025). Sociodemographic factors and vulnerability to multiple extreme weather events: A national study in the US. *Science of the Total Environment, 967*, 178761.

Wood, M. C. (2023). Sky carbon cleanup and biodiversity restoration: Devising regional frameworks. *Vermont Journal of Environmental Law, 25*, 209.

World Bank. (2010). *Cities and climate change: An urgent agenda*. The World Bank.

Wu, W., Zhang, J., Yu, J., Sun, Z., Yu, R., Liu, W., & Fang, M. (2024). Attribution analysis of soil degradation using change vector analysis and the geographical detector from 2010 to 2020 on Hainan Island. *Ecological Informatics, 80*, 102484.

Wunder, S. (2015). Payments for ecosystem services: Some nuts and bolts. *Ecological Economics, 69*, 1202–1211.

Yan, Z., Guo, Y., Sun, B., Gao, Z., Qin, P., Li, Y., & Cui, H. (2024). Combating land degradation through human efforts: Ongoing challenges for sustainable development of global drylands. *Journal of Environmental Management, 354*, 120254.

Yang, Q., Liu, G., Li, H., Santagata, R., & Yang, Z. (2024). Understanding ecological restoration potential: The role of water resources and slope gradient limits. *Science of the Total Environment, 912*, 169001.

Yeremeyev, I., Dychko, A., Remez, N., Kraychuk, S., & Ostapchuk, N. (2021). Problems of sustainable development of ecosystems. In *IOP conference series: Earth and environmental science* (Vol. 628, p. 012014). IOP Publishing. https://doi.org/10.1088/1755-1315/628/1/012014

Yi, Z., Guo, X. M., Han, Y., Jiang, Y. F., Lai, X. H., & Huang, X. Y. (2020). Study on the spatial characteristics and intensity factors of "three-line conflict" in Jiangxi province. *Journal of Natural Resources, 35*, 2428–2443.

Young, T., Petersen, D. A., & Clary, J. J. (2005). The ecology of restoration: Historical links, emerging issues and unexplored realms. *Ecology Letters, 8*, 662–673. https://doi.org/10.1111/j.1461-0248.2005.00764.x

Zhang, D., Wang, X., Qu, L., Li, S., Lin, Y., Yao, R., & Li, J. (2020). Land use/cover predictions incorporating ecological security for the Yangtze River Delta region, China. *Ecological Indicators, 119*, 106841.

Zhang, J., Guan, Q., Shan, Y., Wang, Q., Mi, J., & Pan, L. (2025). Ecological security prediction and land use conflict identification in fragile mountain cities: A case study of Longnan, China. *Journal of Cleaner Production, 497*, 145146.

Zhou, Y., Li, P., Zhang, Q., & Cheng, G. (2025). Socio-economic impacts, challenges, and strategies for whole-region comprehensive land consolidation in China. *Land Use Policy, 150*, 107461.

Revisiting Indigenous Knowledge Systems for Strengthening the Biodiversity Conservation and Management Outlook

Rishav Rawal, Sachin Uniyal, Vikram S. Negi, Ravindra Joshi, and I. D. Bhatt

Abstract This study examines the intricate interconnection between cultural wisdom and ecological knowledge in the context of biodiversity conservation within India's Kailash Sacred Landscape (KSL). Focusing on the Indigenous Rung community in the Byans and Darma valleys, the research highlights how sustainable resource management is deeply embedded in local cultural norms, values, and customary practices. The community has developed complex systems of rules and traditions to protect natural resources and agroecosystems across the region. Their ecological knowledge, intertwined with sacred festivals, customary laws, and strong communal cohesion, functions as a practical mechanism for regulating resource use and preventing environmental degradation. Ethnobotanical knowledge, in particular, plays a vital role in conserving plant diversity, including rare and endemic species as well as local agrodiversity. This study underscores the importance of cultural heritage and traditional governance systems in shaping contemporary conservation strategies and ensuring the sustainable use of biological resources. It advocates for the integration of Indigenous knowledge and ritual practices into formal biodiversity management frameworks.

Keywords Rung community · Ethnobotanical knowledge · Traditional knowledge · Medicinal plants · Indian Himalayan region

R. Rawal
Department of History, Kumaun University, DSB Campus, Nainital, Uttarakhand, India

G.B. Pant National Institute of Himalayan Environment, Almora, Uttarakhand, India

S. Uniyal · R. Joshi · I. D. Bhatt
G.B. Pant National Institute of Himalayan Environment, Almora, Uttarakhand, India

V. S. Negi (✉)
Department of Botany, Hemvati Nandan Bahuguna Garhwal University, B.G.R. Campus, Pauri Garhwal, Uttarakhand, India

S. Siddiqui et al. (eds.), *Frontiers in Sustainability*,
https://doi.org/10.1007/978-3-032-19740-5_12

1 Introduction

Nature is an integral component of culture, and addressing ecological issues requires consideration of both ecosystem structures and the cultural practices of communities interacting with their environments (Wilson, 1992; Negi et al., 2023, 2025). Indigenous communities play a crucial role in assessing environmental conditions, having developed sophisticated knowledge systems through generations of coexistence with nature (Bicker et al., 2003; Negi et al., 2018a, b). These communities often exhibit unique environmental and cultural traits that contribute to high levels of ecological and cultural sustainability. Their dependence on natural resources, combined with traditional knowledge of resource use and management, has long supported biodiversity conservation efforts (Rist & Dahdouh-Guebas, 2006; Kendrick & Manseau, 2008).

In many traditional societies, forests and biodiversity have been conserved through sociocultural and religious practices for centuries (Irakiza et al., 2016; Upreti & Tewari, 2020). For instance, Farooquee et al. (2004) examined natural resource use in different ethnoecological settings and established a link between the local economy and conservation practices among the Rung community in the central Himalaya. Community customs and cultural practices are deeply intertwined with forests and their resources, fostering a lifestyle that sustains ecological balance. One prominent example of community-based conservation is the concept of sacred groves, which exists globally (Onyekwelu et al., 2024). In these areas, biodiversity is protected through religious or cultural rituals practiced by traditional communities (Dar et al., 2022).

The Indian Himalayan Region (IHR) harbors much of the Himalayan biodiversity and is characterized by high endemism, encompassing critical habitats and globally important ecoregions (Rawal et al., 2021, 2024). The IHR is vital to India's ecological and economic security, as recognized in the Indian National Action Plan on Climate Change (Negi et al., 2019; Rawal et al., 2021). The region is also notable for the rich cultural heritage and belief systems of its Indigenous communities, who have conserved biodiversity for centuries. These traditional conservation practices are essential for maintaining the region's ecological integrity and supporting human well-being across both the Himalaya and the adjacent plains. They have contributed significantly to the sustainable use and conservation of biodiversity, including rare and endemic species (Negi et al., 2023; Rawal et al., 2024). These practices are deeply influenced by cultural values and beliefs, which are, in turn, reinforced through the communities' close relationship with nature.

This study seeks to document the traditional ecological knowledge and cultural rituals of Indigenous communities in relation to biodiversity conservation and management. It examines a range of conservation strategies employed by Native communities in the Kailash Sacred Landscape (KSL)–India, with a focus on addressing current sustainability challenges. The study also highlights the profound interconnection between culture and ecology, emphasizing how Indigenous communities maintain a symbiotic relationship with their environment. This relationship has given rise to effective and culturally embedded conservation practices that contribute to the preservation of the region's rich plant diversity and natural resources. In particular, the study underscores the significance of spiritual beliefs and customs in shaping and sustaining biodiversity conservation efforts.

2 Materials & Method

2.1 *Study Area*

The study was conducted in the Kailash Sacred Landscape–India (KSL–India) region, located in the Pithoragarh district of Uttarakhand, India. It encompasses 17 migratory villages in the Darma Valley and nine villages in the Byans Valley, situated across an altitudinal range of 2400 to 3800 meters above sea level in the Western Himalayas. The study area lies between latitudes 28°30′N–31°28′N and longitudes 77°34′E–81°03′E (Fig. 1). KSL–India holds profound cultural and religious significance for diverse ethnic and religious communities and is characterized by complex biological and physical systems. Among these communities, the highland Rung people of Byans, Chaudans, and Darma valleys serve as traditional custodians of the region's forests and plant diversity. Their conservation practices are deeply rooted in ethnoecological knowledge (EEK), combining systematic and ritualistic approaches that contribute significantly to biodiversity conservation and the sustenance of local livelihoods (Negi et al., 2018a).

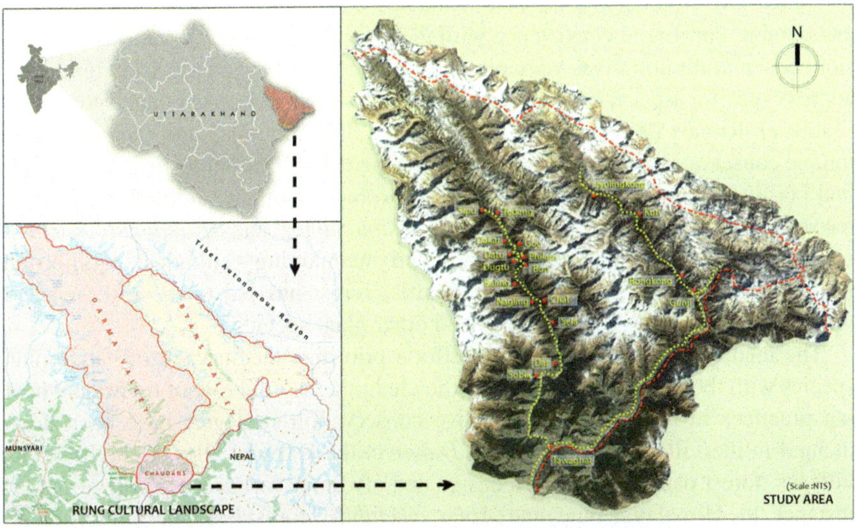

Fig. 1 Study area map of Rung Cultural Landscape (KSL-India), Uttarakhand

2.2 Data Collection and Analysis

This research adopted an ethnographic study approach, following the research design principles of social and cultural anthropology. This involved observing behaviors, rituals, and interactions, providing valuable insights into the community's cultural and ecological practices. We utilized various tools, including in-depth interviews, semi-structured questionnaires, focus group discussions, case studies, observations, and meticulous field note documentation. We implemented strategies to account for potential respondent distortions and inaccuracies to maintain data accuracy and minimize biases. Informal group discussions took place at communal gathering points, focusing on the value of forest products and their consumption. Furthermore, to enrich our investigation, we extensively reviewed historical sources from regional archives and public libraries and published materials in magazines and research papers. Qualitative data gathered from interviews, focus group discussions, and observations were thematically analyzed. The method adopted has contributed valuable insights into the intricate interplay between cultural diversity and biodiversity conservation in the Kailash Sacred Landscape.

3 Results

3.1 Ecological Knowledge of the Rung Community

Darma and Byans Valley in KSL-India is a compelling example of community-driven biodiversity conservation and management practices. Our findings show how residents have transitioned into the dedicated custodians of their natural heritage and exhibit a profound coexistence with the natural environment. Various biodiversity conservation initiatives were observed in the valleys, each precisely tailored to address specific aspects of biodiversity preservation and climate change adaptation (Table 1). Rituals like "miluthim festival" and "bhushe puja" are integral to their unique conservation model. For instance, during the "Miluthim festival ," a ceremonial burning of Juniperus communis serves to repel crop-damaging insects during the season. Similarly, the *"Bhushe puja," "Nabu Samo,"* and *"Kandahali festivals"* play distinct roles in protecting crops and maintaining ecological equilibrium. Protecting Humbalisingh (Abies spectabilis) forest has led to the resurgence of wildlife, including bears, musk deer, and other plant species.

The medicinal plant conservation efforts prioritize the survival of native plant species with therapeutic properties, acknowledging the vital role of traditional healing practices in these valleys. Effective conservation strategies have been documented in the villages of *Baling* and *Dakar* in the Darma Valley. In high-altitude villages, forest patches have been designated off-limits for resource utilization. For instance, the Himalayan blue pine *(Abies spectabilis)* forest patch in Baling, extending uphill to the settlement near *"Byakshi Bugyal,"* is safeguarded with the

Table 1 Cultural festivals and belief systems for biodiversity conservation in KSL, India

Name of festival	Cultural beliefs	Conservation practices
Miluthim	According to deeply rooted local tradition, celebrating the festival in accordance with customary laws is believed to repel crop pests, as the burning of *Juniperus communis* is thought to drive away insects. This belief, passed down through generations, reflects an example of indigenous knowledge systems in pest management	The festival is celebrated in September, coinciding with increased insect activity during crop maturation. As part of the ritual, *Juniperus communis* is burned to produce smoke, which is believed to repel insects from agricultural fields
Bhushe Puja	It is believed that the festival helps protect crops from pests without the need for chemical pesticides. During the associated fasting period, maintaining silence is thought to strengthen one's connection with divinity, reinforcing a spiritual commitment to crop protection	• The festival is celebrated in June in Dugtu village of the Darma Valley, coinciding with a critical stage of crop growth. It is considered important for reducing reliance on chemical pesticides, allowing natural ecological processes to support environmental balance and potentially enhance crop health
Nabu Samo	The festival is called *Nabu Samo*, where *Nabu* means "insects" and *Samo* means "to destroy," clearly reflecting its purpose. During the celebration, insects are collected and destroyed, symbolizing the community's triumph over harmful pests that threaten crop production	This practice serves as a traditional pest control technique aimed at protecting crops from insect-borne diseases. By manually removing harmful insects, the community seeks to improve agricultural yields. During the *Nabu Samo* festival, all family members participate in collecting and destroying pests from the fields, reflecting a collective commitment to crop protection
Kandahali	One of the major festivals of the Chaudans Valley is held every 12 years, coinciding with the full bloom of the *Kandahali* plant (*Strobilanthes kunthiana*). The festival is closely associated with the plant's blooming cycle, which holds deep cultural meaning for the community According to local tradition, a boy once died after applying a paste made from the *Kandahali* plant to a wound. As a result, the community considers the plant harmful and actively prevents it from blooming, integrating this belief into the festival's rituals	This festival also serves as a traditional method for controlling the spread of invasive species through periodic cutting. As part of the ritual, women from the community dress as warriors and are accompanied by men. Together, they ascend to the high ridges where the *Kandahali* plant grows and aggressively cut it down, preventing it from blooming and spreading further
Khirji	Khirji is celebrated in the Boondi village of Byans Valley	Khirji is celebrated in Boondi village of Byans Valley and is similar to the Kandahali festival, which is celebrated to stop the spread of invasive plant species

understanding that this forest cover acts as a protective barrier against glacier-related disasters. In Dakar village, the Bhojpatra patch enjoys complete protection in the name of the deity Larungdev, while a rotational policy governs the utilization of resources in the other nearby forests. Notably, in Dugtu village, the landscape has witnessed a significant increase in forest cover over the past few decades. Dakar village maintains a specific forest patch of *Betula utilis* under the name of the village deity "*Larungdev (Shyankshay)*," which has increased faunal diversity.

3.2 Cultural Laws in Resource Management and Conservation

The Rung community has employed sustainable practices and astute resource management strategies through sociocultural and belief systems (Table 1). A notable dimension of their conservation endeavors lies in establishing a comprehensive community forest governance system, which spans the entire landscape. This governance system, collaboratively overseen by the community and local van panchayats (VPs), rooted in their cultural laws, has implemented stringent measures to curtail excessive resource exploitation within the forested areas. For example, the "Larung Dev" forest in Dakar village, the "Galachingnala forest" in Baling village, and the "Bungj forest" in Kutti village are well managed by the VP council. The Rung community regards the forest not solely as a source of resources but also as a habitat for diverse flora and fauna. Beyond its ecological significance, the forest holds profound cultural and spiritual importance; it is deemed the dwelling place for various types of deities & spirits, including ancestral spirits. Resource extraction adheres to established traditional cultural norms and governance structures in each village, ensuring equitable access for all community members. For instance, a designated 15-day period is allocated for collecting medicinal plants such as *"Pahadi-Jeera," "Atish," and "Kadki"* in Bon village and Dugtu village in Darma Valley. The extraction of resources from the forest, commonly known as "Valleshe" in Tedang village, involves seeking permission from the village authority and, notably, offering a goat as a sacrifice.

The sacred forest patches dedicated to ancestral spirits and deities are meticulously preserved by village communities, often under the governance of temples and their authorities (Table 2). VPs have implemented a yearly rotation system to access resources from other nearby forests. Decisions regarding the use of these sacred groves are typically made through a consensus-driven process guided by the worship of local deities, especially when there is a compelling need or well-defined reasons for harvesting from these sacred sites (Table 2).

The Rung community's conservation practices extend to the careful management of *Bugyals* (alpine meadows), guided by locally established policies that regulate resource extraction and ensure the preservation of ecologically valuable medicinal plants. In several villages, *Van Panchayats* (VPs) collect grazing fees from outsiders who wish to use community lands. For instance, in Baling village, which holds traditional rights over Byakshi Bugyal, the Gaddi community, an Indigenous

Table 2 Sacred forests and protected areas managed by Indigenous communities in the KSL, India

Name of village/ valley	Sacred forests or sites	Associated practices and regulations
Tedang village, Darma	Valleshe forest	In Tedang village, a traditional ritual governs the collection of wood, requiring individuals to obtain explicit permission from village authorities before accessing forest resources. This practice reinforces responsible resource use and helps maintain a balance between human needs and ecological stewardship
Dakar village, Darma	Larungdev (Shyankshay) forest	This practice is specifically associated with the protection of a *Betula utilis* forest patch located above the settlement, where a traditional yearly rotation system is in place. The villagers have conserved and restricted access to this forest under the name of the regional deity Larungdev, highlighting its deep spiritual significance. As a result, the trees within this sacred forest remain untouched and are effectively safeguarded from human intervention
Sipu village, Darma	Mahadev Gufa	*Mahadev Gufa* is a sacred *Betula utilis* forest located near Sipu, dedicated to Lord Shiva. It holds profound spiritual significance for the local community and serves as an important sacred space. As a result, the forest has been protected for generations, with strict prohibitions on resource extraction and grazing that remain in place to this day
Dugtu village, Darma	Sangsuti-Bugyal, Gabbe-Bugyal, Neppa-Bugyal, Ringjangti Tal (Lake of Lord Shiva)	In Dugtu village, several conservation practices are linked to the sacred *bugyals* (alpine meadows), which hold deep spiritual significance. A holy lake, *Ringjangti Tal*—revered as the Lake of Lord Shiva—Is located nearby, where strict rights and restrictions are observed. Resource harvesting from this lake and its surrounding area is entirely prohibited, reflecting the community's commitment to preserving its sacred natural heritage
Kutti village, Darma	Veilshia alpine (rich in medicinal plants)	This practice is linked to restricted areas within Veilshia Alpine, located 10–15 km from Jyolikong, where access is limited due to the abundance of medicinal plant resources and rich biodiversity. The alpine meadow functions as a crucial interface between conservation and local livelihoods, supporting both ecological preservation and traditional health practices. The region is home to a diverse array of medicinal flora, highlighting its ecological and cultural significance
Sela-Rongkong, Darma	*Betula utilis* forest	This forest is considered sacred and devoted to the deity Ghankan near Rongkong. There is a complete ban on resource harvesting from the forest, resulting in the healthy ecological condition of the forest with high species diversity

(continued)

Table 2 (continued)

Name of village/ valley	Sacred forests or sites	Associated practices and regulations
Gunji village, Byans	*Pinus wallichiana* forest	*Pinus wallichiana* forests near Gunji village have been conserved; even the collection of fallen barks and twigs is not permitted from that forest. Villagers constructed a small temple in the forest
Napalchu village, Byans	Hya-Roshe Bugyal (alpine)	This practice is associated with Hya-Roshe Bugyal, an alpine meadow located near the village of Napalchu at 3400 m asl. Here, the villagers have fostered a strong connection between nature and spirituality, reflecting their deep reverence for the landscape. The alpine region serves as a valuable repository of medicinal plant species, playing a vital role in the community's traditional herbal healthcare systems
Kuti village, Byans	Pandav kutti	This is a sacred alpine situated near Kuti village. The harvesting of resources is completely restricted

Himalayan pastoral group, brings their herds to graze during the summer months. In exchange, they pay a grazing tax to the Baling Van Panchayat. This practice not only generates revenue for the VP but also benefits the meadow ecosystem through the natural fertilization provided by livestock manure. Spiritual leaders known as *Dhamis* play a key role in ensuring compliance with community rules and the conservation regulations enforced by the VPs. Their influence supports the sustainable use of natural resources and reinforces cultural adherence to ecological responsibility. Additionally, strict policies govern the harvesting of forest resources in the valleys. Most villages allow only the collection of dry wood, while the extraction of live wood requires special permission from the council. These measures have contributed significantly to forest conservation and biodiversity enhancement. For example, in Baling village, a Himalayan blue pine (*Abies spectabilis*) forest located above the settlement has been completely protected from any resource extraction, even by local residents.

The sacred groves and associated rituals within the scenic Rung landscapes hold immense cultural and ecological significance. These age-old practices designate specific areas of the landscape as sacred, encompassing diverse vegetation, including rare and endangered plant species, distinct life forms, and notable geographical features (see Table 3:). The community's reverence for particular tree species—often adorned with white and red robes as part of unique sociocultural traditions—has contributed to the in situ preservation of a wide range of plant species across the region. A notable example is found in Sipu village, where a sacred site known as *Mahadev Gufa* hosts a *Bhojpatra* (*Betula utilis*) forest patch dedicated to the deity Mahadev (Lord Shiva). Due to its sacred status, the forest is strictly protected from any form of extraction or disturbance. The observance of customary rules and rituals within such sacred spaces not only reflects the community's spiritual values but also reinforces a distinct cultural identity. Importantly, these practices play a critical

Table 3 List of threatened species in the KSL India region

Sr. no.	Species name	Local name	IUCN status
Floral species			
1	*Aconitum violaceum*	Dudhiya bish	VU
2	*Aconitum heterophyllum*	Atish	CE
3	*Allium stracheyi*	Jamboo	EN
4	*Arnebia benthamii*	Ratanjot	CR
5	*Angelica glauca*	Gandreni	EN
6	*Betula utilis*	Bhuj	NT
7	*Dactylorhiza hatagirea*	Hathpanja	EN
8	*Ephedra intermedia*	Ephedra	LC
9	*Fritillaria roylei*	Kakoli/lily	EN
10	*Hippophae salicifolia*	Chuk	NT
11	*Nardostachys grandiflora*	Spikenard/Jatamanshi	EN
12	*Picrorhiza kurroa*	Kutki	EN
13	*Pinus wallichiana*	Kail	LC
14	*Podophyllum hexandrum*	Bankakri	EN
15	*Polygonatum verticillatum*	Ghenauri	VU
16	*Rheum webbianum*	Daal chini	VU
17	*Thymus linearis*	Banaijwain	VU
18	*Dioscorea deltoidea*	Banaloo	CR
19	*Juniperus communis*	Dhoop	LC
Faunal species			
20	*Panthera uncia*	Snow Leopard	VU
21	*Hemitragus jemlahicus*	Himalayan Tahr	NT
23	*Ursus thibetanus*	Asiatic Black Bear	VU
24	*Moschus* spp.	Musk Deer	VU

CR Critically Endangered, *EN* Endangered, *VU* Vulnerable, *NT* Near Threatened, *VU* Vulnerable

role in biodiversity conservation, demonstrating the integration of ecological stewardship with spiritual and cultural heritage in the Rung community.

3.3 *Traditional Agriculture Wisdom*

Despite residing in high-altitude, challenging terrains, the Rung community has mastered the art of farming, ensuring food security and ecological harmony. Cultivating crops in such mountainous terrain demands skill and scientific knowledge. The Rung people successfully grow crops like *Dhaan (Oryza sativa), Madua (Eleusine coracana), Uva (Hordeum himalayense), Chuva (Amaranthus paniculatus),* as well as various seasonal fruits and vegetables. Notably, Fafar *(Fagopyrum esculentum),* Ogal *(Fagopyrum tataricum),* and Uva *(Hordeum himalayense)* are major staples in these high-altitude villages. Traditional farming implements like the "Bhounsi" for digging and "*Ansiya*" for crop harvesting are employed. Rung people practiced *Jhum cultivation (slash and burn)* in the early nineteenth century, burning a portion of the forest and sowing seeds over the ashes to yield high-quality crops. Many villages have long-standing irrigation systems like *Gool* and *Diggiya* to irrigate the crops in the landscape. Labor exchange is a common practice in the village, where community members assist each other with various agricultural operations tasks. However, during the harvest season, some members are compensated for their work in other fields. This is because the labor exchange alone during this season often falls short of meeting the labor demands for the abandonment of agricultural land in a few villages.

3.4 Van Panchayats in Biodiversity Conservation and Management

The community forests or Van Panchayats (VPs) serve as crucial regulators of ecosystem functions and provide essential services that support the local agricultural system, water resources, and overall environmental health in the villages of the landscape (Thakur et al., 2020; Pathak et al., 2021). VPs play a pivotal role in safeguarding biodiversity in the landscape; they achieve this by formulating policies to prevent forest resource degradation and ensure efficient management. VPs have established clear rules and regulations for harvesting forest resources from bugyals (high-altitude meadows) and forests. For example, there has been a designated period from May to July when residents have been permitted to collect resources from "Byakshi-bugyal," such as Keeda jadi *(Ophiocordyceps sinensis)* in Baling village. Similarly, before venturing to "Kolti-bugyal" in Tedang village and "Leppa-bugyal" in Dakar, the community organizes meetings to determine specific times for various activities like grazing and resource extraction. This organized approach has ensured the sustainable use of these valuable natural resources while preventing excessive exploitation and ensuring sustainable utilization of resources.

4 Discussion

Indigenous communities worldwide have profound connections to their environments and cultural heritage, which is intrinsically tied to their ability to maintain sustainability in resource utilization. Their intricate knowledge of local ecosystems enables them to understand resource management practices that are finely attuned to the natural rhythms and ecological functions (Adu-Gyamfi, 2011; Camacho et al., 2016). The Rung community in KSL exhibits a profound understanding of their local ecosystems and sustainable guardianship of resources embedded in their ancestral wisdom and deep cultural heritage. The artistic characteristics of

Indigenous communities emphasize a deep respect for the land, water, and wildlife in the landscape. This reverence fosters a sense of responsibility to protect these vital resources for future generations. Indigenous festivals, rituals, and oral traditions often contain ecological wisdom that guides their interactions with the environment, as documented in previous studies (Datta, 2018; Negi et al., 2018a, 2023, 2025). These cultural inheritances and practices are powerful motivators for sustainable practices, resource utilization, and biodiversity conservation efforts, reinforcing their commitment to maintaining a delicate balance between human activities and nature (Bihari, 2023). The holistic approach of the Rung community exemplifies how a deep understanding of the environment and a strong connection to cultural heritage can contribute to the preservation of ecosystems, making it a valuable lesson for wider global efforts to address environmental challenges.

Strong community bonds and cultural heritage characterize Rung communities in the landscape; this makes their distinct identity in the landscape. Elders pass down traditional practices, ecological wisdom, and cultural values, which sustain their way of life and promote resilience in the changing environment and climate. Past studies also indicated that the knowledge of Indigenous communities plays a vital role in climate resilience and adaptation in the Himalaya (Chaudhary & Bawa, 2011; Negi et al., 2017, 2022; Sharma & Shrestha, 2016) and across the world (Codjoe et al., 2014; Savo et al., 2016; Reyes-García et al., 2019). Cultural continuity is a source of strength, enabling Indigenous communities to adapt to environmental disruptions (Ford et al., 2020) and social changes while preserving their unique identities. The strong sense of community and traditional governance structures of the Rung community ensure collective decision-making and resource management. This plays a pivotal role in responding to environmental fluctuations and external pressures, safeguarding their cultural heritage, and ensuring ecological sustainability. These laws are crucial in preventing resource overexploitation, deforestation, and the degradation of vital ecosystems. This kind of custom worldwide fosters a strong sense of responsibility to safeguard the environment and its biodiversity. The customary laws and cultural beliefs provide a practical conservation framework encouraging long-term commitment to biodiversity preservation (Colchester, 2004; Labadi et al., 2021; Al-Mansoori & Hamdan, 2023). The synergy between environmental knowledge and cultural values creates a sustainable lifestyle that offers an appropriate model for harmonious coexistence with nature.

Indigenous biodiversity-related knowledge is of great significance to food security and livelihoods in the region; this is reflected in the high diversity of crops/landraces and diverse farming practices in the KSL and other parts of the IHR. The sustainable agricultural methods in the landscape, based on crop rotation and agroforestry, ensure food security, mitigate soil erosion, and protect biodiversity in the landscape. Indigenous communities have developed traditional methods of agriculture, which conserve the vast diversity of the farming system (Maikhuri et al., 1997; Negi & Maikhuri, 2013). The traditional methods of maintaining seed stock and cultivars have been critical to the sustenance of indigenous people in the region and play an essential role in in-situ conservation. Agro-diversity maintained by the Indigenous knowledge system in the IHR plays a significant role in (a) increasing productivity per unit area, (b) improving soil fertility by incorporating legumes, (c) minimizing crop loss to insect pest attacks, (d) increasing productivity per unit area, (e) providing diversity to food, (f) conserving soil erosion, (g) minimizing crop loss to pest attacks, and (h) insurance against absolute crop failure (Maikhuri et al., 1997, 2001; Negi et al., 2012).

The Sustainable Development Goals (SDGs) of the UN Agenda 2030 and the Aichi Biodiversity Targets (ABT), established by the Parties to the Convention on Biological Diversity (CBD), have prompted global efforts to advance biodiversity conservation (Navarro et al., 2017). While government-led initiatives play a crucial role, Indigenous and traditional communities—particularly those living in rural and remote areas—have long been managing and conserving biodiversity and natural resources through time-tested practices. Increasingly, Indigenous knowledge systems are being recognized for their alignment with multiple dimensions of the SDGs, highlighting their potential to significantly contribute to the achievement of these global sustainability goals (Negi et al., 2023). The role of Indigenous peoples and local communities in biodiversity management and conservation cannot be overlooked. Indigenous people constitute <5% of the world's population (476 million Indigenous people worldwide) and manage 28% of the world's land surface, which makes them the actual guardians of 80% of the known global biodiversity (Karki et al., 2017; Negi et al., 2018a, 2023). Their traditional knowledge and perspectives are often integrated into conservation strategies to ensure various national and international commitments to climate change mitigation. The local protected areas and sacred groves are considered promising approaches to climate change adaptation, including synergies with climate change mitigation through carbon sequestration and biodiversity conservation.

5 Conclusion

The study concludes that the Rung community of the Kailash Sacred Landscape (KSL) has developed a profound understanding of their local ecosystems through generations of traditional knowledge and spiritual beliefs. These beliefs emphasize the sustainable use of land, water, and biodiversity, fostering a harmonious coexistence with nature. Through community-driven initiatives, the Rung have effectively protected forests, restricted resource extraction in ecologically sensitive areas, and mitigated the risk of glacier-related disasters. Their conservation practices have led to the sustainable use of natural resources, protection of endangered species, restoration of wildlife habitats, and the rejuvenation of critical ecosystems within the landscape. These efforts contribute significantly to the long-term ecological health of the region. KSL serves as an ecologically vital transboundary corridor for wildlife movement, hosting diverse and endangered species across Nepal, India, and China. Preserving this biodiversity is crucial not only for the well-being of local flora and fauna but also for maintaining the resilience of the broader Himalayan ecosystems. Furthermore, the Rung community's unique traditional ecological knowledge plays a pivotal role in both natural resource conservation and the traditional healthcare systems of the region.

Acknowledgments The authors thank Dr. I.D. Bhatt, Director GBP-NIHE, Kosi-Katarmal, Almora, Uttarakhand, for the facilities and encouragement. Partial funding from the Department of Science and Technology (DST), Govt. of India under NMSHE—Task Force 3 (Forest Resources and Plant Biodiversity)—Phase II, and ICIMOD, Kathmandu under KSLCDI-II is gratefully acknowledged. Dr. Vikram S. Negi thanks the Vice Chancellor, HNBGU, and Campus Director, BGR Campus, HNBGU, Pauri Garhwal, for encouragement and support.

References

Adu-Gyamfi, Y. (2011). Indigenous beliefs and practices in ecosystem conservation: Response of the church: Church and environment. *Scriptura: Journal for Contextual Hermeneutics in Southern Africa, 107*(1), 145–155.

Al-Mansoori, F., & Hamdan, A. (2023). Integrating indigenous knowledge systems into environmental education for biodiversity conservation: A study of sociocultural perspectives and ecological outcomes. *AI, IoT and the Fourth Industrial Revolution Review, 13*(7), 61–74.

Bicker, A., Ellen, R., & Parkes, P. (2003). *Indigenous environmental knowledge and its transformations: Critical anthropological perspectives*. Routledge.

Bihari, S. (2023). Cultural heritage and indigenous knowledge: Reviving traditions for future generations. In *Sustainable development goals in SAARC countries: Key issues, opportunities and challenges* (Vol. 1, pp. 24–32).

Camacho, L. D., Gevaña, D. T., Carandang, A. P., & Camacho, S. C. (2016). Indigenous knowledge and practices for the sustainable management of Ifugao forests in Cordillera, Philippines. *International Journal of Biodiversity Science, Ecosystem Services & Management, 12*(1–2), 5–13.

Chaudhary, P., & Bawa, K. S. (2011). Local perceptions of climate change validated by scientific evidence in the Himalayas. *Biology Letters, 7*(5), 767–770.

Codjoe, S. N. A., Owusu, G., & Burkett, V. (2014). Perception, experience, and indigenous knowledge of climate change and variability: The case of Accra, a sub-Saharan African city. *Regional Environmental Change, 14*, 369–383.

Colchester, M. (2004). Conservation policy and indigenous peoples. *Environmental Science & Policy, 7*(3), 145–153.

Dar, J. A., Kothandaraman, S., Khare, P. K., & Khan, M. L. (2022). Sacred groves of Central India: Diversity status, carbon storage, and conservation strategies. *Biotropica, 54*(6), 1400–1411.

Datta, R. (2018). Traditional storytelling: An effective indigenous research methodology and its implications for environmental research. *AlterNative: An International Journal of Indigenous Peoples, 14*(1), 35–44.

Farooquee, N. A., Majila, B. S., & Kala, C. P. (2004). Indigenous knowledge systems and sustainable management of natural resources in a high-altitude society in Kumaun Himalaya, India. *Journal of Human Ecology, 16*(1), 33–42.

Ford, J. D., King, N., Galappaththi, E. K., Pearce, T., McDowell, G., & Harper, S. L. (2020). The resilience of indigenous peoples to environmental change. *One Earth, 2*(6), 532–543.

Irakiza, R., Serge, N. J., Vedaste, M., Elias, B., Nyirambangutse, B., & Marc, N. (2016). Assessment of traditional ecological knowledge and beliefs in the utilisation of important plant species: The case of Buhanga sacred forest, Rwanda. *Koedoe: African Protected Area Conservation and Science, 58*(1), 1–1.

Karki, M., Hill, R., Xue, D., AlanguiX, Ichikawa, K., & Bridgewater, P. (2017). Knowing our lands and resources: Indigenous and local knowledge and practices related to biodiversity and ecosystem services in Asia. In *Knowledges of nature* (Vol. 10, p. 212). UNESCO.

Kendrick, A., & Manseau, M. (2008). Representing traditional knowledge: Resource management and Inuit knowledge of barren-ground caribou. *Society and Natural Resources, 21*(5), 404–418.

Labadi, S., Giliberto, F., Rosetti, I., Shetabi, L., & Yildirim, E. (2021). Heritage and the sustainable development goals: Policy guidance for heritage and development actors. *International Journal of Heritage Studies*.

Maikhuri, R. K., Semwal, R. L., Rao, K. S., Nautiyal, S., & Saxena, K. G. (1997). Eroding traditional crop diversity imperils the sustainability of agricultural systems in Central Himalaya. *Current Science*, 777–782.

Maikhuri, R. K., Rao, K. S., & Semwal, R. L. (2001). Changing scenario of Himalayan agroecosystem: Loss of agrobiodiversity, an indicator of environmental change in central Himalaya, India. *The Environmentalist, 20*, 23–39.

Navarro, L. M., Fernández, N., Guerra, C., Guralnick, R., Kissling, W. D., Londoño, M. C., Muller-Karger, F., Turak, E., Balvanera, P., Costello, M. J., & Delavaud, A. (2017). Monitoring biodiversity change through effective global coordination. *Current Opinion in Environmental Sustainability, 29*, 158–169.

Negi, V. S., & Maikhuri, R. K. (2013). Socio-ecological and religious perspective of agrobiodiversity conservation: Issues, concern and priority for sustainable agriculture, Central Himalaya. *Journal of Agricultural and Environmental Ethics, 26*(2), 491–512.

Negi, V. S., Maikhuri, R. K., & Rawat, L. S. (2012). Paradigm and ecological implication of changing agricultural land-use: A case study from Govind wildlife sanctuary, central Himalaya, India. *Journal of Mountain Science, 9*(4), 547–557.

Negi, V. S., Maikhuri, R. K., Pharswan, D., Thakur, S., & Dhyani, P. P. (2017). Climate change impact in the Western Himalaya: People's perception and adaptive strategies. *Journal of Mountain Science, 14*(2), 403–416.

Negi, V. S., Pathak, R., Sekar, K. C., Rawal, R. S., Bhatt, I. D., Nandi, S. K., & Dhyani, P. P. (2018a). Traditional knowledge and biodiversity conservation: A case study from Byans Valley in Kailash Sacred Landscape, India. *Journal of Environmental Planning and Management, 61*(10), 1722–1743.

Negi, V. S., Kewlani, P., Pathak, R., Bhatt, D., Bhatt, I. D., Rawal, R. S., Sundriyal, R. C., & Nandi, S. K. (2018b). Criteria and indicators for promoting cultivation and conservation of medicinal and aromatic plants in Western Himalaya, India. *Ecological Indicators, 93*, 434–446.

Negi, V. S., Pathak, R., Rawal, R. S., Bhatt, I. D., & Sharma, S. (2019). Long-term ecological monitoring on forest ecosystems in Indian Himalayan Region: Criteria and indicator approach. *Ecological Indicators, 102*, 374–381.

Negi, V. S., Tiwari, D. C., Singh, L., Thakur, S., & Bhatt, I. D. (2022). Review and synthesis of climate change studies in the Himalayan region. *Environment, Development and Sustainability, 24*(9), 10471–10502.

Negi, V. S., Pathak, R., Thakur, S., Joshi, R. K., Bhatt, I. D., & Rawal, R. S. (2023). Scoping the need of mainstreaming indigenous knowledge for sustainable use of bioresources in the Indian Himalayan region. *Environmental Management, 72*(1), 135–146.

Negi, B., Negi, V. S., Rana, S. K., Bhatt, I. D., Manasi, S., & Nautiyal, S. (2025). Role of traditional ecological knowledge in shaping climate resilient villages in the Himalaya. *Journal of Environmental Management, 376*, 124325.

Onyekwelu, J. C., Agbelade, A. D., Stimm, B., & Mosandl, R. (2024). Role of sacred groves in southwestern Nigeria in biodiversity conservation, biomass and carbon storage. *Environmental Monitoring and Assessment, 196*(3), 269.

Pathak, R., Thakur, S., Negi, V. S., Rawal, R. S., Bahukhandi, A., Durgapal, K., Barola, A., Tewari, D., & Bhatt, I. D. (2021). Ecological condition and management status of Community Forests in Indian western Himalaya. *Land Use Policy, 109*, 105636.

Rawal, R., Negi, V. S., & Bhatt, I. D. (2021). Changing outlook on harnessing biodiversity values–A special focus on Indian Himalaya. *Journal of Graphic Era University*, 55–82.

Rawal, R., Negi, V. S., Bhatt, I. D., & Tiwari, L. M. (2024). Rarities pattern of vascular plants in the high-altitude forests of Indian western Himalaya: Conservation implications. *Journal for Nature Conservation, 79*, 126588.

Reyes-García, V., Fernández-Llamazares, Á., McElwee, P., Molnár, Z., Öllerer, K., Wilson, S. J., & Brondizio, E. S. (2019). The contributions of indigenous peoples and local communities to ecological restoration. *Restoration Ecology, 27*(1), 3–8.

Rist, S., & Dahdouh-Guebas, F. (2006). Ethnosciences—A step towards the integration of scientific and indigenous forms of knowledge in the management of natural resources for the future. *Environment, Development and Sustainability, 8*, 467–493.

Savo, V., Lepofsky, D., Benner, J. P., Kohfeld, K. E., Bailey, J., & Lertzman, K. (2016). Observations of climate change among subsistence-oriented communities around the world. *Nature Climate Change, 6*(5), 462–473.

Sharma, R. K., & Shrestha, D. G. (2016). Climate perceptions of local communities validated through scientific signals in Sikkim Himalaya, India. *Environmental Monitoring and Assessment, 188*, 1–11.

Thakur, S., Negi, V. S., Pathak, R., Dhyani, R., Durgapal, K., & Rawal, R. S. (2020). Indicator based integrated vulnerability assessment of community forests in Indian west Himalaya. *Forest Ecology and Management, 457*, 117674.

Upreti, B. M., & Tewari, L. M. (2020). Conservation of rare plant species in thal kedar sacred forest of Pithoragarh, kumaun Himalaya Uttarakhand. *Indian Journal of Natural Science, 10*(60), 19752–19756.

Wilson, D. S. (1992). Complex interactions in metacommunities, with implications for biodiversity and higher levels of selection. *Ecology, 73*(6), 1984–2000.

Protected Areas and their Role in Biodiversity Conservation

Maria Shah, Shujaul Mulk Khan, Jawad Hussain, Qurat Ul Ain, Azra Israr, Fazal Manan, Zareena Batool, Nazir Mohammad, and Shabiha Tuz Zahra

Abstract Protected areas play a pivotal role in the conservation of plant species. These areas are used to protect natural habitats of species, safeguard genetic diversity, and conserve their ecosystems critical for the existence of species. Protected areas serve as shelter for rare, endemic, and endangered plants, protecting them from habitat destruction, climate change, deforestation, overexploitation, and overcollection. These areas also maintain the survival of plant species that impart important ecosystem services such as oxygen production, soil fertility, and carbon sequestration. National parks, game reserves, biosphere reserves, Ramsar sites, and wildlife sanctuaries are some of the protected areas that are involved in the conservation and preservation of plant species. National parks preserve and conserve natural ecosystems, wildlife, and threatened species while allowing controlled public access for research, education, and recreation. Wildlife sanctuaries maintain high levels of biodiversity. Community and game reserves promote conservation through local involvement and sustainable use, while biosphere reserves and Ramsar sites preserve habitat and ecosystems globally. Public awareness via cultivation of native plants, sustainable agroforestry, and ecotourism increases the success of conservation and programs by incorporating local support. GIS and remote sensing tools are essential to track habitat loss, monitor ecosystems, and evaluate biodiversity conservation. In a nutshell, protected areas thus remain indispensable for safeguarding plant biodiversity, assuring ecological balance, and supporting the needs of future generations.

Keywords Protected forests · Sustainable biodiversity · Conservation planning · National parks · Biosphere reserves · Management strategies

---

M. Shah (✉)· J. Hussain · Q. U. Ain · A. Israr · F. Manan · Z. Batool · N. Mohammad
S. T. Zahra
Department of Plant Sciences, Quaid-i-Azam University, Islamabad, Pakistan

S. M. Khan
Department of Plant Sciences, Quaid-i-Azam University, Islamabad, Pakistan

Pakistan Academy of Sciences, Islamabad, Pakistan

S. Siddiqui et al. (eds.), *Frontiers in Sustainability*,
https://doi.org/10.1007/978-3-032-19740-5_13

1 Introduction

Protected areas (PAs) are those particular geographic regions whose primary function is to safeguard and preserve ecosystems. Protected areas are often the only safe spaces where vulnerable and endangered species exist, making their maintenance and restoration even more significant and necessary (Keenleyside, 2012). They play an essential role in the conservation of biodiversity, ecosystem sustainability, enhancing the ecological ecosystem, and the maintenance of the national ecological security (Leverington et al., 2010; McDonald & Boucher, 2011; Zu et al., 2023). According to IUCN (International Union for Conservation of Nature), a protected area is "a geographically defined region that is dedicated, recognized and well managed legally or through other effective ways, in order to accomplish the long-term preservation of nature, along with the ecological services and cultural values" (Dudley, 2008).

More than 12% of the earth's total area or over 100,000 sites have both their spatial and numerical features recorded in the WDPA (World Database on Protected Areas) (Chape et al., 2005). Global biodiversity is declining at an alarming rate (Butchart et al., 2010). According to global research, instead of creating and expanding more protected areas, the focus should be on improving the management of the existing PAs. These efforts can prevent threatened and endangered species from extinction (Leverington et al., 2010). In 147 countries, (Joppa & Pfaff, 2011) used matched pairs techniques to compare both protected and unprotected areas, found that conservation efforts in protected areas significantly decreased habitat conversion in 75% of the cases; however, the impact sizes were smaller when site selection bias was taken into consideration. Additionally, more extensive research highlights that protected areas that are properly controlled and well managed have a higher success rate in maintaining and preserving biodiversity (Geldmann et al., 2019). Protected areas significantly reduced deforestation in core zones, based on a case study carried out in the Himalayan regions by using satellite-derived forest cover trends (Gu et al., 2020). Although private protected areas are the second most commonly reported governance category, only 18,580 protected areas or 6.55% of all sites have been reported globally (Lewis et al., 2023). The number of PAs and their total area coverage across the globe are summarized in Table 1, while PAs and their coverage in Asia, specifically in Pakistan and India, are presented in Table 2.

2 Overview of Biodiversity Conservation and Role of Protected Areas

The diversity of life on Earth, including the various plants, animals, and microorganisms, as well as the genes they possess and the ecosystems they create, is known as biodiversity. It describes species variety (species' numbers), ecosystem variation, and genetic variation within a biome, planet, or region. Human well-being and sustainable development are dependent on the benefits and services provided by

Table 1 Numbers of protected areas and their coverage across different countries

Region	No of PAs	Private PAs	Coverage by PPAs
Australia	7281	1530	–
South Africa	1667	924	55.4%
Canada	2863	1863	–
USA	42,824	11,877	27.73%
Eswatini	13	8	25.0%
Mexico	1185	380	32.06%
Peru	277	95	34.29%
East Asia	1,930,651	–	16.1%
South East Asia	791,681	–	16.0%
Europe	699,761	–	12.4%
South Asia	344,248	–	7.6%

Table 2 Numbers of protected areas and land coverage across Pakistan and India

Protected Areas in Pakistan				
	Number	**Total area (hectare)**	**Coverage % of the land**	**References**
National parks	14	9,170,121 ha	10.4%	Khan et al. (2022), Afza et al. (2016)
Wildlife sanctuaries	99			
Game reserves	96			
Wetlands	9			
Biosphere reserve	19			
Total	237			
Protected Areas in India				
	Number	**Total area (km²)**	**Coverage % of country**	**References**
National parks	104	40,501.03	1.23%	Maan & Chaudhry, (2019), Kala, (2005)
Wildlife sanctuaries	544	118,931.80	3.62%	
Conservation reserves	77	2594.03	0.08%	
Community reserves	46	72.61	0.002%	
Total	771	162,099.47	4.93%	

biodiversity. Biodiversity is essential for organisms to breathe in air, to eat the food, and to drink the water (Rawat & Agarwal, 2015). It is clearly stated that invasive species (Jehangir et al., 2024), overexploitation, climate change, habitat fragmentation, and destruction may significantly contribute to the loss of biodiversity (Prakash & Verma, 2022). In this context, a major role is being played by humans. Anthropogenic pressures are the primary drivers of biodiversity loss, such as deforestation (Manan et al., 2020), and land-use changes. Forty to fifty percent of the ice-free land area has been altered by humans, converting wetlands, prairies, and forest into urban and agricultural systems. In regions where human effects predominate, many species have been eradicated (Chapin Iii et al., 2000). Conservation of biodiversity is essential to sustain the balance in the environment and to maintain the ecosystem services. Sustainable biodiversity is essential for ecological functioning and human well-being (Sala et al., 2000). The conservation of biodiversity is not only essential for the survival of species, but it also contributes to habitat conservation and is likely to reduce climate change (Prakash & Verma, 2022). A balanced strategy is necessary to support sustainable health in the context of its loss (Sharma & Birman, 2024).

To protect and conserve ecological, evolutionary, and biological processes, protected areas are a crucial strategy for the prevention of the current extinction issues regarding biodiversity (Butchart et al., 2012). According to Bacon (2019) protected areas are crucial for the preservation and conservation of biodiversity. These areas provide habitat for rare and threatened species, enhance ecological resilience, and sustain ecosystem services vital for human welfare (Watson et al., 2014). Many researchers stated that protected areas have higher species richness and abundance in contrast to unprotected regions (Achiso, 2020; Gray et al., 2016).

3 Regional and Global Efforts for the Conservation of Biodiversity

Regional and global approaches are very important for the conservation and preservation of biodiversity to reduce the rapid decline of plant species and ecosystems globally. Recent research studies have demonstrated the need for coordination between regional and global biodiversity initiatives to protect the diversity of life (Milner-Gulland et al., 2021; Pierce et al., 2021). Additionally, advancement in technology is continuing to improve preservation strategies, such as machine learning-based models (Estopinan et al., 2024). Digital stewardship using AI and camera traps is an example of innovative strategies that encourage local communities to be involved in conservation. To effectively protect biodiversity, localized initiatives must be integrated with global frameworks, and technology must be utilized (Fergus et al., 2023).

4 Importance of Protected Areas in Plant Conservation

Almost every aspect of life depends on plants, making them an essential part of biodiversity (Ullah et al., 2025). Over 27,000 out of 400,000 known plant species are medicinal, and over 31,000 have been identified as possibly edible for humans (Velazco et al., 2022). Hence, conserving species could also help to preserve the environmental services that they offer. One of the best strategies to protect the ecological environment and natural resources is Protected Areas (Velazco et al.,

2022). To prevent overexploitation of native species, several protected areas restrict or even ban their direct exploitation (de Souza & Prevedello, 2020; Kuempel et al., 2018; Watson et al., 2014). Over the last 20 years, there has been a growing emphasis on the importance of healthy ecosystems in sustaining human societies. Important ecosystem services, including water, food security (Zeb et al., 2026), management of wild stocks, protection of crop relatives, and carbon storage, can be obtained from well-managed protected areas (Postel & Thompson Jr, 2005; Scharlemann et al., 2010; Soares-Filho et al., 2010; Watson et al., 2014). To conserve more habitat, for instance forest cover, protected areas may support higher plant population (Fahrig, 2013). PAs are the best possible ways to protect and conserve rare, endemic and threatened plant species and to mitigate climate change. There are several strategies through which plants can be conserve and preserve, such as in situ conservations strategies.

5 Types of Protected Areas

In situ conservation emphasizes the conservation of species in their natural habitats (Öztürk et al., 2022). This strategy attempts to sustain the ecological integrity of the habitats and the natural processes vital for the support of biodiversity. Key components of in situ conservation involve the creation of protected areas, like wildlife sanctuaries, national parks, sacred groves, and biosphere reserves (Fig. 1).

5.1 National Parks and Wildlife Sanctuaries

National parks are founded to preserve wild species (Ozturk et al.), and their natural habitats alongside the provision of controlled tourism in the interest of public recreation. To avoid environmental decay, practices such as grazing, hunting, and logging are mostly prohibited. Parks are usually allocated to preserve certain species alongside securing the preservation of whole ecosystems (Mahanayak, 2024). National parks and wildlife sanctuaries maintain high levels of biodiversity, which is important for conservation (Asmat et al., 2022). National parks increase plant density, and contributing to carbon sequestration (Ali et al., 2022). In Chitral Gol National Park, even with varied medicinal plants, conservation is insufficient (Asmat et al., 2021). Wildlife sanctuaries are set aside as protected habitats where hunting, shooting, and capturing wildlife are strictly prohibited unless under special permits. Unlike national parks, the sanctuaries can allow human activities, for example, forestry, as long as they are not harmful to wildlife. They both act as crucial habitats for threatened species and help preserve ecological balance (Mahanayak, 2024).

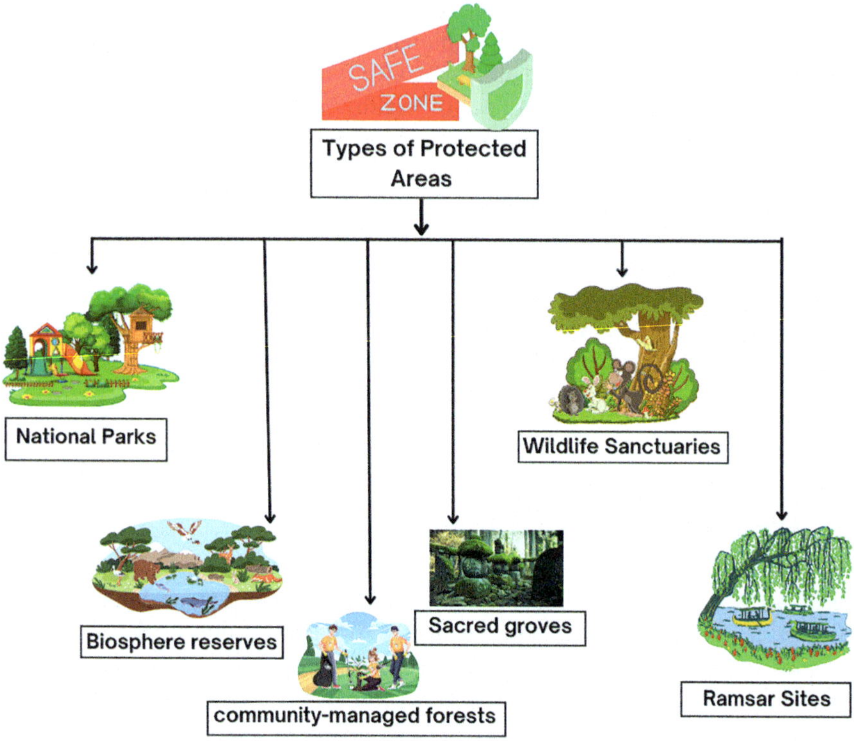

Fig. 1 Different types of protected areas

5.2 *Biosphere Reserves*

Biosphere reserves are specially marked sites that maintain distinct ecosystems and biodiversity while supporting research and sustainable development. They harmonize human activities with conservation, supporting sustainable use of resources and community participation (Mahanayak, 2024). Biosphere reserves exhibit three primary roles: i) protecting natural and semi-natural ecosystems, ii) facilitating sustainable economy and human development, and iii) facilitating research, monitoring, and education. These roles are organized through zonation, including core zones (strict protection), buffer zones (sustainable management), and transition zones where local people practice sustainable livelihoods (Dabard et al., 2024; Stoll-Kleemann & O'Riordan, 2017).

5.3 *Sacred Groves and Community-Managed Forests*

Sacred groves, which are forest patches prized for their cultural or religious values. Being sacred, the groves are maintained free from destruction and act as havens for numerous species (Kamran et al., 2020). The most intact habitats are those found in the form of sacred groves, which play a vital role in preserving endemic, threatened, and endangered species (Mahanayak, 2024). Sacred groves are conserved forest fragments, either primary or secondary, that support high biodiversity and have been protected for centuries because of their indigenous, religious, and sociocultural importance. Nevertheless, ignorance of their long-term ecological values has resulted in their degradation. Such factors as religious changes, growing pilgrim, tourist activities, and utilization of plants for economic purposes have played a crucial role in the deterioration of these previously flourishing ecosystems (Kandari et al., 2014). Community-managed forests, on the other hand, empower the local population to manage their resources sustainably, preserving centuries-old practices simultaneously. Bangladesh's Chittagong Hill Tracts (CHTs) natural forests are undergoing degradation, whereas Village Common Forests (VCFs) provide successful people-managed forests. This research in Komolchori VCF, Khagrachari, recorded 921 tree plants in 25 randomly selected plots, which reflects its diversity (Chowdhury et al., 2019).

5.4 *Ramsar Sites*

According to a recent study, Ramsar-designated wetlands support diverse and well-organized plant communities that are essential for ecological sustainability and preservation. For instance, a systematic survey of Boubyan Island in Kuwait, which serves as a Ramsar Site, discovered about 33 plant taxa belonging to 31 genera and 14 families, with Asteraceae, Amaranthaceae, and Poaceae being the prevalent groups. About 73% of species included grasses and annual forbs, and 5 different plant communities were identified through cluster analyses, demonstrating ecological variability and conservation importance of that site (Omar, 2025).

6 Ecosystem Services Provided by Protected Areas

The advantages that humans receive from ecosystems, either directly or indirectly, are known as ecosystem services (ESs) (Wang et al., 2021). Protected areas provide many services that support both human well-being and nature (Fig. 2).

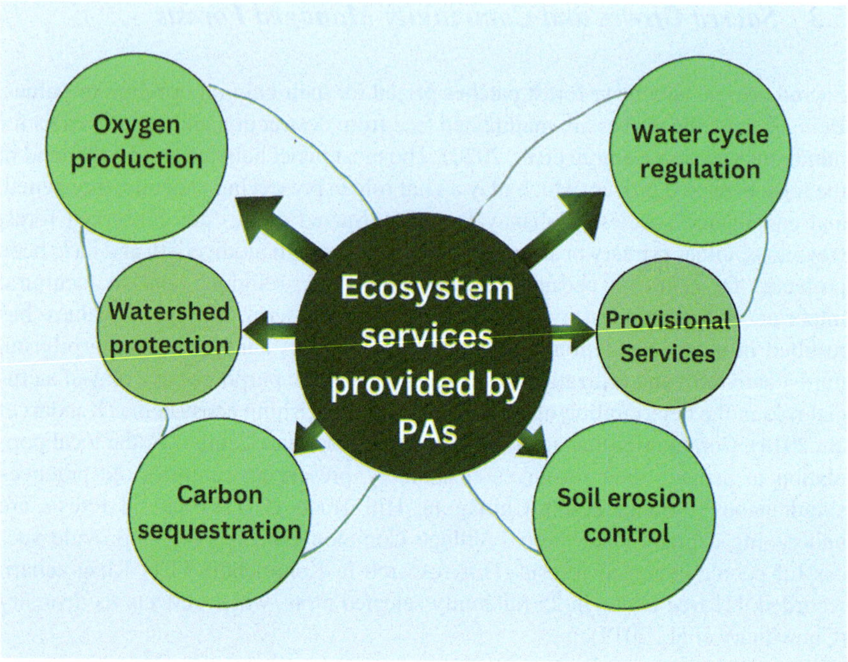

Fig. 2 Ecosystem services provided by the protected areas

6.1 Oxygen Production and Carbon Sequestration

Forests are essential for halting climate change and advancing sustainable growth (Ali et al., 2023). They are essential for controlling the planet's temperature and sustaining life since they absorb carbon dioxide and release oxygen. This is why they are frequently referred to as the lungs of the Earth (Ansari et al., 2021). According to research, higher productivity is expected to support some services, such as carbon storage and carbon sequestration (Hulvey et al., 2013). 96% of the world's tree species are found in tropical forests, which also store 25% of the world's carbon (Poorter et al., 2015). In this regard, national parks are the primary source of increased plant diversity and abundance, and thus carbon sequestration (Ali et al., 2022; Asmat et al., 2022). This helps to lessen the consequences of global warming (Kumar et al., 2017). It has been estimated that 15% of the global terrestrial carbon stock is found in protected areas (Lopoukhine et al., 2012). According to other research estimates, protected areas currently store 0.5 Pg C yearly, or roughly one-fifth of the carbon stored by all land ecosystems combined (Melillo et al., 2016).

6.2 Soil Erosion Control

Protected areas are crucial because it minimizes the rate of soil erosion (Borrelli et al., 2021). Soil erosion rates are greatly accelerated by human activities (Khan et al., 2025), including intensive plowing, inappropriate agricultural methods, and excessive grazing (Panagopoulos et al., 2019), deforestation (Borrelli et al., 2016), and related land use changes. The soil loss rate values vary by category, with prospective rangelands, including grasslands, thinly vegetated areas, and sclerophyllous vegetation, showing the greatest rates (Stefanidis et al., 2022). In terms of storing carbon and retaining nutrients, ILs (Indigenous Lands) outperformed protected areas; almost two-thirds of ILs were very successful in providing these ecosystem services (Zanin et al., 2024). Recognizing the importance of soil resources for sustainable development, the Sustainable Development Goals (SDGs) of the UN promote their preservation in order to achieve the lofty target of zero land degradation by 2030 (Keesstra et al., 2016).

6.3 Water Cycle Regulation and Watershed Protection

Protected Areas (PAs) are essential tools for conserving natural resources and preserving or improving ecosystem services, like watershed management (Zanin et al., 2024). About one-third of the world's, 105 biggest metropolitans drinking water provided by protected areas (Lopoukhine et al., 2012). About 80% of the downstream local community users acquire water from upstream protected areas (Postel & Thompson Jr, 2005) and 20% of the world's total, or around forty thousand km^3 per year, PAs supply a large portion of continental runoff (Harrison et al., 2016). In order to preserve watersheds and the availability and quality of freshwater supplies, protected forests are essential. By stabilizing the soil, tree roots prevent erosion and preserve the purity of the water. Forested regions frequently remove sediments and contaminants from runoff by acting as natural filters (Rajasugunasekar et al., 2023). This is supported by other research showing that compared to those outside protected areas, white-water and clear-water river reaches inside protected areas showed lower levels of suspended and dissolved sediments. This is due to the fact that forest cover that has been properly maintained reduces erosion (Zanin et al., 2024). Additionally, wetland protected areas also offer beneficial water storage services (Stolton et al., 2015).

6.4 Provisional Services

It has been shown that pleasant surroundings are beneficial for both physical and psychological therapy (Leverington et al., 2010). Protected areas benefit the local population and incorporate local activities into their management. Protected areas (PAs) contribute to public health as well as in economy by supplying medicinal herbs and genetic resources for pharmaceutical businesses (Stolton et al., 2015).

7 Threats to Biodiversity in Protected Areas

Protected areas play a crucial role in the preservation of biodiversity (Leverington et al., 2010). However, biodiversity is still declining even while conservation efforts both inside and outside protected areas are getting more attention (Butchart et al., 2010). Even in the best-managed PAs, the invasion of alien species is one of the factors that threaten biodiversity directly ((Foxcroft et al., 2013). Climate change is also a driver that threatens the conservation of biological diversity where environmental conditions are changing at a rate that makes it impossible for species to adapt and migrate (Füssel, 2007; Hussain et al., 2025). Global ecosystem functions are known to be threatened by climate change, which has also been demonstrated to have the ability to quickly alter the geographic range, diversity, and abundance of species (Harrison et al., 2006; Holmes et al., 2013).

8 Conservation Strategies and Management Approaches

8.1 Habitat Restoration

Restoration programs are crucial for improving plant biodiversity within PAs. In recent times, the most pressing issues faced by the Earth's surface and ecosystem are anthropogenic activities (Gairola et al., 2023). Native plant species also face challenges from distinctive climatic conditions, which can restrict their adaptability and ecosystem functions (Shah et al., 2025). Ecological restoration plays a major role in revitalizing these affected areas. Habitat restoration is recognized as a critical strategy that focuses on structure, function, and biodiversity, thus aiming to restore these damaged ecosystems to their previous state and reinstate ecosystem services (Dong et al., 2020). Practically, this can be brought about by creating habitat for diverse native species and providing natural solutions to the global climate crisis (Cheng & Li, 2024).

8.2 *Reforestation and Afforestation Initiatives*

It is crucial to have comprehensive insight into the extent to which a plant community has endured degradation before initiating restoration efforts, acknowledging that the ecosystem may degrade in a short period of time or otherwise gradually, upon straining to critical extremities (Koch et al., 2021). In recent times, the regenerative potential of forests in man-made tropical landscapes has been gaining recognition exponentially (Chazdon & Guariguata, 2016), it emerged as a powerful tool in forest restoration toward carbon sequestration, biodiversity preservation, and benefiting local human communities (Aguirre-Gutiérrez et al., 2023).

8.3 *Sustainable Agroforestry*

Species and ecosystems that are essential for the maintenance of human life and for the health of our earth are declining at a rapid pace, causing serious threats to food security, climate change, and biodiversity (Batool et al., 2025). Hence, there is a need for quick action to establish viable approaches for biodiversity conservation. Policy-makers and researchers are discovering the importance of agroforestry in the sustaining of biodiversity in both temperate and tropical areas (Harvey et al., 2006; Jose, 2009; McNeely, 2004; Schroth et al., 2013). Generally, agroforestry contributes to the conservation of biodiversity in major ways: (a) it lowers the rate of natural habitat conversion to agricultural land by offering sustainable and productive alternatives; (b) it helps in the preservation of germplasm of insecure species; (c) it provides many ecosystem services, including water refill, loss of habitat, and control of erosion that helps in the preservation of biodiversity; and (d) it provides suitable habitat for those species that cannot live in disturbed places. To develop and manage an agroforestry (Khan et al., 2024), framework for preservation, one would need to evaluate the whole terrain and apply easy and affordable agricultural techniques to optimize the environmental advantages (Jose, 2009).

8.4 *Public Awareness and Ecotourism Initiatives*

In protected areas (PAs), plant species conservation efforts and initiatives are significantly supported by public awareness programs and ecotourism efforts. It was discovered that an area's degree of protection was positively correlated with its tourism appeal (Xue & Tisdell, 2001). Every year, tourists spend US$1.5 × 109 in the densely populated Lake District National Park in the United Kingdom (Baldi et al., 2017). Ecotourism helps in the establishment of a state economy by generating revenue that can help in the sustainability of biodiversity and ecosystems (Buckley, 2011). To conserve ecosystems and to maintain restoration projects, forest conservation areas and national parks are fully dependent on revenues generated by ecotourism. Ecotourism also provides an opportunity to educate tourists about native plants, their role in ecology, conservation efforts, and their threats.

8.5 Public Engagement and Educational Initiatives

Educational programs in research centers are crucial to increase public awareness regarding biodiversity and plant conservation problems. Ballantyne & Packer (2011) stated that participation in environmental stewardship may rise due to these educational initiatives. Incorporating indigenous knowledge into conservation strategies is essential, as Vandebroek & Balick (2012) stated that the involvement of community can lead to sustainable and effective results.

8.6 Community Involvement in Conservation Initiatives

Protected areas have been intentionally isolated from human use (Vodouhê et al., 2010). Local communities' participation in the conservation of biodiversity in protected areas provides economic benefits for their livelihoods (Lepetu et al., 2008), while ensuring the preservation of biodiversity in protected areas by lowering illegal activities (Mbaiwa, 2004; Vodouhê et al., 2010), including fires, hunting, and excessive harvesting of natural resources.

9 Future Directions in Protected Area Management

9.1 Role of Technology (GIS, Remote Sensing) in Biodiversity Monitoring

Remote sensing is about data collection without physical interaction with earth (Joseph et al., 2011). Remote sensing and geographic information systems (GIS) technologies are now essential tools for evaluating biodiversity conservation and ecological services (Nizamani et al., 2023). Remote sensing provides opportunities to monitor biodiversity in a spatially extensive, reliable, and systematic way for earth observation (Pettorelli et al., 2016). Through remote sensing techniques, communities and ecosystems can be observed to monitor the climate drivers of biodiversity change and demonstrate the ecological context (Reddy et al., 2021). The amount of accessible remote sensing data is rising. Additionally, GIS tools are continuously storing in situ data, and new observations are frequently generated (Vihervaara et al., 2012, 2013). To monitor forest degradation and deforestation, the only effective way in real time is through satellite data (Abbas et al., 2013; Petrou et al., 2015). A potential way to solve problems in biodiversity monitoring is to combine in situ and remotely sensed data in modeling (Vihervaara et al., 2017). These techniques are essential for sustainable conservation strategies because they make it possible to determine critical areas and track ecological changes.

9.2 Climate Adaptation Strategies for the Conservation of Plant

The conservation of plants is facing serious threats due to climate change (Ahmad et al., 2025). Assisted migration is one such strategy, which entails moving plant species to regions with more suitable climates in order to guarantee their survival when their natural environments become unstable (Vitt et al., 2010). To reduce ecological threats, this technique requires appropriate and targeted species and location (McLachlan et al., 2007).

Connecting fragmented patches of habitat with corridors is the best way to prevent and restore scattered populations. Ecological studies have shown that corridors can aid in migration between patches and can improve species abundance and species richness (Gilbert-Norton et al., 2010). It has also been shown that corridor design, which can vary substantially between locations, can influence dispersal behavior, migration rates, and community composition (Christie & Knowles, 2015).

9.3 Expand Coverage of Protected Area

Adopt the Kunming-Montreal Global Biodiversity Framework's "30 by 30" target, which demands the protection of 30% of marine and land regions by 2030. In order to optimize conservation objectives, this expansion should give priority to ecologically appropriate areas.

Targets include restoring 30% of the planet's degraded inland water, terrestrial, marine, and coastline ecosystems (Target 2), encouraging private companies to monitor their biodiversity impacts, and increasing protection of inland water, terrestrial, and marine regions to at least 30% from the previous 17% terrestrial and 10% marine (Target 3) (Watson et al., 2023).

9.4 Promote Collaborative Management

Implementation of collaborative management partnerships (CMPs) such as involvement of governments, NGOs, local communities, and other private sectors can strengthen the role of protected areas. These partnerships have always remained effective in minimizing deforestation rates and strengthening the conservation of biodiversity in different regions. The globe is now facing huge biodiversity loss and climate change challenges; CMPs can be used as a template for future conservation efforts around the world. The funding and capacities of existing PAs can be escalated through strong CMPs between governmental and non-governmental organizations (Baghai et al., 2018; Desbureaux et al., 2025).

9.5 Foster International Collaborations

Addressing transboundary protected areas conservation (TBPAs) issues requires international collaboration. Coordinated efforts can improve the management of ecosystems and multinational migratory species (Pressey et al., 2015). Transboundary protected areas have been shown to encourage biodiversity conservation and ecological connections when different countries utilize shared governance techniques (Vasilijević et al., 2015).

9.6 Secure Sustainable Financing

Securing a number of funding sources, such as international funding, public budgets, and private sector investments, is significant for long-term conservation sustainability (Waldron et al., 2017). Numerous conservation initiatives have gained advantages from private-public partnerships, which increase management capability and offer financial security (Žuvela et al., 2023). Sustainable approaches to financing protected areas have been recommended, including biodiversity offsets and conservation trust funds (Doinjashvili et al., 2021).

10 Conclusion

Protected areas (PAs) play a significant role in the conservation of biodiversity by preserving genetic diversity, protecting natural habitats and conserving rare and threatened plant species. PAs provide important ecosystem services, such as water regulation, control of soil erosion, and carbon sequestration. These ecosystem services contribute to the maintenance of biodiversity and ecosystems. However, despite their importance, protected areas are facing some challenges like climate change, overexploitation of plant species, and habitat loss. Strengthening protected areas management through habitat restoration and afforestation, ecotourism, educational initiatives, and sustainable agroforestry can improve the conservation of biodiversity while supporting and helping local communities. Advanced technology like remote sensing and GIS further enhances adaptive management strategies and biodiversity monitoring for better conservation plans. Furthermore, policy framework and international collaboration are crucial for managing conservation problems and ensuring sustainable funding for protected areas.

References

Abbas, A., Khan, S., Hussain, N., Hanjra, M. A., & Akbar, S. (2013). Characterizing soil salinity in irrigated agriculture using a remote sensing approach. *Physics and Chemistry of the Earth, Parts A/B/C, 55*, 43–52.

Achiso, Z. (2020). Biodiversity and human livelihoods in protected areas: Worldwide perspective—A review. *SSR Institute of International Journal of Life Sciences, 6*(3), 2565–2578.

Afza, R., Ahmad, H., Saqib, Z., Ali, N., & Khan, J. (2016). Phytodiversity of Ayubia National Park Pakistan: Conservation and management issues. *Journal of Biodiversity and Environmental Sciences, 8*(2), 327–336.

Aguirre-Gutiérrez, J., Stevens, N., & Berenguer, E. (2023). Valuing the functionality of tropical ecosystems beyond carbon. *Trends in Ecology & Evolution, 38*(12), 1109–1111.

Ahmad, Z., Khan, S. M., Wani, Z. A., Pant, S., & Bhat, J. A. (2025). Editorial: Conserving plants in a changing climate. *Frontiers in Conservation Science, 6*. https://doi.org/10.3389/fcosc.2025.1743981

Ali, S., Khan, S. M., Siddiq, Z., Ahmad, Z., Ahmad, K. S., Abdullah, A., et al. (2022). Carbon sequestration potential of reserve forests present in the protected Margalla Hills National Park. *Journal of King Saud University-Science, 34*(4), 101978.

Ali, S., Khan, S. M., Ahmad, Z., Siddiq, Z., Ullah, A., Yoo, S., Han, H., & Raposo, A. (2023). Carbon sequestration potential of different forest types in Pakistan and its role in regulating services for public health. *Frontiers in Public Health, 10*. https://doi.org/10.3389/fpubh.2022.1064586

Ansari, N. A., Agus, C., & Nunoo, E. K. (2021). Foundations of 'SDG15–LIFE on land': Earth, forests and biodiversity. In *SDG15–life on land: Towards effective biodiversity management* (pp. 7–48). Emerald Publishing Limited.

Asmat, S., Khan, S. M., Ahmad, Z., Abdullah, A. M., Noor, R., et al. (2021). Study of medicinal plants of the Chitral Gol National Park (CGNP) with special emphasis on future conservation and sustainability. *Ethnobiology of Mountain Communities in Asia*, 71–85.

Asmat, S., Khan, S. M., Ahmad, Z., Manan, F., Noor, R., Zaman, I. U., & Abdullah. (2022). Role of chitral gol national park in maintaining and conserving plant diversity of the region. In *Biodiversity, conservation and sustainability in Asia: Volume 2: Prospects and challenges in south and middle Asia* (pp. 199–217). Springer.

Bacon, E., Gannon, P., Stephen, S., Seyoum-Edjigu, E., Schmidt, M., Lang, B., Sandwith, T., Xin, J., Arora, S., Adham, K. N., Espinoza, A. J. R., Qwathekana, M., Prates, A. P. L., Shestakov, A., Cooper, D., Ervin, J., de Souza Dias, B. F., Leles, B., Attallah, M., Mulongoy, J., & Gidda, S. B. (2019). Aichi Biodiversity Target 11 in the like-minded megadiverse countries. *Journal for Nature Conservation, 51*, 125723. https://doi.org/10.1016/j.jnc.2019.125723

Baghai, M., Miller, J. R., Blanken, L. J., Dublin, H. T., Fitzgerald, K. H., Gandiwa, P., et al. (2018). Models for the collaborative management of Africa's protected areas. *Biological Conservation, 218*, 73–82.

Baldi, G., Texeira, M., Martin, O. A., Grau, H. R., & Jobbágy, E. G. (2017). Opportunities drive the global distribution of protected areas. *PeerJ, 5*, e2989.

Ballantyne, R., & Packer, J. (2011). Using tourism free-choice learning experiences to promote environmentally sustainable behaviour: The role of post-visit 'action resources'. *Environmental Education Research, 17*(2), 201–215.

Batool, Z., Khan, S. M., Abdullah, A., Hussain, J., Khan, M. S., Hussain, I., Abbass, Z., Wu, Z.-Y., & Balslev, H. (2025). Foraging wild food plants in the dry gorges of Pakistan's high Karakorum mountains: A generational perspective. *Economic Botany, 79*(4), 529–550. https://doi.org/10.1007/s12231-025-09656-y

Borrelli, P., Panagos, P., Langhammer, J., Apostol, B., & Schütt, B. (2016). Assessment of the cover changes and the soil loss potential in European forestland: First approach to derive indicators to capture the ecological impacts on soil-related forest ecosystems. *Ecological Indicators, 60*, 1208–1220.

Borrelli, P., Alewell, C., Alvarez, P., Anache, J. A. A., Baartman, J., Ballabio, C., et al. (2021). Soil erosion modelling: A global review and statistical analysis. *Science of the Total Environment, 780*, 146494.

Buckley, R. (2011). Tourism and environment. *Annual Review of Environment and Resources, 36*(1), 397–416.

Butchart, S. H., Walpole, M., Collen, B., Van Strien, A., Scharlemann, J. P., Almond, R. E., et al. (2010). Global biodiversity: Indicators of recent declines. *Science, 328*(5982), 1164–1168.

Butchart, S. H., Scharlemann, J. P., Evans, M. I., Quader, S., Arico, S., Arinaitwe, J., et al. (2012). Protecting important sites for biodiversity contributes to meeting global conservation targets. *PLoS One, 7*(3), e32529.

Chape, S., Harrison, J., Spalding, M., & Lysenko, I. (2005). Measuring the extent and effectiveness of protected areas as an indicator for meeting global biodiversity targets. Abstract. *Philosophical Transactions of the Royal Society B: Biological Sciences, 360*(1454), 443–455. https://doi.org/10.1098/rstb.2004.1592

Chapin Iii, F. S., Zavaleta, E. S., Eviner, V. T., Naylor, R. L., Vitousek, P. M., Reynolds, H. L., et al. (2000). Consequences of changing biodiversity. *Nature, 405*(6783), 234–242.

Chazdon, R. L., & Guariguata, M. R. (2016). Natural regeneration as a tool for large-scale forest restoration in the tropics: Prospects and challenges. *Biotropica, 48*(6), 716–730.

Cheng, C., & Li, F. (2024). Ecosystem restoration and management based on nature-based solutions in China: Research progress and representative practices. *Nature-Based Solutions, 100176*.

Chowdhury, M. A., Islam, K. N., Hafiz, N., & Islam, K. (2019). Diversity of trees in a community managed forest: The case of Komolchori VCF, Khagrachari, Bangladesh. *Geology, Ecology, and Landscapes, 3*(2), 95–103.

Christie, M. R., & Knowles, L. L. (2015). Habitat corridors facilitate genetic resilience irrespective of species dispersal abilities or population sizes. *Evolutionary Applications, 8*(5), 454–463.

Dabard, C. H., Gohr, C., Weiss, F., von Wehrden, H., Neumann, F., Hordasevych, S., et al. (2024). Biosphere reserves as model regions for transdisciplinarity? A literature review. *Sustainability Science, 19*(6), 2065–2081.

de Souza, A. C., & Prevedello, J. A. (2020). The importance of protected areas for overexploited plants: Evidence from a biodiversity hotspot. *Biological Conservation, 243*, 108482.

Desbureaux, S., Kabore, I., Vaglietti, G., Baghai, M., Lindsey, P., Robson, A., et al. (2025). Collaborative management partnerships strongly decreased deforestation in the most at-risk protected areas in Africa since 2000. *Proceedings of the National Academy of Sciences, 122*(1), e2411348121.

Doinjashvili, P., Méral, P., & Andriamahefazafy, F. (2021). Sustaining protected areas through conservation trust funds: a review. *International Journal of Sustainable Development & World Ecology, 28*(3), 193–202.

Dong, S., Shang, Z., Gao, J., & Boone, R. B. (2020). Enhancing sustainability of grassland ecosystems through ecological restoration and grazing management in an era of climate change on Qinghai-Tibetan plateau. *Agriculture, Ecosystems & Environment, 287*, 106684.

Dudley, N. (2008). *Guidelines for applying protected area management categories*. Iucn.

Estopinan, J., Servajean, M., Bonnet, P., Joly, A., & Munoz, F. (2024). AI-based mapping of the conservation status of orchid assemblages at global scale. *arXiv preprint arXiv:2401.04691*.

Fahrig, L. (2013). Rethinking patch size and isolation effects: The habitat amount hypothesis. *Journal of Biogeography, 40*(9), 1649–1663.

Fergus, P., Chalmers, C., Longmore, S., Wich, S., Warmenhove, C., Swart, J., et al. (2023). Empowering wildlife guardians: An equitable digital stewardship and reward system for biodiversity conservation using deep learning and 3/4G camera traps. *Remote Sensing, 15*(11), 2730.

Foxcroft, L. C., Pyšek, P., Richardson, D. M., & Genovesi, P. (2013). Plant invasions in protected areas. *Patterns, problems and challenges. Invading nature—Springer series in invasion ecology, 7*, 3–18.

Füssel, H.-M. (2007). Adaptation planning for climate change: Concepts, assessment approaches, and key lessons. *Sustainability Science, 2*, 265–275.

Gairola, S. U., Bahuguna, R., & Bhatt, S. S. (2023). Native plant species: A tool for restoration of mined lands. *Journal of Soil Science and Plant Nutrition, 23*(2), 1438–1448.

Geldmann, J., Manica, A., Burgess, N. D., Coad, L., & Balmford, A. (2019). A global-level assessment of the effectiveness of protected areas at resisting anthropogenic pressures. *Proceedings of the National Academy of Sciences, 116*(46), 23209–23215.

Gilbert-Norton, L., Wilson, R., Stevens, J. R., & Beard, K. H. (2010). A meta-analytic review of corridor effectiveness. *Conservation Biology, 24*(3), 660–668.

Gray, C. L., Hill, S. L., Newbold, T., Hudson, L. N., Börger, L., Contu, S., et al. (2016). Local biodiversity is higher inside than outside terrestrial protected areas worldwide. *Nature Communications, 7*(1), 12306.

Gu, C., Zhao, P., Chen, Q., Li, S., Li, L., Liu, L., & Zhang, Y. (2020). Forest cover change and the effectiveness of protected areas in the Himalaya since 1998. *Sustainability, 12*(15), 6123.

Harrison, P., Berry, P., Butt, N., & New, M. (2006). Modelling climate change impacts on species' distributions at the European scale: Implications for conservation policy. *Environmental Science & Policy, 9*(2), 116–128.

Harrison, I. J., Green, P. A., Farrell, T. A., Juffe-Bignoli, D., Sáenz, L., & Vörösmarty, C. J. (2016). Protected areas and freshwater provisioning: A global assessment of freshwater provision, threats and management strategies to support human water security. *Aquatic Conservation: Marine and Freshwater Ecosystems, 26*, 103–120.

Harvey, C. A., Gonzalez, J., & Somarriba, E. (2006). Dung beetle and terrestrial mammal diversity in forests, indigenous agroforestry systems and plantain monocultures in Talamanca, Costa Rica. *Biodiversity and Conservation, 15*, 555–585.

Holmes, K. R., Nelson, T. A., Coops, N. C., & Wulder, M. A. (2013). Biodiversity indicators show climate change will alter vegetation in parks and protected areas. *Diversity, 5*(2), 352–373.

Hulvey, K. B., Hobbs, R. J., Standish, R. J., Lindenmayer, D. B., Lach, L., & Perring, M. P. (2013). Benefits of tree mixes in carbon plantings. *Nature Climate Change, 3*(10), 869–874.

Hussain, J., Khan, S. M., Khan, M. S., Saqib, Z., Ahmad, Z., Shah, M., et al. (2025). Distribution modelling and future prediction of a threatened species-Heracleum candicans Wall. ex DC.; Within the framework of biotic and abiotic interactions. *Journal of Environmental Management, 387*, 125818.

Iyendo, T. O., Uwajeh, P. C., & Ikenna, E. S. (2016). The therapeutic impacts of environmental design interventions on wellness in clinical settings: A narrative review. *Complementary Therapies in Clinical Practice, 24*, 174–188.

Jehangir, S., Khan, S. M., Ejaz, U., Qurat-ul-Ain, Zahid, N., Rashid, N., Noshad, Q., Din, Z. U., & Shoukat, A. (2024). Alien flora of Pakistan: Taxonomic composition invasion status geographic origin introduction pathways and ecological patterns. *Biological Invasions, 26*(8), 2435–2451. https://doi.org/10.1007/s10530-024-03311-8

Joppa, L. N., & Pfaff, A. (2011). Global protected area impacts. *Proceedings of the Royal Society B: Biological Sciences, 278*(1712), 1633–1638.

Jose, S. (2009). *Agroforestry for ecosystem services and environmental benefits: An overview*. Springer.

Joseph, S., Murthy, M., & Thomas, A. (2011). The progress on remote sensing technology in identifying tropical forest degradation: A synthesis of the present knowledge and future perspectives. *Environmental Earth Sciences, 64*, 731–741.

Kala, C. P. (2005). A multifaceted review on the biodiversity conservation of the Valley of Flowers National Park, India. *The International Journal of Biodiversity Science and Management, 1*(1), 25–32.

Kamran, S., Khan, S. M., Ahmad, Z., Rahman, A. U., Iqbal, M., Manan, F., Haq, Z. U., & Ullah, S. (2020). The role of graveyards in species conservation and beta diversity: A vegetation appraisal of sacred habitats from Bannu. *Pakistan Journal of Forestry Research, 31*(4), 1147–1158. https://doi.org/10.1007/s11676-019-00893-1

Kandari, L. S., Bisht, V. K., Bhardwaj, M., & Thakur, A. K. (2014). Conservation and management of sacred groves, myths and beliefs of tribal communities: A case study from North-India. *Environmental Systems Research, 3*, 1–10.

Keenleyside, K. (2012). *Ecological restoration for protected areas: Principles, guidelines and best practices* (Vol. 18). iucn.

Keesstra, S., Bouma, J., Wallinga, J., Tittonell, P., Smith, P., Cerdà, A., et al. (2016). The significance of soils and soil science towards realization of the United Nations sustainable development goals. *The Soil, 2*(2), 111–128.

Khan, A. N., Khan, H. A., & Ammar, E. (2022). A review: Protected areas of Pakistan management and current issues. *American Journal of Biological and Environmental Statistics, 8*(4), 112–116.

Khan, I. A., Khan, S. M., Jahangir, S., Ali, S., & Tulindinova, G. K. (2024). Carbon storage and dynamics in different agroforestry systems. In A. Raj, M. K. Jhariya, A. Banerjee, R. K. Jha, & K. P. Singh (Eds.), *Agroforestry* (pp. 345–374). Wiley.

Khan, M. S., Khan, S. M., Abdullah, Liu, J., Zeng-Yuan, W., Hussain, J., Zeb, S. A., Mohammad, N., Batool, Z., Saqib, Z., Afza, R., Manan, F., & Ali, S. (2025). Ecological assessment of Iris hookeriana across subalpine and alpine regions of the Hindu-Himalayas. *Frontiers in Forests and Global Change, 8*. https://doi.org/10.3389/ffgc.2025.1539025

Koch, J., Pearson, D. E., Huebner, C. D., Young, M. K., & Sniezko, R. A. (2021). Restoration of landscapes and habitats affected by established invasive species. *Invasive species in forests and rangelands of the United States*, 185.

Kuempel, C. D., Adams, V. M., Possingham, H. P., & Bode, M. (2018). Bigger or better: The relative benefits of protected area network expansion and enforcement for the conservation of an exploited species. *Conservation Letters, 11*(3), e12433.

Kumar, V., Sharma, K., Sharma, V., Arya, V. M., Kumar, R., Singh, V., et al. (2017). Soil quality refurbishment through carbon sequestration in climate change: A review. *International Journal of Current Microbiology and Applied Science, 6*, 1210–1223.

Lepetu, J., Makopondo, R. O., & Darkoh, M. B. (2008). Community-based natural resource management and tourism partnership in Botswana: Which way forward? *Botswana Notes and Records*, 113–124.

Leverington, F., Costa, K. L., Pavese, H., Lisle, A., & Hockings, M. (2010). A global analysis of protected area management effectiveness. *Environmental Management, 46*, 685–698.

Lewis, A. H., Gottlieb, B., Wilson, B., Sutton, J., Lessmann, J., Delli, G., et al. (2023). Coverage and beyond: How can private governance support key elements of the global biodiversity framework's target 3? *Frontiers in Conservation Science, 4*, 1303801.

Lopoukhine, N., Crawhall, N., Dudley, N., Figgis, P., Karibuhoye, C., Laffoley, D., et al. (2012). Protected areas: Providing natural solutions to 21st century challenges. SAPI EN. S. Surveys and Perspectives Integrating Environment and Society, (5.2).

Maan, J., & Chaudhry, P. (2019). People and protected areas: Some issues from India. *Animal Biodiversity and Conservation, 42*(1), 79–90.

Mahanayak, B. (2024). Ex-situ and in-situ conservation of wild life. *World Journal of Biology Pharmacy and Health Sciences, 18*, 277–282.

Manan, F., Khan, S. M., Ahmad, Z., Kamran, S., Haq, Z. U., Abid, F., Iqbal, M., & Abdullah. (2020). Environmental determinants of plant associations and evaluation of the conservation status of Parrotiopsis jacquemontiana in Dir the Hindu Kush Range of Mountains. *Tropical Ecology, 61*(4), 509–526. https://doi.org/10.1007/s42965-020-00109-2

Mbaiwa, J. E. (2004). The success and sustainability of community-based natural resource management in the Okavango Delta, Botswana. *South African Geographical Journal, 86*(1), 44–53.

McDonald, R. I., & Boucher, T. M. (2011). Global development and the future of the protected area strategy. *Biological Conservation, 144*(1), 383–392.

McLachlan, J. S., Hellmann, J. J., & Schwartz, M. W. (2007). A framework for debate of assisted migration in an era of climate change. *Conservation Biology, 21*(2), 297–302.

McNeely, J. (2004). Nature vs. nurture: Managing relationships between forests, agroforestry and wild biodiversity. *Agroforestry Systems, 61*(1), 155–165.

Melillo, J. M., Lu, X., Kicklighter, D. W., Reilly, J. M., Cai, Y., & Sokolov, A. P. (2016). Protected areas' role in climate-change mitigation. *Ambio, 45*, 133–145.

Milner-Gulland, E., Addison, P., Arlidge, W. N., Baker, J., Booth, H., Brooks, T., et al. (2021). Four steps for the earth: Mainstreaming the post-2020 global biodiversity framework. *One Earth, 4*(1), 75–87.

Nizamani, M. M., Zhang, Q., Muhae-Ud-Din, G., Awais, M., Qayyum, M., Farhan, M., et al. (2023). Application of GIS and remote-sensing technology in ecosystem services and biodiversity conservation. In *Deep learning for multimedia processing applications* (pp. 284–321). CRC Press.

Omar, S. A. (2025). Floristic diversity and vegetation structure of Boubyan Island, a RAMSAR site in Kuwait. *Wetlands Ecology and Management, 33*(2), 34.

Öztürk, M., Khan, S. M., Altay, V., Efe, R., Egamberdieva, D., & Khassanov, F. O. (2022). *Biodiversity conservation and sustainability in Asia. Volume 2: Prospects and challenges in south and middle Asia.* Springer.

Panagopoulos, Y., Dimitriou, E., & Skoulikidis, N. (2019). Vulnerability of a Northeast Mediterranean Island to soil loss. Can grazing management mitigate erosion? *Water, 11*(7), 1491.

Petrou, Z. I., Manakos, I., & Stathaki, T. (2015). Remote sensing for biodiversity monitoring: A review of methods for biodiversity indicator extraction and assessment of progress towards international targets. *Biodiversity and Conservation, 24*, 2333–2363.

Pettorelli, N., Wegmann, M., Skidmore, A., Mücher, S., Dawson, T. P., Fernandez, M., et al. (2016). Framing the concept of satellite remote sensing essential biodiversity variables: Challenges and future directions. *Remote Sensing in Ecology and Conservation, 2*(3), 122–131.

Pierce, J. R., Drill, S., Halder, M. D., Tan, M. M. J., Tiwari, A., & López Guijosa, P. A. (2021). Scaling biodiversity conservation efforts: An examination of the relationship between global biodiversity targets and local plans. *Frontiers in Conservation Science, 2*, 752387.

Poorter, L., van der Sande, M. T., Thompson, J., Arets, E. J., Alarcón, A., Álvarez-Sánchez, J., et al. (2015). Diversity enhances carbon storage in tropical forests. *Global Ecology and Biogeography, 24*(11), 1314–1328.

Postel, S. L., & Thompson, B. H., Jr. (2005). Watershed protection: Capturing the benefits of nature's water supply services. *Natural Resources Forum, 29*(2), 98–108. https://doi.org/10.1111/j.1477-8947.2005.00119.x

Prakash, S., & Verma, A. K. (2022). Anthropogenic activities and biodiversity threats. *International Journal of Biological Innovations, IJBI, 4*(1), 94–103.

Pressey, R. L., Visconti, P., & Ferraro, P. J. (2015). Making parks make a difference: Poor alignment of policy, planning and management with protected-area impact, and ways forward. *Philosophical Transactions of the Royal Society B: Biological Sciences, 370*(1681), 20140280.

Rajasugunasekar, D., Patel, A. K., Devi, K. B., Singh, A., Selvam, P., & Chandra, A. (2023). An integrative review for the role of forests in combating climate change and promoting sustainable development. *International Journal of Environment and Climate Change, 13*(11), 4331–4341. https://doi.org/10.9734/ijecc/2023/v13i113614

Rawat, U. S., & Agarwal, N. K. (2015). Biodiversity: Concept, threats and conservation. *Environment Conservation Journal, 16*(3), 19–28.

Reddy, C. S., Kurian, A., Srivastava, G., Singhal, J., Varghese, A., Padalia, H., et al. (2021). Remote sensing enabled essential biodiversity variables for biodiversity assessment and monitoring: Technological advancement and potentials. *Biodiversity and Conservation, 30*, 1–14.

Sala, O. E., Stuart Chapin, F., Armesto, J. J., Berlow, E., Bloomfield, J., Dirzo, R., et al. (2000). Global biodiversity scenarios for the year 2100. *Science, 287*(5459), 1770–1774.

Scharlemann, J. P., Kapos, V., Campbell, A., Lysenko, I., Burgess, N. D., Hansen, M. C., et al. (2010). Securing tropical forest carbon: The contribution of protected areas to REDD. *Oryx, 44*(3), 352–357.

Schroth, G., da Fonseca, G. A., Harvey, C. A., Gascon, C., Vasconcelos, H. L., & Izac, A.-M. N. (2013). *Agroforestry and biodiversity conservation in tropical landscapes*. Island press.

Shah, M., Khan, S. M., Manan, F., Hussain, J., Din, Z. U., Shah, I., et al. (2025). The role of plant nurseries in the spread of invasive alien plant species in Pakistan's Subtropical Region; a threat for urban greening. *Urban Forestry & Urban Greening, 128688*.

Sharma, I., & Birman, S. (2024). Biodiversity loss, ecosystem services, and their role in promoting sustainable health. In *The climate-health-sustainability nexus: Understanding the interconnected impact on populations and the environment* (pp. 163–188). Springer.

Soares-Filho, B., Moutinho, P., Nepstad, D., Anderson, A., Rodrigues, H., Garcia, R., et al. (2010). Role of Brazilian Amazon protected areas in climate change mitigation. *Proceedings of the National Academy of Sciences, 107*(24), 10821–10826.

Stefanidis, S., Alexandridis, V., & Ghosal, K. (2022). Assessment of water-induced soil erosion as a threat to Natura 2000 protected areas in Crete Island, Greece. *Sustainability, 14*(5), 2738.

Stoll-Kleemann, S., & O'Riordan, T. (2017). The challenges of the Anthropocene for biosphere reserves. *Parks, 23*(1), 89–100.

Stolton, S., Dudley, N., Avcıoğlu Çokçalışkan, B., Hunter, D., Ivanić, K.-Z., Kanga, E., et al. (2015). Values and benefits of protected areas. In *Protected area governance and management* (pp. 145–168).

Ullah, S., Khan, F. M. S., Khan, S. M., Manan, F., Shah, M., & Ullah, T. (2025). Ecological variation and geographic distribution of vegetation in Karak northwestern Pakistan. *Pakistan Journal of Botany, 57*(6). https://doi.org/10.30848/PJB2025-6(11)

Vandebroek, I., & Balick, M. J. (2012). Globalization and loss of plant knowledge: Challenging the paradigm. *PLoS One, 7*(5), e37643.

Vasilijević, M., Zunckel, K., Mckinney, M., Erg, B., Schoon, M., Rosen Michel, T., & Niccolini, F. (2015). The challenge of developing a shared vision: Marittime Alps-Mercantour. In *Transboundary conservation: A systematic and integrated approach* (pp. 68–68). IUCN.

Velazco, S. J. E., Bedrij, N. A., Rojas, J. L., Keller, H. A., Ribeiro, B. R., & De Marco, P. (2022). Quantifying the role of protected areas for safeguarding the uses of biodiversity. *Biological Conservation, 268*, 109525.

Vihervaara, P., Kumpula, T., Ruokolainen, A., Tanskanen, A., & Burkhard, B. (2012). The use of detailed biotope data for linking biodiversity with ecosystem services in Finland. *International Journal of Biodiversity Science, Ecosystem Services & Management, 8*(1-2), 169–185.

Vihervaara, P., D'Amato, D., Forsius, M., Angelstam, P., Baessler, C., Balvanera, P., et al. (2013). Using long-term ecosystem service and biodiversity data to study the impacts and adaptation options in response to climate change: Insights from the global ILTER sites network. *Current Opinion in Environmental Sustainability, 5*(1), 53–66.

Vihervaara, P., Auvinen, A.-P., Mononen, L., Törmä, M., Ahlroth, P., Anttila, S., et al. (2017). How essential biodiversity variables and remote sensing can help national biodiversity monitoring. *Global Ecology and Conservation, 10*, 43–59.

Vitt, P., Havens, K., Kramer, A. T., Sollenberger, D., & Yates, E. (2010). Assisted migration of plants: Changes in latitudes, changes in attitudes. *Biological Conservation, 143*(1), 18–27.

Vodouhê, F. G., Coulibaly, O., Adégbidi, A., & Sinsin, B. (2010). Community perception of biodiversity conservation within protected areas in Benin. *Forest Policy and Economics, 12*(7), 505–512.

Waldron, A., Miller, D. C., Redding, D., Mooers, A., Kuhn, T. S., Nibbelink, N., et al. (2017). Reductions in global biodiversity loss predicted from conservation spending. *Nature, 551*(7680), 364–367.

Wang, B., Zhang, Q., & Cui, F. (2021). Scientific research on ecosystem services and human Well-being: A bibliometric analysis. *Ecological Indicators, 125*, 107449.

Watson, J. E., Dudley, N., Segan, D. B., & Hockings, M. (2014). The performance and potential of protected areas. *Nature, 515*(7525), 67–73.

Watson, J. E., Venegas-Li, R., Grantham, H., Dudley, N., Stolton, S., Rao, M., et al. (2023). Priorities for protected area expansion so nations can meet their Kunming-Montreal global biodiversity framework commitments. *Integrative Conservation, 2*(3), 140–155.

Xue, D., & Tisdell, C. (2001). Valuing ecological functions of biodiversity in Changbaishan Mountain biosphere reserve in Northeast China. *Biodiversity and Conservation, 10*, 467–481.

Zanin, P. R., Cavalcante, R. B. L., Fleischmann, A. S., Peres, C. A., Ferreira, D. M., de Oliveira Serrão, E. A., & Pontes, P. R. M. (2024). Do protected areas enhance surface water quality across the Brazilian Amazon? *Journal for Nature Conservation, 81*, 126684. https://doi.org/10.1016/j.jnc.2024.126684

Zeb, S. A., Abdullah, A., Ahmad, Z., Haq, Z. U., Ali, S., Zeb, T. A., Afza, R., Ahmad, I., Jatt, T., & Khan, S. M. (2026). Soil nutrients control the ecological characteristics of the riparian vegetation. *Pakistan Journal of Botany, 58*(6). https://doi.org/10.30848/PJB2026-6(7)

Zu, K., Lenoir, J., Fang, J., Tang, Z., Shen, Z., Ji, C., et al. (2023). Elevational shift in seed plant distributions in China's mountains over the last 70 years. *Global Ecology and Biogeography, 32*(7), 1098–1112.

Žuvela, A., Dragija, M. Š., & Jelinčić, D. A. (2023). Partnerships in heritage governance and management: Review study of public–civil public–private and public–private–community partnerships. *Heritage, 6*(10), 6862–6880. https://doi.org/10.3390/heritage6100358

Correction to: Frontiers in Sustainability: Biodiversity Conservation in a Changing Climate

Sazada Siddiqui, Mohd Tariq, Zishan Ahmad Wani, and Pankaj Kumar

Correction to:
S. Siddiqui et al. (eds.), *Frontiers in Sustainability*, https://doi.org/10.1007/978-3-032-19740-5

The original version of the book was inadvertently published with incorrect affiliations for editors Mohd Tariq and Pankaj Kumar. We have now updated the book with the correct affiliation details.

The updated version of this book can be found at
https://doi.org/10.1007/978-3-032-19740-5

S. Siddiqui et al. (eds.), *Frontiers in Sustainability*,
https://doi.org/10.1007/978-3-032-19740-5_14

Index

S. Siddiqui et al. (eds.), *Frontiers in Sustainability*,
https://doi.org/10.1007/978-3-032-19740-5

The manufacturer's authorised representative in the EU is Springer Nature Customer Service Centre GmbH, Europaplatz 3, 69115 Heidelberg, Germany. If you have any concerns regarding our products, please contact ProductSafety@springernature.com

Printed and bound by CPI Group (UK) Ltd, Croydon, CR0 4YY
14/10/2026
02233569-0001